Bernhard Varenius

(1622–1650)

Brill's Studies in Intellectual History

VOLUME 159

Bernhard Varenius (1622–1650)

Edited by

Margret Schuchard

BRILL

LEIDEN • BOSTON
2007

On the cover: Detail from the engraved title *Geographia Generalis*, 1650. It probably shows Varenius on the left with two Mayors of Amsterdam, Klaes Corver and Ridder Dr. Banning Cocq. (See also Fig. 2 on page xxii.)

This book is printed on acid-free paper.

Library of Congress Cataloging-in-Publication Data

A C.I.P. record for this book is available from the Library of Congress.

ISSN 0920 8607
ISBN 978 90 04 16363 8

PRINTED IN THE NETHERLANDS

CONTENTS

II

Varenius as a Transmitting Agent of Knowledge—Japan / Varenius als Wissensvermittler einer fremden Welt—Japan

III

Geographia Generalis and Its Legacy / Die *Geographia Generalis* und ihre Nachwirkungen

PREFATORY NOTE

This book is dedicated to Bernhard Varenius, whose geographical books influenced the history of the discipline to such a degree that Isaac Newton, Alexander von Humboldt, Tsar Peter the Great and many North American universities referred to him. A research project of the National Center for Geographic Information and Analysis called itself after Varenius in 1997.

Varenius achieved this renown as a result of his *Descriptio regni Japoniae*, 1649, the first comprehensive description of Japan from a European perspective (which was published as one of the 'republics' by Elzevier in Amsterdam). But his real influence derives from his general geography (*Geographia Generalis*, 1650) in which he describes geography as such, without special regional features. He classifies it as a mathematical science and gives geography a theoretical foundation as a discipline in its own right with a systematic and methodological approach of its own.

The contributions in this book originated in a symposium at the Herzog August Bibliothek in Wolfenbüttel and were extended by further articles. Thus a multifaceted view of Varenius studies describes the young scholar in the context of his time and suggests the enduring effects of his works. Each German article is prefaced by an abstract in English (and vice versa).

In the first part of the book the focus is on biographical studies along with Varenius' education and family history as well as the theological framework in the era of confessional disputes. Varenius' book on Japan is discussed in the second part. Here the emphasis is on the transmission of knowledge acquired without any personal acquaintance with the country described and exclusively based on a diversity of sources. His critical use of source materials and the literary form of his presentation are remarkable. The third part is devoted to his main work, *Geographia Generalis* of 1650. The author's attitude toward Aristotelianism and Cartesianism as well as his terminology are examined. The final articles show how his work was distributed and how it spread to other European countries and beyond to North America.

Ernst-Christian Volkmann, the German translator (1974) of *Descriptio regni Japoniae*, very kindly contributed his translation of Varenius' dedication and chapter 1 of *Geographia Generalis*, now printed for the first time in the Appendix. It is juxtaposed with the Latin text taken from the 1650 edition.

A final word of special thanks is addressed to my friend, Kathleen Rabl, lector of English at the University of Munich. She gave invaluable help with translations at the Wolfenbüttel conference and then very graciously brought her editing expertise to bear on English texts written by non-native speakers. I am immensely grateful for her unfailing sense of style and the meticulous care and sensitivity with which she worked.

The Editor

VORBEMERKUNG

Der vorliegende Band ist Bernhard Varenius gewidmet, dessen Bücher zur Geographie die Wissenschaftsgeschichte so nachhaltig beeinflusst haben, dass sich Isaac Newton, Alexander von Humboldt, Zar Peter der Große und viele nordamerikanische Universitäten bis heute auf ihn bezogen haben. Ein Forschungsprojekt des National Center for Geographic Information and Analysis in den USA nannte sich 1997 nach Varenius.

Gewirkt hat Varenius durch die erste umfassende Beschreibung von Japan aus europäischer Sicht (*Descriptio regni Japoniae*, 1649), die im Rahmen der Länderbeschreibungen von Elzevier in Amsterdam publiziert wurde. Doch sein eigentlicher Einfluss beruht auf seiner allgemeinen Geographie (*Geographia Generalis*, 1650), in der die Geographie als solche, also ohne regionale Besonderheiten zu berücksichtigen, betrachtet wird. Er ordnet sie der mathematischen Wissenschaft zu; als eigenständiges Fach wird sie theoretisch fundiert und in systematischer und methodischer Ausrichtung vorgestellt.

Die Beiträge dieses Bandes gehen auf ein interdisziplinäres Symposium an der Herzog August Bibliothek in Wolfenbüttel zurück und wurden um weitere Artikel ergänzt. So ist ein facettenreiches Bild von Varenius-Studien entstanden, die den jungen Gelehrten im Kontext seiner Zeit und in der Nachwirkung seines Werkes zeigen. Vor jedem englischen Beitrag steht eine deutsche Zusammenfassung (und umgekehrt).

Untersuchungen zur Biographie und seiner Bildungs- und Familiengeschichte, den theologischen Rahmenbedingungen im Zeitalter des Konfessionalisierungsstreits und bibliographische Fragen bilden den ersten Teil. Das Japan-Buch steht im Mittelpunkt des zweiten Teils; hier wird sein Verfahren von Wissensvermittlung ohne eigene Anschauung von dem beschriebenen Land allein auf Grund von Quellen unterschiedlicher Art dargelegt, sein kritischer Umgang mit den Quellen und die literarische Darstellungsform. Im dritten Teil geht es um das Hauptwerk von Varenius, die 1650 erschienene *Geographia Generalis*. Untersucht werden im Besonderen des Autors Haltung zwischen Aristotelismus und Cartesianismus und seine Begrifflichkeit. Studien zur Verbreitung des Buches in Europa, Nordamerika und Russland bilden den Schlussteil.

Ernst-Christian Volkmann, dem wir die Übersetzung ins Deutsche (1974) von Varenius' *Descriptio regni Japoniae* verdanken, hat uns freundlicherweise seine bisher ungedruckte Übersetzung des Widmungsschreibens und des 1. Kapitels aus *Geographia Generalis* zur Verfügung gestellt. Sie ist in der Appendix dem lateinischen Text der Ausgabe von 1650 gegenübergestellt.

Zum Schluss ein Wort des Dankes an meine Freundin Kathleen Rabl, Lektorin für Englisch an der Universität München. Sie hat die Entstehung dieses Buches überaus stilsicher, sensibel und sorgfältig begleitet mit ihrer langen Erfahrung im Edieren von englischen Texten, die nicht von Muttersprachlern verfasst wurden. Dafür bin ich ihr (und ihrer geduldigen Familie) unendlich dankbar.

Die Herausgeberin

CHRONOLOGICAL TABLE—THE LIFE OF BERNHARD VARENIUS

1622		Born in Hitzacker (Elbe), Germany, as one of probably four children. Father: Heinrich Varenius (1595–1636), court preacher of Duke August of Braunschweig-Lüneburg) and Lutheran pastor. Mother: Anna (Freder)
1627		Late that year moved to Uelzen on his father's appointment as praepositus and superintendent
1635		Death of his father
1636	Jan. 10	Matriculated at Helmstedt University: "Bernhard Varenius, Hitzackerensis", left in 1640 for Hamburg
1640	Apr. 1	Matriculated at Hamburg Academic Gymnasium (under Jungius)
1642	Nov. 16	Disputation *De Definitione motus Aristotelica*
1643	July 4	Matriculated at Königsberg University, left in 1645 for Dutch States
1645	Apr. 30	Visited John Pell in Amsterdam
	May 6	Matriculated at Leiden University
1646	autumn	Moved to Amsterdam (Blue House, Heerengracht) to work as tutor for a year
1649	June 22	Medical disputation at Leiden for D. med. (*De febri in Genere*)
	July 1	*Descriptio Regni Japoniae* dedicated to consuls and senate of Hamburg, dated in Amsterdam
	Aug. 1	*Tractatus... de Iapone religione* dedicated to Christina of Sweden, dated in Amsterdam
1650	Aug. 1	*Geographia Generalis* dedicated "ex meo musaeo" to mayor, trustees of the gymnasium, and city treasurer of Amsterdam
		Probably died in 1650; nothing is known of the circumstances of his death

BERNHARD VARENIUS—FAMILY BACKGROUND

I.

Lorenz Smedes (Schmidt)
(matric. 1517, †1555)
Alderman, Rostock
∞ Margarete Bercholt
(†1565/69)

MATTHÄUS KOCHHAF (CHYTRAEUS)
(1495–1559)
Lutheran pastor, Ingelfingen u. Menzingen
∞ Barbara Neuburger alias Nellberg

II.

[Joachim Kron
Alderman, Rostock
(1521–1597)
∞ Agneta Smedes
(1530–1619)]

JOHANNES FREDER
(1510–1562)
∞ Anna Falk

DAVID CHYTRAEUS
(1530–1600)
Prof. Theol., Rostock
1) ∞1553 Margaretha Smedes
(1531–1571)
[2) ∞ 1572 Margarethe Pegel
(† after 1600)]

Nathan Chytraeus
(1543–1598)
Prof. Latin, Rostock,
Headmaster, Bremen

III.

[Bartholom. Klinge
(Clingius) (1535–1610)
Prof.iur.&log., Rostock
2) ∞ Margareta Kron
(1555–1618)]

JOHANN VARENIUS
Brewer and citizen,
Herford
∞ Margaretha Plüstermann

JOHANNES FREDER ∞ 1573 Margaretha Chytraeus
(1544–1604) (1555/8–1600)
Prof. & Superintendent,
Rostock

IV.

[Bernhard Klinge
(1584–1648)
Mayor 1631, Rostock]
3) ∞ 1633 Ursula Dieck (†after 1652)

HEINRICH VARENIUS ∞ Anna Freder
(1595–1635) († late in 1623)
Lutheran pastor, Hitzacker,
Praepositus & superintendent, Uelzen

V.

AUGUST VARENIUS
(1620–1684)
Prof. Hebr.,Theol., Rostock
∞ Charitas Heckenberg, geb.
Kregel (1624–1672)

BERNHARD VARENIUS
(1622–1650)

Sources: Funeral sermons on D. Chytraeus, J. Freder (1604), A. Varenius; Rohrbach, Möhlmann, Thüringer

BERNHARD VARENIUS—STEMMA OF EDITIONS

(I)

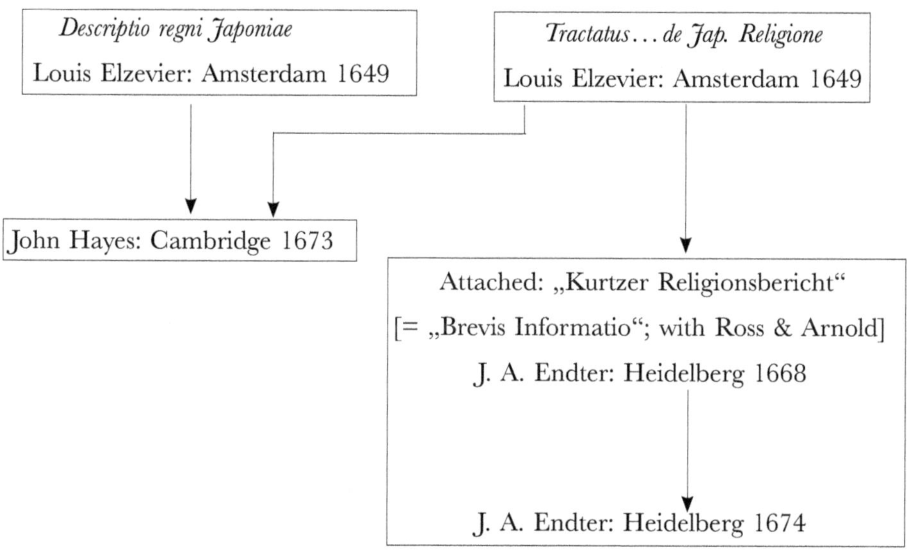

STEMMA OF EDITIONS

(II)

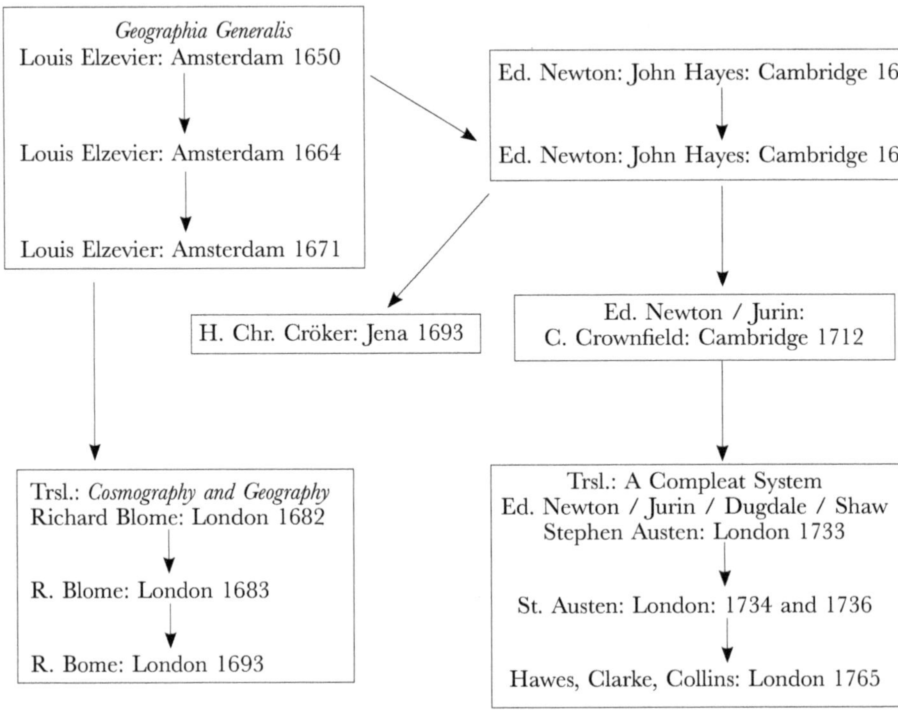

LIST OF ILLUSTRATIONS

Figures 1–15

Acknowledgments:
Figs. 1, 2, 4, 12–15 Landesbibliothek Eutin; 7 from S. Klocke Daffa (ed.), *Engelbert Kaempfer*, 2003, S. Fischer Verlag; 8, 9, 10, 11 Peter Kapitza, Judicium Verlag; 3, 5, 6 Herzog August Bibliothek Wolfenbüttel.

Fig. 1. *Descriptio Regni Japoniae*, 1649. Title-page of the first edition

Fig. 2. *Geographia Generalis*, 1650. Title-page of the first edition

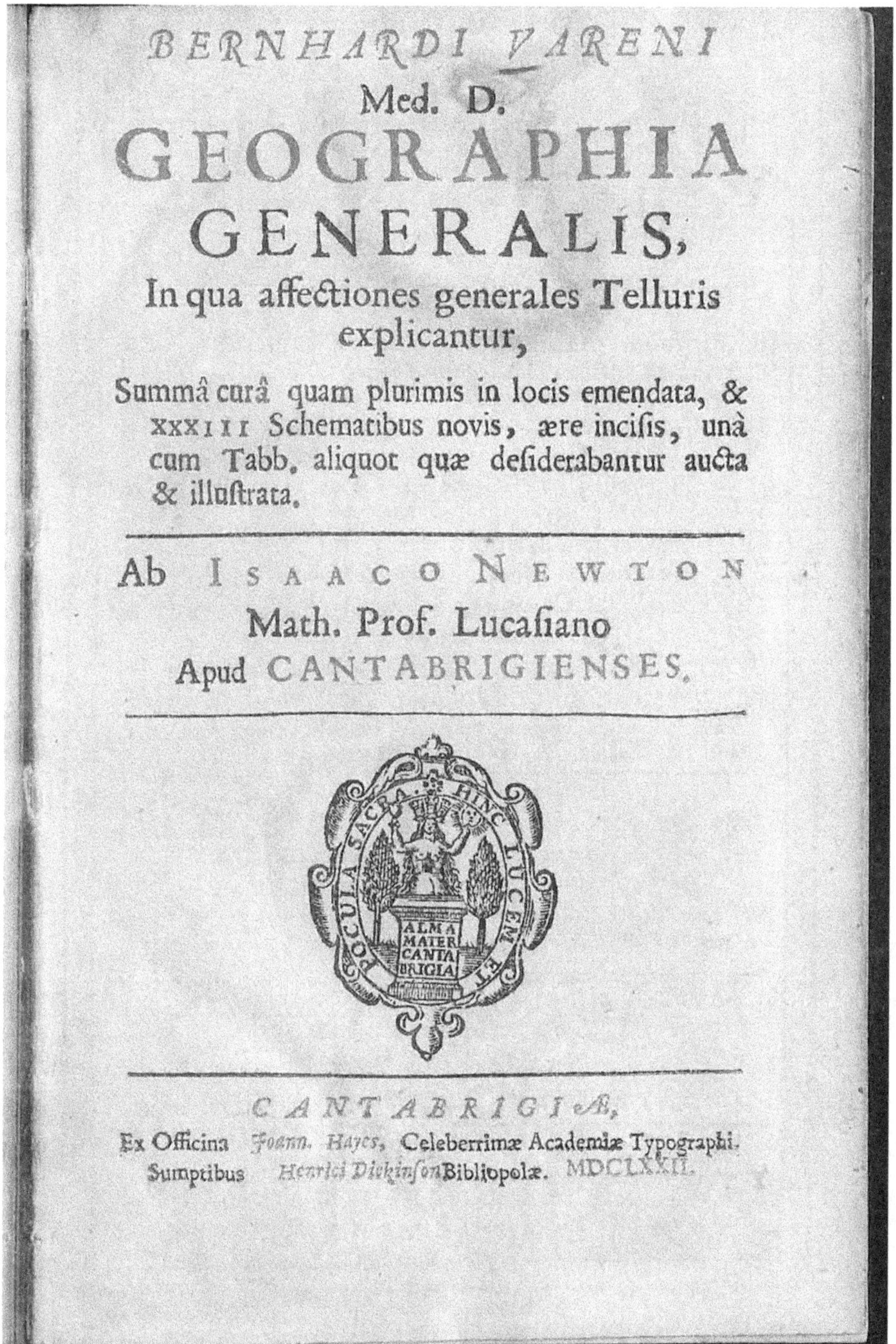

BERNHARDI VARENI
Med. D.
GEOGRAPHIA
GENERALIS,
In qua affectiones generales Telluris
explicantur,
Summâ curâ quam plurimis in locis emendata, &
XXXIII Schematibus novis, ære incisis, unà
cum Tabb. aliquot quæ desiderabantur aucta
& illustrata.

Ab ISAACO NEWTON
Math. Prof. Lucasiano
Apud CANTABRIGIENSES.

CANTABRIGIÆ,
Ex Officina *Joann. Hayes*, Celeberrimæ Academiæ Typographi.
Sumptibus *Henrici Dickinson* Bibliopolæ. MDCLXXII.

Fig. 3. *Geographia Generalis*, 1672. Newton's first edition of Varenius

A COMPLEAT

SYSTEM

OF

General Geography:

EXPLAINING

The Nature and Properties of the EARTH;

VIZ.

It's Figure, Magnitude, Motions, Situation, Contents, and Diviſion into Land and Water, Mountains, Woods, Deſarts, Lakes, Rivers, *&c.*

With particular Accounts of the different Appearances of the Heavens in different Countries; the Seaſons of the Year over all the Globe; the Tides of the Sea; Bays, Capes, Iſlands, Rocks, Sand-Banks, and Shelves.

The State of the Atmoſphere; the Nature of Exhalations; Winds, Storms, Tornados, *&c.*

The Origin of Springs, Mineral-Waters, Burning Mountains, Mines, *&c.*

The Uſes and Making of Maps, Globes, and Sea-Charts.

The Foundations of *Dialling*; the Art of *Meaſuring Heights and Diſtances*; the Art of Ship-Building, *Navigation*, and the Ways of *Finding the* LONGITUDE AT SEA.

Originally written in *LATIN*

By *BERNHARD VARENIUS*, M. D.

Since Improved and Illuſtrated

By Sir ISAAC NEWTON and Dr JURIN;

And now Tranſlated into *Engliſh*; with additional *Notes*, *Copper-Plates*, an *Alphabetical Index*, and other Improvements.

Particularly uſeful to Students in the *Univerſities*; Travellers, Sailors, and all thoſe who deſire to be acquainted with *Mix'd Mathematics*, *Geography*, *Aſtronomy*, and *Navigation*.

By Mr *DUGDALE*.

The whole Reviſed and Corrected by PETER SHAW, M. D.

In TWO VOLUMES.

LONDON: Printed for STEPHEN AUSTEN, at the *Angel* and *Bible*, in *St Paul's Church-Yard*. 1733.

Fig. 4. *A Compleat System*, 1733. Dugdale's translation of the edition by Newton, Jurin and Shaw

INTRODUCTION

This volume collects contributions on Bernhard Varenius which attempt to reconsider his place in the history of scientific thought within the context of his age, the early seventeenth century, and after. It is not an easy matter, and since this book owes its origin to a symposium and the availability and willingness of scholars to contribute to interdisciplinary work, not all the fields one would like to see investigated could be covered. For any exhaustive study of Varenius the fact that there is still no complete translation available of his *Geographia Generalis* into a modern vernacular is a major impediment. Such work would require the cooperation of Latinists and historians of mathematics, natural philosophy and geography among others, not to mention the necessary funding. A beginning is, however, made by providing a translation by Ernst-Christian Volkmann into German of the dedication and the first chapter in the Appendix, since these parts of the book convey Varenius' purpose in writing, his theoretical considerations and his overall design. For Varenius it was a matter of course (as it was for scholars from Descartes to Newton) to express what was deeply important to him in the Latin language he had studied from childhood. Not only were school boys in his time steeped in the classics, they, at least the more gifted ones among them, could write elegant prose or gracefully composed poems in the traditional styles of poetic diction. Latin was also the language of serious theological study and eloquent debate; it could reveal or conceal by not trading controversial and complex issues openly in the market square, and yet reach the wide audience of the Latinate public sphere. It was still a matter of ritual that letters in a certain social context had to be written in Latin, witness Bernhard's letters to his former teacher Jungius or the letters his father and his brother wrote to Duke August.

In the first part of this collection the focus is on biographical studies along with Varenius' education and family history as well as the theological framework in the era of confessional disputes. Varenius was a natural philosopher firmly rooted in Christian theology, as Margret Schuchard points out. In a line of five generations of Lutheran theologians (among them the eminent David Chytraeus, who was educated by Melanchthon), he was the first to depart from the family tradition and to choose secular fields of enquiry, medicine, mathematics and natural philosophy. Yet in the taxonomy of knowledge he presented in his works, he never alienated himself from theology. This is most obvious in his extensive treatise on the religion of the Japanese and his survey of the religions of diverse peoples, and Varenius found Schouten's description of the religion of the Siamese interesting enough to add it to the volume on Japan when space allowed him to do so.

Johann Anselm Steiger's exhaustive and meticulous study of Varenius' father's intrepid defence of Arndt shows the background of confessional strife in which young Varenius grew up at the time of the Thirty Years' War. Johann Arndt, the influential superintendent in Celle, had died in 1621, and two years later Osiander of Tübingen charged Arndt with presenting false doctrine of a spiritualistic-enthusiastic kind in a widely read book. Heinrich Varenius, court preacher to Duke August, took sides at once in his book of 1624, doubtless with his noble patron's backing, to defend Arndt against Osiander's accusation of his being outside Lutheran orthodoxy. Heinrich Varenius displays impressive theological learning but sometimes gets carried away and overstates his case rather sweepingly; as does Osiander. Steiger selects three aspects of the controversial issues for discussion—the spiritual birth of Christ in man, the mystic union of the believer with God, and the relation between verbum internum and verbum externum. In his careful and penetrating analysis he suggests that neither party is unbiased, yet both parties are right, because Arndt's text ultimately does not allow for an unambiguous reading. Varenius' spirited vindication of Arndt paved his way to becoming a superintendent himself so that Bernhard Varenius grew up in Uelzen where his father Heinrich took up the duties of his new post.

Both Lehmann and Schuchard attempt to add new pieces to the puzzle of Bernhard Varenius' life of which so little is known. To start with a marginal but obvious point of curiosity discussed by Blanke: We would like to know what Varenius looked like but cannot be certain whether the young man on the engraved titles to his description of Japan and to his geography book (see figs. 1 and 2) is even the same person, let alone whether it is Varenius. In all likelihood, and in the absence of evidence to the contrary, most scholars agree that here we are indeed looking at Varenius. Biographical sources are scarce enough, unfortunately, and what is available we owe to a large extent to Bernhard's brother August's well documented life. The other important source of information is the correspondence between Jungius and

Varenius already mentioned. Between them, Schuchard and Lehmann explain how Varenius was brought up, where he was educated and who his teachers were. Klaus Lehmann focuses on teaching methods at school and university, and for the first time the formative influence of his teachers at the University of Helmstedt is discussed here. Joachim Jungius, head of the Academic School in Hamburg, trained Varenius in disputation and, as Lehmann points out, appeared to have been his living example of how one might collect information and put it into some kind of order. In the case of Jungius' notes labelled "Geographia generalis", Staffhorst, who studied them in detail, was able to show conclusively that Varenius could not have profited from them for his own work. Jungius never worked these notes into shape, and they were evidently not intended for preparing an eventual publication.

Varenius' laborious road to authorship in the Dutch Republic is traced by Schuchard, and his efforts to support himself through his own work and patronage are recorded. When he did find his publisher in Louis Elzevier of Amsterdam, who accepted his book on Japan as the last of the 'republics', which his cousins in Leiden evidently had no intention of adding to their series, he could begin to make a name for himself as a scholar, but his short life ended before he could achieve what he had set out to do. He came to Holland in high hopes of finding intellectual stimuli at the renowned University of Leiden which had attracted an elite of learned men, and he witnessed the spreading of Descartes' ideas and the fierce battles over them. Rienk Vermij describes the world of learning in the 'golden age' of the Dutch republic ruled by urban oligarchies whose prime interest lay in advancing Dutch trading power and dominating the seas. So Varenius, even as a stranger in the country, enjoyed greater factual freedom, civic and religious, than he could have found under the absolutist government of a German prince. Vermij explores the reasons why Varenius failed in his application for the chair of mathematics as John Pell's successor and why the demand for instruction in practical mathematics for navigation purposes, which he could easily have met in order to support himself, was not an alternative open to him or compatible with his talent and ambition. Vermij finds it reasonable to assume that Varenius, educated as a humanist like most of his contemporaries, must have practiced a rather old-fashioned kind of mathematics, still largely based on the study of classical authors rather than the new analytical techniques developed by Descartes. Since those unpublished papers have not survived, this point cannot be settled, but Staffhorst's findings in the third part of this book suggest a treatment that includes a great awareness of Cartesian method and yet does not dispense with the ancients.

Varenius' description of Japan is discussed in the second part of this book. Here the emphasis is on the transmission of knowledge acquired without any personal acquaintance with the country described and exclusively based on a diversity of sources. The wide horizon of a seafaring nation had drawn the attention of Varenius to what lies beyond the European coast lines. In the context of travel reports and discovery, Reichert's study of Varenius' sources describes how the European image of Asian countries was formed. He begins with Varenius' critical use of Marco Polo, who established Japan in the European mindscape (Polo called it Zipangu and classified it as an Indian island). Varenius also used Jesuit authors, less fabulous than Polo's reports from hearsay, because as Varenius was well aware, the Jesuit mission work brought authentic information about Japan to Europe, and he supplemented his findings with what he could discover from those who worked for the United East-India Company (Verenigde Oostindische Compagnie V.O.C.), particularly Caron. Varenius' critical treatment of source materials and the literary form of his presentation are significant. Düchting's approach is that of a philologist, thoroughly familiar with bibliographical issues and the Neo-Latin manner of writing. He notes that Varenius makes use of the traditionally open form of such descriptions in the style of a chorography and examines his terminology with the greatest care (not without looking at Chytraeus' description of his native region for comparison and contrast). Düchting's study is supplemented by Blanke, who enquires into the way Varenius acquired and structured knowledge and how this knowledge was filtered and then transferred. Here the practice of apodemic literature, at the time an established genre of teaching the art of travelling with quite a list of topics to be considered, appears to have had a guiding influence on Varenius in writing about Japan for a wider readership than the select circle of the learned. His *Geographia Generalis* is a different matter, if not a change of paradigm, in Blanke's view, and it marks an intellectual breakthrough in Varenius' attempt to systematise knowledge in well considered categories.

The third part is of course devoted to Varenius' main work, *Geographia Generalis* of 1650 and its legacy. The author's attitude toward Aristotelianism and Cartesianism as well as his terminology are also explored. This section begins with Frank Richter's assessment that Varenius cannot really be assigned to any one school of natural philosophy. There are some leanings toward Cartesianism without his entirely abandoning Aristotelian approaches. Richter, in critical engagement with Warntz and Bowen, raises once again the question of where to locate Varenius within the various channels of transmitting scientific thought from the Greeks to Renaissance scientists and on to Descartes, Newton, Wolff, Kant and Humboldt, only to find Varenius between all philosophical stools, because definitive answers remain elusive. Nor can his careful scrutiny of Varenius' key terminology as to 'affectiones' or 'accidentia' establish a well-founded basis for his affiliations with any one view of the world. Staffhorst, when raising some of the same issues and after a detailed exposition of how these key terms are used by Varenius in context and by authors he knew, nevertheless comes to the conclusion that Varenius, although he refers to old and new authors equally, is quite close to Descartes and to ancient atomism. Thanks to Staffhorst's study of the *Geographia Generalis* as a whole, we now know that Varenius supported the heliocentric view of the world. In the preliminary sections Varenius expressed himself cautiously and, as he often did, reported different opinions in juxtaposition, which led to the misunderstanding that he found the Ptolemaic view more convincing than the Copernican, but this is not true; and it is the more remarkable, because none of his predecessors, whether Merula, Keckermann, Alstedt, Göllnitz or Christiani, reached this conclusion. Varenius thought for himself and did not uncritically repeat ancient or church authorities on trust, which is also evident in his exhibiting signs of reasoning by means of causal connections.

The final articles deal with how the *Geographia Generalis* was distributed and how it spread. Varenius picked his way through his source material with dexterity, and his concept of introducing the subject in a systematic manner to make it suitable for teachers and learners attracted Newton's attention so that his initiative in selecting Varenius for republication eventually turned it into an export article. The Cröker edition of 1693, based on Newton, was on the syllabus of the Gymnasium Illustre of Karlsruhe in Germany, as Staffhorst discovered. By then the main centre of activities had long moved from the Netherlands to England, the chief focal point for both Schuchard and Mayhew in their complementary discussions of the sequence of editions and the circumstances that brought them about. Blome's hackwork translation into English of 1682 deserves attention only in passing. What is at stake is Newton's edition and its afterlife, closely connected with the Royal Society, and here Mayhew presents a highly convincing argument based on the context of Bentley's involvement in the project so that the Varenius text, promoted by this scholar of classical literature in Cambridge, who was also a Newtonian, acquired the status of a classic. Jurin and afterwards Shaw (and the translator Dugdale), when they edited, annotated and revered it, treated it as a canonical text at the time when the greatest editorial care was in the same manner devoted to Shakespeare (Nicholas Rowe, Alexander Pope). Mayhew enriches our understanding of the *Geographia Generalis* by drawing our attention to it not only for its contribution to scientific thought but to its textual status, aware of the interplay of linguistic signs, thus meeting with Düchting's and Staffhorst's fine sense of the textual complexities in Varenius the writer.

In the colonial period in North America the editions based on Newton made their way into the colleges as a prescribed text book for geography and in Pennsylvania even for theology, as Schuchard mentions. Jedidiah Morse, particularly interested in special geography, spread Varenius' ideas, and Jefferson is also bound to have come across them. Schuchard and Rebok explore the influence of Varenius on Jefferson and Humboldt. Jefferson was very knowledgeable about geographical questions, and both his *Notes on the State of Virginia* and his preparation of the Lewis and Clark expedition are mile stones in the development of the discipline. Humboldt met Jefferson in person after his explorations in Spanish America. Rebok notes that in the indexes and search tools, which are available at present, explicit references to Varenius in Jefferson are not so far listed; the same is true of Humboldt with the exception of his long comment in *Kosmos* that he wrote late in his life, but a striking similarity of their scientific concepts to what Varenius proposed is very noticeable.

Varenius also made his way to Russia. When Peter the Great began to tackle the huge task of reforming Russia to modernise his empire with a view to developments in western Europe, he commissioned the translation of the *Geographia Generalis* into Russian (published in 1718), and since he desired it to be widely understood, he explicitly wanted it in plain Russian. Denis J.B. Shaw explains the context for the tsar's needs to unify and centralise his country by making use of the necessary skills developed elsewhere for mapmaking, land inventory and cadastral surveys. Here Varenius' book was valuable by way of disseminating learning in geography, a geography that was solidly founded on modern European science, and was instrumental in securing the tsar's power and advancing trade and commerce.

At the end of this survey of the themes and subjects dealt with in this book, there is a note on a regrettable absence. Wolfgang Griep, the former director of the Eutin State Library, gave a paper on "Lost Traces: News of the Varenius Scholar Carl Rohrbach (1861–1932) and the Varenius Collection in the Eutin State Library" at the Wolfenbüttel symposium. Unfortunately it is missing from this collection, because new obligations hindered Griep from seeing it into print. A century ago a huge mass of source material on Bernhard Varenius, published and unpublished, was collected by a school master in Gotha, Professor Dr. Carl E.M. Rohrbach (1861–1932), educated in the humanist tradition and graced with the honorary title of Geheimrat. His huge collection of Varenius' printed books and of archival material remained virtually unknown until the Eutin Landesbibliothek could buy it from the Rohrbach estate in 2000. It is now available to researchers besides a wealth of other sources, especially in the field of travel literature and travel studies (see: www.bibliothek-eutin.de/lb/). Rohrbach focused on biographical and bibliographical aspects with scrupulous attention to detail, on the geographical and historical sources of *Descriptio Regni Japoniae* and *Geographia Generalis*, and he began to translate the *Geographia Generalis*. Griep described Rohrbach's labours as "vergebliche Wissenschaft" (scholarship to no avail), because although Rohrbach had made it his life's work to prepare a book on Varenius, his labour of love tragically never saw the light of publication; only a few articles made it into print. His collection now at Eutin is, however, of the first importance to Varenius scholars.

This renewed examination of Varenius' works and re-evaluation of his achievements brings to mind what Hanno Beck wrote in 1982 when he called Varenius "definitely the greatest geographer since antiquity". Even if we do not go as far as that, Varenius' part in shaping the subject and advancing its standing within the development of geography as a scientific discipline is decisive, and this endeavour remained interesting to scholars of later generations throughout the process of forging the 'modern mind'. The legacy of the scientific revolution has, however, been radically questioned in our time, and critics (among them feminists and ecologists like Margarita Bowen and Carolyn Merchant) have challenged the dominant concern of scientists to objectify our knowledge of the earth. Instrumentalisation and exploitation are on one side of the coin; the other displays a disintegration of certainties as regards the reality of the material world. In our present struggles with problems of virtuality concomitant with the electronic age, we are once again beginning to realise what Varenius knew in the very different context of his era—how high the price for our knowledge can be.

M.S.

I

VARENIUS' FORMATIVE YEARS IN THE CONTEXT OF HIS TIME

VARENIUS IM BILDUNGSHORIZONT SEINER ZEIT

VARENIUS AND HIS FAMILY: A DYNASTY DEDICATED TO SCHOLARSHIP AND ROOTED IN CHRISTIAN PHILOSOPHY

Margret Schuchard

Abstract

Varenius und seine Familie: Eine Dynastie im Dienst der Wissenschaft und verwurzelt in der christlichen Philosophie

Varenius entstammte einer Familie von Pfarrern, die sich bis in die erste Reformationszeit zurückverfolgen lassen. Soziologisch betrachtet sehen wir an ihnen ein familiäres Beziehungsnetz, geknüpft aus gezielter Begabungsförderung und geschickter Heiratspolitik (Einheiratung); zum Profil dieser enzyklopädisch gebildeten Gelehrtenfamilien gehört ihre lutherische Konfession sowie der geographische Bezugsraum ihrer Karriereverläufe in den Hansestädten. Über fünf Generationen hinweg bekleideten sie führende Ämter in Universität und Kirche. Besonders ist hier Varenius' Urgroßvater David Chytraeus zu nennen, ein Schüler Melanchthons, der zum bedeutendsten Professor der Universität Rostock in der frühen Neuzeit wurde. Auch sein Großvater Johannes Freder d. J. und sogar noch sein Bruder August hatten Lehrstühle der Theologie an der Universität Rostock. Bernhard Varenius ist der erste, der angetrieben von wissenschaftlicher Neugier einen anderen Weg ging und aus der Traditionslenkung durch Theologie ausscherte.

An der Schnittstelle zwischen späthumanistischer Polymathie und sich ausdifferenzierender Wissenschaft blieb er theologischem Denken eng verbunden, wenn er durchaus quellenkritisch Berichte über fremde Religionen zusammenzufassen suchte. Im Zusammenhang mit seinen Japan-Studien veröffentlichte er den *Tractatus in quo agitur de Iaponiorum religione.* Der Abschnitt "Brevis informatio de diversis gentium religionibus" hat eine komplizierte Publikationsgeschichte; er wurde 1668 von Endter in der Übersetzung von Finx als *Varenii kurtzer Religionsbericht von mancherley Völkern* im Anhang zu Alexander Ross' *Pansebeia* publiziert. In seinen beiden Hauptwerken, die in der weltkundigen Stadt Amsterdam entstanden, machte Varenius das Wissen seiner Zeit in einer systematisierten Ordnung zugänglich.

Varenius was born into a family of Lutheran clerics, where in the direct line of five generations we find six theologians of whom four were also professors of theology. Bernhard Varenius dared to interrupt the tradition when it was still taken for granted that family practices must be passed on from one generation to the next, whether in the guilds of craftsmen and merchants or in the academic professions. The function was to stabilise the social order by such channelling, and in the seventeenth century this was particularly felt to be necessary in the midst of religious, social, and scientific upheavals. Familial networks guaranteed professional connections, and in the case of Bernhard Varenius he profited from them in his early and formative years. There was never a doubt, whatever hardships might have to be faced, that like his ancestors he should have an excellent education. Although young Bernhard did not enjoy much direct continuity in his family because of the early death of his parents, he dedicated himself to scholarship like his forebears, but, as it turned out, not to theology, the prime and most serious field of enquiry. He chose secular subjects where he thought his individual talents lay, mathematics, medicine, natural philosophy, and he had the courage to leave the established track and to look for something new without yet being certain where this route would take him. When tracing the familial chain of his forebears down to Bernhard, his own achievements stand out the more clearly.

Bernhard Varenius[1] was born in Hitzacker some time in 1622 where his father Heinrich was pastor primarius and chaplain to Duke August of Brunswick-Luneburg (Braunschweig-Lüneburg), a descendent of the younger house of the Welfs. Bernhard was one of probably four children. There were his older brother August, born 20 September 1620, a sister whose name is unknown, and possibly a younger brother called Georg Heinrich. The children lost their mother in infancy; she died in 1623, and in 1628 the Varenius family left Hitzacker for good; because until his death, Heinrich Varenius held the post of praepositus (Propst, Hauptpastor) at St. Mary's church in Uelzen and was the superintendent for the surrounding parishes.

After 1635 the children were without parental care and had to be brought up by others. Who their guardian was remains unknown. Soon after their father's death the rectory would have been occupied by his successor. The children are likely to have been separated, they certainly were when it was time for the boys to be sent to school and

[1] For biographical information the best source (beside his letters to Jungius) is the funeral sermon preached when his brother August died in 1684; it was afterwards published together with tributes to August. See Justus Christoph Schomerus, *Die Gewissens-Freudigkeit im Tode* [Leichenpredigt August Varenius], Rostock 1684.

to be boarded with their teachers as was customary. Yet both August and Bernhard continued to feel attached to Uelzen as their home town in later life. In 1635 they must have been just old enough to remember their father's stories, within the span of the usual four generations of oral tradition, and their teachers cannot have been unfamiliar with the boys' background. Also there were the books by their ancestors to inform them further as they grew up.

The dominant family lineage and the one most remembered was on Bernhard's mother's side, for his father's family did not make its mark on posterity. All we know is that Heinrich Varenius came from Herford where his father was a citizen and a brewer. The matrilineal line goes back to Matthäus Kochhaf of Brackenheim, who studied theology in Tübingen and took sides with Luther's followers; and Kochhaf was thus among the first Lutheran preachers who dared to get married. Kochhaf had ten children, and it was in Menzingen where Bernhard Varenius' great-grandfather grew up. He was David Kochhaf. David Kochhaf, later called Chytraeus,[2] became a leading figure in the history of the later phase of the reformation in Germany. He was such a gifted child that he was given the best possible education. When not quite ten years of age, his Latin was already good enough for him to be matriculated at Tübingen University, and in the course of his studies he was taught Greek by Melanchthon's friend Joachim Camerarius. David Kochhaf received his M.A. at fourteen, and on the recommendation of Melanchthon's brother travelled to Wittenberg where Melanchthon, in order to test this extremely young Master's Greek, asked him to translate a text by Thucydides. Kochhaf passed the test with flying colours and not only translated fluently but also explicated the passage in Latin,—whereupon Melanchthon took him into his house and brought him up like a son. (Tu merito es Magister, et tu eris mihi filii loco.)[3] With his other pupils who boarded with him, Melanchthon himself attended to his education for about five years. David heard Luther and Melanchthon lecture, but his mentor suggested

---

[2] See also the long entries for each of these theologians in *Biographisch-Bibliographisches Kirchenlexikon (BBKL)*, ed. Friedrich Wilhelm Bautz, 1990– : Johann Arndt, David Chytraeus, Nathan Chytraeus, Johannes Freder the Elder, Joachim Jungius. No entry for Heinrich Varenius.

[3] Funeral sermon by Lucas Bacmeister on the death of Chytraeus. Quoted in Stefan Rhein, „Ein Gruß aus Wittenberg: David Chytraeus und die Hausschule Melanchthons," *David Chytraeus (1530–160): Norddeutscher Humanismus in Europa*, eds. Karl-Heinz Glaser and Steffen Stuth, Ubstadt-Weiher 2000, 13.

that besides theology he should also devote himself to history, medicine and astronomy. Meanwhile, David Kochhaf had taken his cue from Melanchthon, who like other humanists had changed his name into Greek (from Schwartzerdt), and so Kochhaf became Chytraeus.

On Melanchthon's recommendation Chytraeus went to Rostock, and in 1550 the Duke of Mecklenburg appointed him a professor of the university. He retained this position throughout his life and declined nine calls to other universities. For half a century Chytraeus took it on himself to continue to propagate Melanchthon's academic line in Lutheran Protestantism as one of its chief protagonists,[4] and he became the most renowned scholar at Rostock University in the second half of the sixteenth century. He taught classical and biblical exegesis and gave encyclopaedic lectures. Besides theological themes, he also dealt with historical, philological and methodological subjects, and he never denied the differences between his esteemed and beloved teachers, Luther and Melanchthon; yet he tried to avoid discord, since he saw his responsibility in reconciling conflicting views and in reassuring his readers in the orthodox Lutheran faith.[5] As a gifted organiser he was in great demand in many areas. He was responsible for reforming the University of Rostock where he served five terms as rector, and he was one of the founders of a university in the Duchy of Brunswick-Luneburg, which was established in Helmstedt in 1576.[6] He declined a call to Helmstedt, the very university where later his great-grandsons Varenius would be matriculated.

Among David Chytraeus' most important students was the younger Johannes Freder, who was his friend's son, and he educated him in his own house. Freder eventually became his successor, and also his son-in-law, when he married his daughter Margareta in 1573. Freder accompanied Chytraeus on his many travels. In the year before his marriage into the Chytraeus family he became a professor at Rostock University, where he then served as rector several times and edited the papers of his father-in-law. Here we see the familial network in full operation.

---

[4] Walter Thüringer, „David Chytraeus (1530–1600): Aus dem Leben eines dem Kraichgau verbundenen Theologen und Historikers," *Kraichgau* 6 (1979): 164.

[5] Rudolf Keller, *Die Konfus Augustana im theologischen Wirken des Rostocker Professor David Chytraeus (1530–1600)*, Göttingen 1994, esp. ch. 5

[6] Chytraeus' role in the establishment of Helmstedt University and its early history are described in detail by Peter Baumgart, *Universitäten im konfessionellen Zeitalter: Gesammelte Beiträge*, Münster 2006, 103–295.

Yet in 1600 there is a decisive break in the continuity of the Chytraeus—Freder families, because in that year both David Chytraeus and his daughter Margarete Freder died, and only four years later Johannes Freder was buried in the grave next to Chytraeus.[7] Thus the Varenius family knew of the background of the Freders by hearsay only. When Heinrich Varenius came to Rostock University as a student in 1613, where he obtained his MA in 1617, not a single member of the older generation who had so vigorously shaped the Reformation and the university in Rostock was still alive.

As we come to consider the life of Heinrich Varenius (1595–1636), we have to turn to the role of Duke August in the history of this family. Duke August of the house of Brunswick-Luneburg was fifteen when he left his home town of Dannenberg for Rostock. He was a student there in 1594, and honorary rector of the university at the same time, witnessing the battles between the Reformation and Counterreformation. After the long travels of his grand tour, he settled in Hitzacker in 1604, a small territory to rule indeed, but one which he held for thirty years, as it turned out, until he could better himself by succeeding to the Wolfenbüttel line of the Welfen (Guelfs) in 1635. When still in Hitzacker, he appointed Heinrich Varenius, who had been a student in Rostock since 1613 and had just been conferred his MA, as his chaplain in 1617, and in this role Heinrich served him until late in 1627. In Hitzacker Heinrich Varenius was also pastor primarius of St. John's church in town, only a stone's throw away from the ducal chapel, and it was in Hitzacker that his children August (1620) and Bernhard (1622) were born. The Duke graciously condescended to grant his preacher's humble request and became godfather to August, who was named after him.

It would be interesting to discover how the Duke came to appoint Varenius to his court. Since Varenius had not yet published anything, he must have been recommended to the Duke by his mentors in Rostock, Hamburg or Helmstedt. One of these recommendations is certain to have been his connection with the Chytraeus family, because Chytraeus, whom the student duke had met in Rostock, was a man he highly respected and with whom he continued to correspond afterwards.

[7] Funeral sermon by Valentin Schacht., quoted in Otto Krabbe, *Die Universität Rostock im fünfzehnten und sechzehnten Jahrhundert*, Rostock 1854, 668.

Heinrich Varenius was not yet thirty and still lived in Hitzacker when in 1624 he publicly raised his voice to take up the defence of Johann Arndt[8] against the allegations of Osiander that Arndt was not a true Lutheran Christian, but an enthusiast. Heinrich Varenius dedicated the second part of his book to Duke August on 2 July 1624. Heinrich Varenius' public involvement in warning true Lutherans against "the false teachings of Papists, Calvinists, Photinians, Weigelinians and similar hordes"[9] came to the attentions of his superiors, and so at only 33 years of age, he was endowed in 1628 with a very important benefice, ranking only second in the principality.[10] He became praepositus (Propst) and superintendent in Uelzen. He therefore left Hitzacker and Duke August, who graciously let him go after eleven years, and entered the service of August's cousin, Duke Christian the Elder, who began his rule in 1611 and resided in Celle. We learn details about this transition from the farewell sermon he gave in Hitzacker and the introductory sermons in Uelzen; the collection of these sermons (*Trifolium sacrum*) with prefatory material was again printed by Johann and Heinrich Stern in Lüneburg in 1628.

It is quite possible that Varenius' sons August and Bernhard were at these services, although they were still very young in 1628. After their father's death in Uelzen, they continued to draw a small income from what he had left them until the dreadful day when three fifths of the town were destroyed by fire on 20 September 1646. A drunken man who was attempting to dry flax over an open fire caused the catastrophe. Heinrich Varenius' successor, Magister Heino Cregel (1594–1663) also suffered, and with his whole house lost also his cherished theological library.[11] Maybe this was the very house (with a part of Varenius' library) where the Varenius family had lived. Incidentally, Cregel later became August's father-in-law. Perhaps not unexpectedly, because in all the generations of this family we can observe how the network was tightened by a carefully contrived marriage policy.

---

[8] See Steiger's paper below.

[9] Farewell sermon with „Valediction", 58.

[10] Thus Rohrbach in unpublished typescript on Heinrich Varenius in the Eutiner Landesbibliothek, p. 14.—Funeral sermon by Johannes Saubert about Ernst Cregel (Altdorf 1674), the son of Varenius' successor and brother-in-law of August Varenius. Here we learn that his father Heino Cregel was responsible for about twenty parishes in the vicinity.

[11] Funeral sermon on Ernst Cregel, 27f.

The Varenius Brothers

In the fifth generation of the family, I should like to sketch August's life before that of his younger and (to us) so much more interesting brother. For August's biography the sermon preached at his funeral is a mine of information which allows us to trace the main stages of his life in details that are sadly absent in the case of Bernhard.

August went to school in Uelzen, Hannover and Braunschweig where he learnt Hebrew, and in 1634, at fourteen, he was matriculated at Helmstedt University which he had to leave in 1637 because the war made it impossible for him to stay. As there was no home to return to, he made his way to Hamburg and ran into a kind customs officer from Hitzacker who looked after him until his death, after which August returned to Hamburg. This time he found employment with a Dutch merchant and could thus continue his education and take part in disputations. In January 1640, in his twentieth year, he delivered a disputation against the Jews in the Hebrew language, to the admiration of his learned listeners.[12] From Hamburg he moved on to Königsberg in 1640 and was fortunate enough to be noticed by the Royal Polish delegate who appreciated his linguistic competence, particularly in Latin, and, as his secretary, August made himself useful when travelling with him to the Diet at Warsaw and to other places in Poland. From Königsberg, he soon moved on again and ended up in Rostock.

At last the family clan functioned in his favour, in the person of mayor Klinge,[13] who was the grandchild of Chytraeus' sister-in-law, in other words, his deceased mother's second cousin. To live in the house of a mayor and senator of a Hanseatic town, himself the son of a professor of law and logic at Rostock, was an education in itself. Klinge had been at the Diet in Regensburg and was versed in the political affairs of the municipal and the wider world; for a while he took on the responsibility for his young cousin. Then August was apparently passed on to Schnobel and Lütkemann, who both taught at the University. Joachim Lütkemann, who became professor of physics and metaphysics in 1641, left Rostock in 1649 when Duke August, meanwhile in Wolfenbüttel, appointed him to take office as the highest ranking theologian in his principality. As August Varenius worked hard, he managed to make his

[12] Schomerus 51.
[13] Schomerus 53.

way in Rostock and directed all his energies to passing examinations in order to be eligible for a secure position. He duly received his M.A., his licentiate with the venia legendi in 1648 and his doctorate in 1650. As early as 1643, Duke Adolf Friedrich of Mecklenburg made him professor of Hebrew on no other qualification than an M.A. Further degrees were often acquired a long time after the appointment to a professorship, and August Varenius is no exception.

August Varenius was obviously proud of his earlier disputation of 28 March 1648. The young professor of the Hebrew language at Rostock lost no time in sending it to Duke August on 8 April 1648 with a long covering letter in Latin,[14] reminding him of the time when the young Duke was rector of Rostock University in 1594. Varenius dedicated his work to Duke August in memory of his maternal grandfather Johannes Freder, his father and his own childhood; this explicit reference to his childhood in Hitzacker may be an allusion not only to the Duke in his role as godfather but perhaps he had received favours in the past. Strangely enough, we have no evidence that Duke August played a supporting role in the lives of the brothers Varenius even though the Duke was the most powerful prince of the Welf dynasty after 1635 when he left Hitzacker for Braunschweig and Wolfenbüttel.

August Varenius established himself in the town of his grandparents and stayed there for the rest of his life. He soon held other important posts and declined calls to Lund and Greifswald. August was graced with living to old age and being mourned by his family, his children and grandchildren; when he was taken to his grave in 1684, innumerable mourners took part in a ceremonial act of the university and the town. August, who was highly esteemed in his day, has been forgotten. In the discourse of Protestant theology he has not found his way into the standard works of reference today, because he never was a leading figure in the period of confessionalisation in the seventeenth century, only a participant.

How different it was with Bernhard. He never had a home of his own, and he died young in 1650, in obscurity. Bernhard never enjoyed a secure position and could not reap the fruits of what he set out to achieve, yet his contributions have won him the lively interest of today's scholars investigating the history of science. Parallel lives? Yes, at first. As in the case of August, Bernhard enjoyed a good education. Like his

[14] Herzog August Bibliothek Wolfenbüttel, 56 Extravagantes f. 284r.

brother he went to school in Uelzen and was fourteen when he was matriculated at Helmstedt.[15] As he had just lost his father, it may have been a consolation for him to be with his brother for a time. Incidentally, both brothers belonged to the minority who entered a university before they were fifteen (eight out of 321) when the average age for beginners was 18.9 years in the period between 1618 and 1648.[16] Four years later, on 1 April 1640, Bernhard became a student at the Hamburg Academic School, whose director Joachim Jungius became his most important teacher and mentor. It was Jungius, who perfected his skill in disputation, because under his supervision and following his concept of refuting Aristotle's principles and definitions of natural philosophy, Varenius had his first disputation published in 1642 in which he analysed and repudiated Aristotle's definition of *motus*. Bernhard, one of the best students Jungius had ever taught, left in 1643 and continued writing afterwards. This correspondence is one of the few significant sources we have of Bernhard's thoughts beyond the factual information of archival listings and, of course, the evidence of his publications. Bernhard's next stop was Königsberg where August had been three years earlier, but Bernhard did not much like it, because mathematics was not taught and only medicine turned out to be somewhat useful to him; besides, it was too expensive, and so he decided to move on, but not to Hamburg or Rostock. He travelled to Leiden in 1645, and at this point the brothers parted for good. No more parallel lives, and certainly on Bernhard's part nothing that was modelled on his older brother. When he chose to study in Holland, he did so because intellectual pursuits, the family legacy, were given first priority; but as it turned out, without the familial network there was no way for him to support himself in Holland.

Theological Concerns

Although Varenius was born into a family of theologians and renowned academics, he departed from this tradition. Yet this is not to say that even where he tackled scientific problems in his chosen subjects,

---

[15] For his education in Uelzen, Helmstedt and Hamburg see Lehmann below.

[16] This study by Forschungsstelle für Personalschriften, Marburg University, was based on funeral sermons.

especially in geography, he had, despite an increasing secularisation of learning, alienated himself from theology. If there was a bipolarity there, it certainly was not a conflict. This is evident in his continual engagement with religions and religious differences.

When working on Japan, Varenius had collected so much material on the religion of the Japanese that he planned a separate edition for it to be published a few weeks after the *Descriptio regni Japoniae*, but soon his *Tractatus in quo agitur de Iaponiorum religione* was bound with the *Descriptio*. The treatise contains 11 chapters of different length on 215 pages. Varenius begins by listing his authorities (as he had done in the *Descriptio*) and notes immediately that reliable sources for Japan are not easy to find and that they are often contradictory, and to this he will draw attention. Here as elsewhere he organizes his presentation in three parts: the first part of 80 pages dealing with the "false religion" of the Japanese (chs. 1–4) is about the priests, the temples (especially those in Nara), holiday feasts and the cult of the dead. The next 78 pages are about the introduction of Christianity by the Jesuits to the Japanese islands (chs. 5–10) and include a digression (in ch. 6) on the mission work of the Portuguese in Asia, the various forms of resistance to Christianisation, but also its success with Japanese Christians. The third and concluding part tells us in a long and gruesome chapter (ch. 11) of 57 pages how in the 1620s Christians were most cruelly persecuted, tortured, beheaded and burnt, their creed was being forbidden.[17] Nowhere else, Varenius says, did they have to suffer as much. But he takes the long view of mission work and wants to make his treatise available for better times when missionaries will be able to go back and begin again.

Many years later one section of the book following the *Tractatus* had an interesting afterlife. The half-title reads, "Brevis informatio de diversis gentium religionibus". Its publication history is somewhat confusing and not easily disentangled, but to cut a long story short: It was rediscovered by the Nuremberg publisher Johann Andreas Endter (1625–1670),[18] who published it in 1668 in a German translation as part of a compilation of three texts, each by a different author, of 1152 pages in all (see fig. 5). The first book was written by Alexander Ross, a

---

[17] For the historical facts see Reichert below.

[18] Funeral sermon in *Archiv für die Geschichte des Buchwesens* 10 (1970), cols. 1399–1414. For the Endter family see Severin Corsten et al., eds., *Lexikon des gesamten Buchwesens*, vol. 2, Second ed. Stuttgart 1989; and Josef Benzing, *Die Buchdrucker des 16. und 17. Jahrhunderts im deutschen Sprachgebiet*, Wiesbaden 1982.

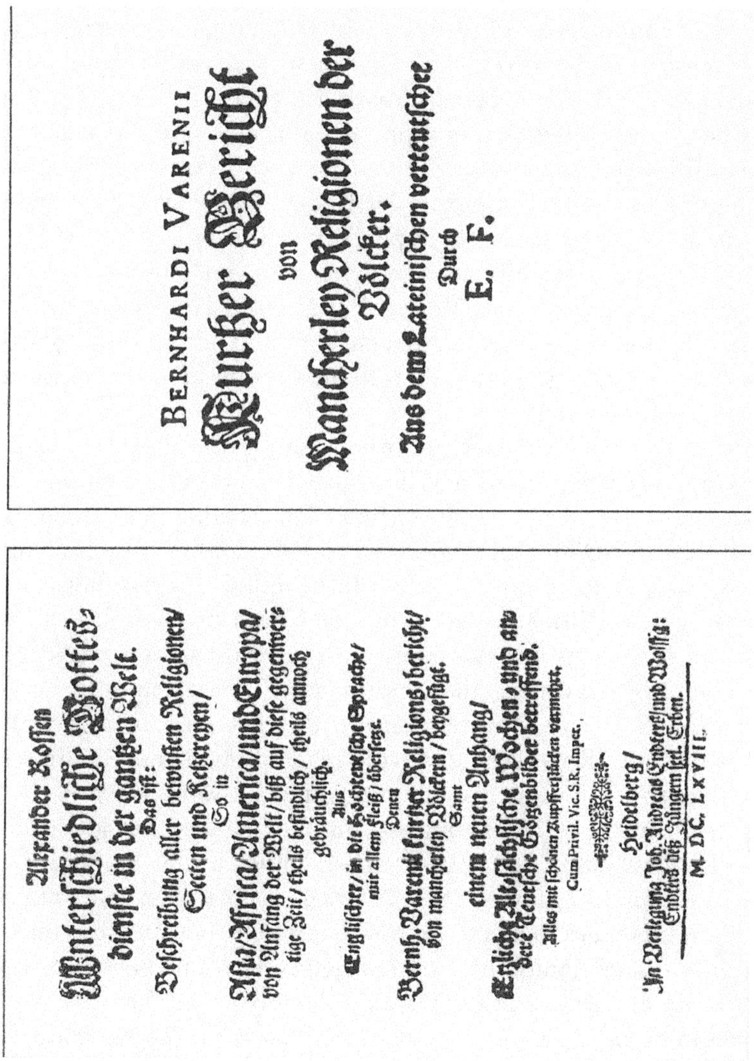
Alexander Rossen
Unterschiedliche Gottes-
dienste in der gantzen Welt.
Das ist:
Beschreibung aller bewusten Religionen/
Secten und Ketzereyen/
So in
Asia/Africa/America/und Europa/
von Anfang der Welt/biß auf diese gegenwär-
tige Zeit/theils befindlich/theils annoch
gebräuchlich.
Aus
Englischer/in die Hochteutsche Sprache/
mit allem Fleiß/übersetzt.
Denen
Bernh. Varenii kurtzer Religions-bericht/
von mancherley Völckern/beygefügt.
Samt
einem neuen Anhang/
Etliche Alt-sächsische Wochen- und an-
dere Teutsche Götzenbilder betreffend.
Alles mit schönen Kupfferstücken vermehrt.
Cum Privil. Vic. S.R. Imper.
Heidelberg/
In Verlegung Joh. Andreas Endters/und Wolffs:
Endters deß Jüngern sel. Erben.
M. DC. LXVIII.

Bernhardi Varenii
Kurtzer Bericht
von
Mancherley Religionen der
Völcker.
Aus dem Lateinischen verteutschet
Durch
E. F.

Fig. 5. Varenius' "Kurtzer Bericht" within *Unterschiedliche Gottesdienste* by Ross, 1668

Scottish theologian. His *Pansebeia: Or a View of all Religions in the World*, originally published in 1653, had become an international bestseller, and Endter helped himself to an existing translation into German by Albert Reimar of Lübeck, which Joachim Nosche had published and printed in Amsterdam the year before, in 1667, without giving the translator's name. Endter commissioned Erasmus Finx (1627–1094) for the translation of Varenius' *Brevis informatio* into German and added annotating comments to the text. Finx called himself Erasmus Francisci after his father's first name. Francisci, a non-practicing lawyer and the son of a distinguished lawyer, was one of the first men who supported himself by writing. He was often employed by Endter. Since such work was beneath him, the translator only appears with his initials E.F.[19] Only Christoph Arnold's appendix on heretics has no credentials to hide. The Nuremberg publisher had the book printed in Heidelberg rather than at home, which was not his usual custom, and all this deviousness may point less to outright piracy than to an attempt to avoid religious censorship in Nuremberg.

Alexander Ross (1591–1654), the main author, had been chaplain to King Charles I before the English Civil War, and in his writings he inveighed against the Copernican system of astronomy and violently disagreed with Sir Thomas Browne, Thomas Hobbes and William Harvey, which fails to recommend him to us today. (He did, however, provide the first English translation of the Koran in 1649, based on a French version.) Ross and Varenius must have turned over in their graves to be thus thrown together and forced between the same book covers by Endter. There is more than just the difference of a generation between them. Ross turns to a wide reading public and addresses them in English in the form of questions and answers. He preaches and edifies, instructs or lashes out, for instance against the Jews in a vile manner. Ross writes about the many religions in the world, as he explains in the preface, not to make his readers love them but rather because he wants them to recognize how detestable they are, thus making the light of Christianity shine more brightly. He quite believes what he finds in his sources about griffins, the phoenix or witches.

Varenius in his *Brevis Informatio* (pp. 225–308) takes a different approach in using the Latin language of scholars and sceptically examining his

[19] See *Biographisch-Bibliographisches Kirchenlexikon*, ed. Friedrich Wilhelm Bautz as well as the early biographies in *Deutsche Biographie* on microfiche.

sources. As in his other works he begins with a systematic part and classifies the religions of the contemporary world; he then turns to the different religions in Europe and the other continents, but deliberately does not use the customary division into believers in God and atheists, because that would present problems of where to place the heathens. When he writes about the heathens, he concentrates on the religions of the present, not of the past (as Vossius and Ross had done so extensively). Varenius does not accept as true that people everywhere believe in some divinity (here he differs from Vossius), and by drawing upon the reports of travellers he relies on "experience" for support rather than "reason", quite in the empirical spirit. As to Jews, Christians and 'Mahometans,' they are classified as revealed religions; of this group, he claims, only the Christians are divided among themselves. Quite in the spirit of his father, he mentions the "infamous" Anabaptists and Socianists along with other less known sects (paragraph 7, p. 229). But otherwise his report is restrained and not as adorned with extra epithets of value judgement as in Francisci's additions to the actual translation (he refers to heathens as "schädliches Unkraut" or "Greuel" and speaks of "Irrglaube" where Varenius simply says "Mahometanorum Religio", 259). Among Varenius' acknowledged sources are Maffeus (Giovan Pietro Maffei) on the Abessinians with his book *Historia Indiae Orientalis* (252), Johann Heinrich Alstedt about Mahometans (264), the Jesuit Nicolaus Trigautius (Trigault) on China (266), the Portuguese Petrus (Pedro) Texeira on India (290), the head of the Dutch trading settlement in India Johannes à Twist (294), and several Dutchmen who had just returned from Angola, as Varenius was writing this (307). Varenius cites Leo Africanus for Africa, but leaves out Japan and Siam, since their religions had been discussed separately.

Although Ross's book was written in the vernacular for a general readership and reprinted so many times and in different languages, this is not true of Varenius' booklet, because *Varenii kurtzer Religionsbericht von mancherley Völkern* was only attached to Endter's version of 1668 and its reprint of 1674 and so did not have the wide distribution of Ross's work that Gottfried Lange assumed.[20]

Varenius' *Brevis informatio* takes us back to our initial considerations, Varenius' coming of age within a theological discourse of conflicting

---

[20] Gottfried Lange, „Das Werk des Varenius: Eine kritische Gesamtbibliographie," *Erdkunde* 15 (1961): 9, 15–18.

claims of orthodoxy, which he absorbed from his family and witnessed everywhere he went. This immense interest of the period in religion had led to comparisons of religions, past and present, which included distant parts of the world. Such studies began to flourish. The Humanists had long been interested in the religions of ancient Egypt, Babylon, Persia, Greece, Rome and Italy; the study of the Oriental languages of the Bible also promoted the ambitious production of polyglot bibles. Meanwhile new information had come from travellers who reported from China, India and the American Indians. Many authors tried therefore to answer the demand for a taxonomy of knowledge,[21] and Varenius, the son of a theologian, rose to the occasion. He undoubtedly also pursued this interest with the earlier achievement of Vossius in mind, for in 1641 Vossius' *De theologia gentili et physiologica Christiana* had become available in two volumes and was afterwards reprinted several times;[22] later Newton studied Vossius' work intensively and added his annotations to it.[23] Vossius listed and classified polytheistic religions and argued that the heathen deities were really derived from biblical prototypes and that the original revelation was lost through misunderstandings and depreciation. He wanted to reveal the errors of the heathen religions. His huge compendium of mythologies was also meant to show that behind any religion one can find a kind of "natural religion", only knowable through reason.[24] Ross expressed similar opinions, but Varenius does not comment on the historical origins of other religions. He surveys the world religions of his day.

Varenius and the seventeenth-century writers on these subjects integrated non-Christian religions into the mythological history of Jews and Christians as participating in this one history of mankind or at least anticipating it. In Varenius' time it was not yet possible to appreciate other religions as having an independent history of their own, for any view other than Eurocentric was still outside this framework. Just as the European culture was aligned with Christianity, so it was assumed

[21] Richard H. Popkin, "The Crisis of Polytheism and the Answers of Vossius, Cudworth, and Newton," *Essays on the Context, Nature, and Influence of Isaac Newton's Theology*, eds. James E. Force and Richard H. Popkin, Dordrecht 1990, 9 .

[22] C.S.M. Rademaker, *Life and Work of Gerardus Joannes Vossius (1577–1649)*, Assen 1981, 368f.

[23] Popkin, "Polytheism, Deism, and Newton," *Essays on the Context, Nature, and Influence of Isaac Newton's Theology*, eds. James E. Force and Richard H. Popkin, Dordrecht 1990, 30.

[24] Popkin, "The Crisis", 11f; "Polytheism", 28f.

that other geographical parts of the world were equally associated with their respective religions. This simple analogy remained in currency for a long time.[25]

Certainly Varenius was not a man to be satisfied with the compiling and editing of what was known already as an end in itself. However much he thought knowledge important for its practical use (an aspect he never neglected), he yet wanted to enquire into its general meaning beyond the singular and individual. Trained in mathematics and logic, he was looking for a theoretical concept in geography which would systematise his material. In that way it was intended to become a useful textbook.

Nature's book (like the Bible) continued to be read as the work of the divine artificer, and this explains how different enquiries could cohere, where we would see separate disciplines. Varenius, quite within a line of tradition, attempted to apply the method of reducing its manifold phenomena to mathematical principles. (In Galilei's words, nature's book is written in the language of mathematics.) Mathematics in conjunction with science is absolute, not subject to time and historical processes of variability. Since Melanchthon the cosmological context of divine providence had remained undisputed. In his *Loci* of 1559 Melanchthon explained that God's providence was visible in nature's ordered and systematic formation; we experience it, for instance, in the change of the seasons and in the revolving stars, and this shows that God in his great care has ordered all this to be of use to man. Although Melanchthon did not accept the Copernican heliocentric view of the world, he promoted the publication of his hypotheses of motion. Melanchton's Aristotelianism in its synthesis of Christian-Aristotelian thought dominated the Varenius family's traditional universities, Rostock and Helmstedt, through Chytraeus, of course. Keckermann was its ardent advocate as well. In the course of the seventeenth century the idea develops that where God's divine agency and man's actions concur (allowing for contingency), we find *providentia generalis* in all creation, *providentia specialis* in all living beings (and in believers, in particular, we see *providentia specialissima*). According to the Lutheran view, God's first creational act and secondary causes are ever present in simultaneous

[25] See Bernhard Maier, "Religionsgeschichte" (580f.) and Gregor Ahn, „Religion I" (514) in *Theologische Real-Enzyklopädie* 28, Berlin 1997.

influx.[26] Varenius was captivated by his generation's fascination with mathematics, and his curiosity was equally roused by the 'special' phenomena accessible through experience. He was a natural philosopher firmly rooted in Christian theology.

[26] Deuser 309f.—See other articles in *Theologische Realenzyklopädie,* esp. „Naturwissenschaft I" (E. Wölfel, 24: 189–192), „Welt/Weltanschauung/Weltbild IV/4" (W. Sparn, 35: 595–598).

HEINRICH VARENIUS' *RETTUNG* VON JOHANN ARNDTS *WAHREM CHRISTENTUM*

Johann Anselm Steiger

Abstract

Heinrich Varenius' Rettung *[Defence] in Vindication of Johann Arndt's* Wahres Christentum *[True Christianity]*

Johann Arndt's *Vier Bücher von wahrem Christentum* (1605/1610) (*BWC*) [Four Books of True Christianity] are among the most successful and repeatedly published devotional writings of the post-reformation Lutheran period. The *BWC* were translated into Latin and several European languages. It was deservedly a bestseller, which influenced religious life, and devotional poetry, not only in German speaking areas but in various parts of Europe (among them Scandinavia, the Baltic countries and Russia). The *BWC* exercised a multi-faceted influence inside the Lutheran church and beyond, particularly in separatist Pietism and in the context of mystic spiritualism. Arndt's *BWC* had both an ecclesiastical-orthodox and a heterodox impact. Soon after its publication, opinion was divided as to whether Arndt's position placed him inside or outside Lutheran orthodoxy, and during his life time there were diverse disputes over how to interpret the *BWC* adequately.

The most extensive literary attack against Arndt's *BWC* was written by the Tübingen professor of theology, Lukas Osiander the Younger (1571–1638) and was entitled *Theologisches Bedencken*...(1623). Osiander exposes the *BWC* as a concoction composed mainly of false doctrine of a spiritualistic-enthusiastic kind, beside some orthodox teachings. Only a year after Osiander's philippic an extensive defence of Arndt was published: *Christliche / Schrifftmässige / wolgegründete Rettunge der Vier Bücher vom wahren Christenthumb*...(1624). The author was Bernhard Varenius' father, Heinrich Varenius (1595–1635), who was the Hitzacker chaplain of Duke August of Braunschweig-Lüneburg, and his work is not only the most extensive but also the most effective Arndt apology.

Varenius, who was familiar with the sources, points out quite correctly that the commonplace of Christ's spiritual birth in the heart of the believer, which Osiander censured, also plays an important role with Luther and Lutheran-orthodox theologians. Thus Varenius can demonstrate that Osiander's criticism of Arndt is much too sweeping and lacks the necessary critical interpretation. At the same time, however, it becomes clear that in his defence of Arndt Varenius does not always differentiate clearly either, especially when we look closely at the sources he used to prove Arndt's orthodoxy. From his precipitous discovery of identical commonplaces used by both Arndt and orthodox Lutherans,

Varenius deduces Arndt's supposedly unshakable Lutheran orthodoxy. This is not always convincing, as a closer comparative analysis of the Lutheran texts that Varenius stated were identical does not confirm his claims. Nevertheless, Varenius rightly points out how the *BWC* should be taken *ad meliorem partem*: he reads Arndt through the perspective of prominent orthodox theologians and contributes a great amount of different material on orthodox Arndtianism.

In the controversy between Osiander and Varenius, it was Varenius who was the victor, because the Stuttgart consistorium criticised Osiander for not having submitted his book for censorship before publication, and when Osiander wanted to reply to Varenius in print, the consistorium refused permission.

Despite all their inadequacies, Osiander and Varenius' books make two things clear: after having read Osiander's *Bedencken*, nobody can deny that Arndt's *BWC* are not only open to a spiritualistic interpretation but that to a great extent such a reading even suggests itself. On the other hand, whoever has read Varenius will have thoroughly learned to appreciate the intricate difficulty of granting Arndt an *interpretatio orthodoxa*. In their very different approaches both Osiander and Varenius were right, as we can see from the way Arndt's works were discussed in the 17th, 18th and 19th centuries, and we find this confirmed in the research on Arndt today.

Johann Arndts *Vier Bücher von wahrem Christentum* (1605[1]/1610[2]) (*BWC*) gehören zu den erfolgreichsten und auflagenstärksten Erbauungsschriften

---

[1] Hinweis: Das Literaturverzeichnis zu dieser Studie ist am Ende dieses Aufsatzes angeführt und nicht in der Bibliographie am Ende des Buches enthalten.—Vom Erstdruck des ersten Buches der *BWC*, das bei Jonas Rosa und Nikolaus Hoffmann d. Ä. erschien, sind nach dem derzeitigen Kenntnisstand nur zwei Exemplare überliefert. Johann Arndt: *Von wahrem Christenthumb/ heilsamer Busse/ wahrem Glauben/ heyligem Leben vnd Wandel der rechten wahren Christen. Das erste Buch* [...]. Frankfurt a. M. 1605 (Herzog August Bibliothek [fortan: HAB] Wolfenbüttel Th 82 und Yv 1111.8° Helmst. [Titelblatt fehlt im zweiten Exemplar]). Vgl. zum erstgenannten Druck die von der Arndt-Forschung lange Zeit vergessene bzw. übersehene (vgl. Blaufuß. 216f und 6, 378) Notiz von Milchsack. Vgl. jetzt die Neuedition Arndt 2005.

[2] In diesem Jahr erschien die erste komplette Ausgabe der vier Bücher bei Johann Francke in Magdeburg. Mir lag eine Mischauflage vor: Johann Arndt: *Vier Bücher Von wahrem Christenthumb/ Heilsamer Busse/ Hertzlicher Rewe vnd Leid vber die Sünde vnd wahrem Glauben: auch heiligem Leben vnd Wandel der rechten wahren Christen. Derer Jnhalt nach dem Titel zu finden. Das Erste Buch Jetzo auffs newe vbersehen vnd gebessert* [...]. Magdeburg 1610. *Das Ander Buch Vom wahren Christenthumb/ Wie Christi Menschwerdung/ Liebe/ Demuht/ Sanfftmuht/ Gedult/ Leyden Sterben/ Creutz/ Schmach vnd Todt/ vnser Artzney vnd Heylbrunnen/ Spiegel/ Regel vnd Buch vnsers Lebens sey* [...]. Magdeburg 1615. *Das Dritte Buch vom wahren Christenthumb/ vom Jnwendigen Menschen. Wie Gott den höchsten Schatz: sein Reich in des Menschen Hertz geleget hat/ als einen verborgenen Schatz im Acker: Als ein Göttliches Liecht der Seelen/ vnd wie dasselbe in vns zu erwecken vnd zu suchen [...].* Magdeburg 1614. *Das Vierdte Buch vom wahren Christenthumb/ Liber Naturae. Wie das grosse Weltbuch der Natur/ nach Christlicher Außlegung/ von GOtt zeuget/ vnd zu Gott führet/ wie auch alle Menschen Gott zu lieben/ durch die Creaturen gereitzet/ vnd durch jhr eigen Hertz vberzeuget werden* [...]. Magdeburg 1615 (HAB Wolfenbüttel 955–956 Theol. [1]).

des nachreformatorischen Luthertums. Auch die Tatsache, dass „bis zum Beginn des 19. Jahrhunderts [...] über 200 Drucke (scil. der *BWC*) nachweisbar“[3] und die *BWC* ins Lateinische und in mehrere europäische Sprachen sowie ins Varugische und (teilweise) ins Tamulische[4] übersetzt worden sind, zeigt, wie sehr dieses Werk verdient, ein 'Bestseller'[5] genannt zu werden, von dem das Frömmigkeitsleben nicht nur im deutschsprachigen Raum, sondern in verschiedenen Teilen Europas wichtige Anstöße erhalten hat—u.a. in Skandinavien,[6] im Baltikum (Jarlert), in Russland (Tschizewskij) und Ungarn (Nagy, Zoltán). Auch die geistliche Dichtung haben die *BWC* beeinflusst.[7] Was die Auflagenhöhe und die internationale Wirkung betrifft, können den *BWC* allenfalls noch Johann Gerhards *Meditationes Sacrae* an die Seite gestellt werden.[8] Die tatsächliche Breitenwirkung der *BWC* indes, die z.B. durch Visitationsakten[9] und Nachlassverzeichnisse dokumentiert wird (Fritz), ist nicht zuletzt darum nicht abschließend bewertbar, da eine vollständige Arndt-Bibliographie fehlt—sie ist ein dringliches Desiderat[10] und müsste sämtliche Neuauflagen der Opera Arndts bis ins 20. Jahrhundert nachweisen. Die Erfolgsgeschichte der *BWC* lässt sich u.a. anhand der Vielzahl der mit emblematischen Darstellungen gezierten Ausgaben ausmachen, deren erste 1678/79 in Riga erschien und damit eine lange Reihe von Neuauflagen initiierte, die erst im Jahre 1930 zum Abschluss kam.[11]

Die *BWC* zeitigten eine höchst facettenreiche Wirkung sowohl innerhalb der lutherischen Kirche als auch außerhalb derselben, näherhin im separatistischen Pietismus und im Kontext des mystischen Spiritualismus. Arndts *BWC* haben sowohl eine kirchlich-orthodoxe als auch eine heterodoxe Wirkungsgeschichte. Schon früh war man geteilter Meinung darüber, ob Arndt innerhalb der lutherischen Orthodoxie zu verorten

[3] Schneider 1995: 89.

[4] Bach-Nielsen 17.

[5] Wallmann 1984 in: Breuer 53: „Ich wüßte außer Bibel, Gesangbuch und Katechismus kein Werk, für das man im protestantischen Deutschland des 17. Jahrhunderts eine ähnliche Zahl von Druckorten nennen könnte“.

[6] Vgl. Tiililä, Pleijel, Jarlert.

[7] Krummacher 1976 u. 1964, van Ingen.

[8] Vgl. *Bibliographia Gerhardina*.

[9] Vgl. zur bislang zu wenig bearbeiteten Frage, welche Bücher Gegenstand des Studiums und der Lektüre von Pfarrern gewesen sind, die instruktive Untersuchung von Ernst Koch. Bei den von ihm untersuchten Quellen rangiert Arndt allerdings 'nur' an dritter Stelle hinter Johann Gerhard und Christian Scriver (290f).

[10] So schon Wallmann 1980: 18.

[11] Dazu Greschat, Müller-Mees, Peil 1977 u. 1978, Schilling.

sei oder außerhalb derselben. Die unterschiedlichen Richtungen innerhalb der heutigen Arndt-Forschung spiegeln letztlich diese historische Kontroverse. Schon zu Lebzeiten Arndts hat es vielfältige Streitigkeiten darüber gegeben, wie die *BWC* angemessen zu interpretieren seien—so etwa in Braunschweig, dem damaligen Wirkungsort Arndts selbst, wie Walch im 3. Band seiner Studie berichtet (191–195), aber auch in Danzig, wo die 1619 ausgebrochene Arndt-Kontroverse (197–202) in die Debatten um die theologische Position von Hermann Rahtmann (1585–1628) mündete.[12] Nicht nur Arndt selbst war darum bemüht, die Rechtgläubigkeit seiner *BWC* unter Beweis zu stellen oder zumindest zu beteuern. Vielmehr wurden auch andere literarisch-apologetisch für ihn tätig, so z.B. der Danziger Pfarrer Daniel Dilger (1572–1643), der im Jahre 1620 eine ausführliche Verteidigungsschrift für Arndt (gegen Johann Corvinus) publizierte.[13]

Die wohl umfangreichste literarische Attacke gegen Arndts *BWC* stammt aus der Feder des Tübinger Theologie-Professors Lukas Osiander d. J. (1571–1638).[14] Schon der Titel dieser 1623 erstmals gedruckten und im Folgejahr wiederaufgelegten Schrift lässt die Stoßrichtung erahnen: *Theologisches Bedencken/ Vnd Christliche Treuhertzige Erinnerung/ welcher Gestalt Johann Arndten genandtes Wahres Christenthumb/ nach Anleitung deß H. Worts Gottes/ vnd der reinen Evangelischen Lehr vnd Bekandtnussen/ anzusehen und zuachten seye. Allen Gottseligen Christen/ denen jetzgemeldtes Christenthumb zur Hand kommen/ zu notwendiger Nachrichtung* [...].[15] Kaum ein gutes Haar lässt Osiander an Arndt und enttarnt die *BWC* als ein Machwerk, das neben einigen orthodoxen Lehren hauptsächlich spiritualistisch-enthusiastischen Irrglauben enthalte.

---

[12] Vgl. Halverscheid, Steiger 1998 u. 2004.

[13] Daniel Dilger: *Des Ehrwürdigen/ Achtbaren vnd Hochgelarten Herrn Johannis Arndes, im Löblichen Fürstenthumb Lüneburg General Superintendenten, Richtige/ vnd in Gottes Wort wolgegründete Lehre/ in den vier Büchern vom wahren Christenthumb: Jn etlichen Puncten auß dringenden/ Nothwendigen vrsachen/ die in der Vorrede angezogen/ repetiret vnd wiederholet* [...]. Stettin 1620 (HAB Wolfenbüttel 1152 Theol. [5]). Zu Dilgers Biographie vgl. *ADB* 5, 223 und *DBA* 238, 432–435.

[14] Zu Osiander s. Wallmann 2003.

[15] Bayerische Staatsbibliothek [fortan: BSB] München Polem. 2024. Im folgenden wird nach der zweiten Auflage Tübingen 1624 zitiert: *Theologisches Bedencken/ Vnd Christliche Treuhertzige Erinnerung/ welcher Gestalt Johann Arndten genandtes Wahres Christenthumb/ nach Anleitung deß H. Worts Gottes/ vnd der reinen Evangelischen Lehr vnd Bekandtnüssen/ anzusehen und zuachten seye. Allen Gottseligen Christen/ denen jetzgemeldtes Christenthumb zur Hand kommen/ zu notwendiger Nachrichtung* [...], Tübingen 1624 (HAB Wolfenbüttel Alv. Ac 551 [1]). Ein weiterer Druck dieser Schrift erschien 1624 in Stettin (Universitäts- und Landesbibliothek Halle/S. AB 38 2/i, 17).

Osianders Hauptsorge war es, „daß sich der Spiritualismus eines Paracelsus, Weigel oder Schwenckfeld weiter verbreiten könnte".[16] Nicht lange auf sich warten ließ indes eine noch umfänglichere Schutzschrift für Arndt, die schon ein Jahr nach der Veröffentlichung von Osianders Philippica die Druckerpresse verließ. Sie war abgefasst von Heinrich Varenius (1595–1635),[17] der als Hofprediger in Hitzacker in den Diensten Herzog Augusts zu Braunschweig-Lüneburg stand,[18] und trägt den Titel *Christliche/ Schrifftmässige/ wolgegründete Rettunge der Vier Bücher vom wahren Christenthumb/ des seligen/ vmb die Kirche Christi wolverdienten recht Lutherischen Evangelischen Theologi H. Johannis Arndten* [...] (siehe Fig./Abb. 6).[19] Diese Schrift darf als die nicht nur umfangreichste, sondern wohl auch eingehendste Arndt-Apologie betrachtet werden und wurde 1624 bei den Gebrüdern Stern in Lüneburg in Verlag genommen. Die Publikation dieser Schrift war keine leichte Geburt, da der Rat der Stadt Lüneburg eine solche verboten hatte. Es ist der Intervention Herzog Christians zu verdanken, dass Varenius' Werk dennoch das Tageslicht erblickte.[20] Dass die Sterne, die ohnehin eine große

---

[16] Sommer 1999: 248.

[17] Vgl. *DBA* 1: 1301, 293.

[18] 1628 verließ Varenius Hitzacker und trat das Amt des Propsten und Superintendenten in Uelzen an. Vgl. Heinrich Varenius: *Eine Antrits=Predigte Von Dem Ampte aller getrewen Lehrer vnnd fleissigen Zuhörer/ voll schöner Lehr vnd Erinnerunge/ Aus Dem ordentlichen Evangelio Dominica Sexagesimae, Jn der Kirchen zu S. Marien in Vltzen gehalten* [...]. Lüneburg 1628 (HAB Wolfenbüttel Yj 50.12° Helmst. [1]).

[19] 2 Teile. Lüneburg 1624 (HAB Wolfenbüttel Alv. Ac 550). Im Jahre 1689 wurde das Werk nochmals in Lüneburg gedruckt (HAB Wolfenbüttel Yv 1202.8° Helmst.). Varenius' Schrift erfuhr eine Erwiderung durch den mecklenburgischen Hofprediger Georg Rost (1582–1629): Georg Rost, *Amica ac fraterna Admonitio Super Controversijs DE VERO DN. JOANNIS ARNDTEN generalis in Ducatu Luneburgico Superintendentis, p. m. Christianismo, inter D. D. Lucam Osiandrum & M. Henricum Varenium, Dn. Paulum Eggardum, aliosque Theologos & politicos ortis* [...]. Rostock 1626 (HAB Wolfenbüttel Yv 1113.8° Helmst. [1]).—Hierauf antwortete Heinrich Varenius mit: *BREVIS CONSIDERATIO ADMONITIONIS GEORGII ROSTII SUPER CONTROVERSIIS DE VERO B. ARNDI CHRISTIANISMO ORTIS. In qua B. ARNDI innocentia ejusdemque Christianismi orthodoxa veritas asseritur, & a φλυαρίαις & corruptelis G. Rostij vindicatur, quaestionesque motae dextrè & solidè explicantur* [...]. Lüneburg 1626 (HAB Wolfenbüttel Yv 1113.8° Helmst. [2]).—Rost reagierte nochmals mit folgender Schrift: Georg Rost, *EXAMEN BREVIS CONSIDERATIONIS VARENIANAE APOLOGETICUM, Seu Justa ac modesta defensio amicae ac fraternae Admonitionis super controversijs de vero ARNDI p. m. Christianismo ortis, IN QVA ARNDIANI CHRIstianismi orthodoxia approbatur, heterodoxia fraternè emendatur, & quaestiones ac controversiae motae solidè ac nervosè discutiuntur, & à Φλυαρίαις, convitijs & corruptelis M. HENRICI VARENII vindicantur*. Rostock 1628 (HAB Wolfenbüttel Yv 1113.8° Helmst. [3]). Vgl. Walch 3/1, 207f.

[20] Wallmann (1980: 31 Anm. 79) zitiert aus dem Schreiben Herzog Christians an den Rat der Stadt Lüneburg vom 14.4.1624, laut Kopialbuch der Gebrüder Stern, das

Christliche/
Schrifftmässige/wolgegründete
Rettunge der Vier Bücher vom wahren
Christenthumb/des seligen/ umb die Kirche Christi
wolverdienten recht Lutherischen Evangelischen Theologi
H. Johannis Arndten/weiland general Superintendenten der Kirchen im löblichen Fürstenthumb Lüneburg/ und Pfarrherrn
in Zell.
Verfertiget/und D. Lucæ Osiandri
Theologischem Bedencken entgegen
gesetzt/
Von
M. Henrico Varenio,
H. W. Pastoren in Hitzacker.
Erster Theil.
Gedruckt zu Lüneburg/ bey und in Vorlegunge Johann und Heinrich Sternen/
Buchführer daselbsten.
Anno M. DC. XXIV.

Ander Theil und Beschluß
Der Christlichen Rettunge der
Vier Bücher vom wahren Christenthumb/
Des sehligen und umb die Kirche Christi wol verdienten Theologi
H. Johannis Arndten/weiland general Superintendenten der Kirchen im löblichen Fürstenthumb Lüneburg/ und Pfarrherrn
in Zell.
Verfertiget/und D. Lucæ Osiandri
Theologischem Bedencken entgegen
gesetzt/
Von
M. Henrico Varenio,
Pastorn und Prediger der reinen Evangelischen Warheit in Hitzacker.
Gedruckt zu Lüneburg/ bey und in Vorlegunge Johann und Heinrich Sternen/
Buchführer daselbsten.
Anno M. DC. XXIV.

Fig. 6. Heinrich Varenius' *Rettunge*, 1624

Anzahl von Arndtiana im Verlagsprogramm hatten, mitunter ein vitales wirtschaftliches Interesse an der Rettung ihres Hausautoren gehabt haben werden, wird sich nicht von der Hand weisen lassen und dürfte auch der Grund dafür gewesen sein, dass bei den Sternen im selben Jahr auch des Nortorfer Pfarrers Paul Egard (ca. 1580–1655)[21] *Ehrenrettung Johannis Arndten*[22] und im Jahr darauf eine anonyme Schutzschrift für Arndt[23] gedruckt wurden. Letztere Schrift stammt von Melchior Breler (†1627), seines Zeichens Leibarzt Herzog Augusts zu Braunschweig und Lüneburg, der wie Varenius in Hitzacker tätig war und bereits unmittelbar nach Arndts Tod mit Verteidigungsschriften für seinen Freund von sich reden gemacht hatte.[24] Somit stehen Varenius, Egard und Breler im Gefolge Johann Valentin Andreaes, der schon früher—zunächst mit seinem Christianismus genuinus (1615) als „Propagandist Arndts" hervorgetreten ist.[25] Auffällig ist, dass mit Varenius und Breler zwei Personen zur Verteidigung Arndts auf den Plan treten, die „in den unmittelbaren Umkreis Herzog Augusts gehören",[26] der zu Recht als

sich im Bestand des Niedersächsischen Staatsarchivs Wolfenbüttel befindet (VI Hs 11 Nr. 229): „...wir werden glaubhaft berichtet, daß Ihr nicht zugeben oder verstatten wollet, daß die Sterne Magistri Henrici Varenii Scriptum, darin Doctor Lucae Osiandri wider unseres gewesenen General Superintendenten Herrn Arndt (unleserlich) refutiret, drucken sollen... begehren demnach hiermit gnediglich, ihr wollet gedachten Stern an Druck und Verfertigung solcher refutation keine Verhinderung thun".

21 Vgl. Jöcher Teil 2, Sp. 285f und Sträter in *RGG* 4. Aufl.

22 Paul Egard, *Ehrenrettung Johannis Arndten/ Das ist/ Christliche vnd in Gottes Wort wolgegründete Erinnerung/ was von D. Lucae Osiandri, Theologiae Professoris zu Tübingen Vrtheil vnd Censur, uber Johan Arndten wahres Christenthumb/ sey zu halten. Allen Frommen vnd Gottfürchtigen zu nothwendiger Nachrichtung/ aus Liebe der Warheit beschrieben vnnd ans Liecht gegeben* [...]. Lüneburg 1624 (HAB Wolfenbüttel 857 Theol. [3]).

23 [Melchior Breler:] *Warhafftiger/ Glaubwirdiger vnd gründlicher Bericht von den vier Büchern vom Wahren Christenthumb Herrn Johannis Arndten/ auß den gefundenen brieflichen Vrkunden zusammen getragen. Darauß Sonnenklar zu beweisen ist/ daß gedachte Bücher vom Wahren Christenthumb mit der Papisten/ Calvinisten/ Schwenckfeldts/ Weigelij/ Enthusiasten/ vnd dergleichen Schwärmern Jrrthumen/ zur vngebür bezüchtiget vnd außgeruffen werden. Nebenst Herrn Johann Arndten kurtzen Bedencken vber V. Weigelij Dialogum de Christianismo. Entgegen gesatzt. Den hin vnd wieder außgesprengeten/ gedachter Bücher/ vnwarhafften Beschüldigungen/ falschen Auflagen vnnd Verdacht/ Jnsonderheit aber dem vntheologischen Bedencken D. Lucae Osiandri Professorn zu Tübingen. Vnd GOtt zu Ehren/ zu Offenbarung der Warheit/ zur Nachrichtung vieler tausend Christen an den tag gegeben [...].* Lüneburg 1625 (HAB Wolfenbüttel G 412.8° Helmst.). Die sog. Schutzbriefe wurden 1688 (HAB Wolfenbüttel Xb 1508 [4]) und 1706 in erweiterter Form unter dem Titel 'Apologetica Arndiana [...]' (Bibliothek der Franckeschen Stiftungen Halle/S. 45 I 7 [6]) neu aufgelegt. Vgl. Wallmann 1980: 19 Anm. 24.—Zu Breler vgl. Jöcher 1: 1360.

24 Walch 3/1: 202f.

25 Brecht 2002: 139.

26 Wallmann 1980: 24.

fürstlicher Schutzherr des Arndtschen Erbes und seiner Verteidiger bezeichnet worden ist.[27] Dieses „Patrocinium" für Arndt, den ehemaligen Generalsuperintendenten zu Celle, zu übernehmen, hatte Varenius das „gantze hochlöbliche Fürstliche Hauß Lüneburg" (1. fol. c2r) in seiner Vorrede zu seiner *Rettung* gebeten.

Unbestreitbar ist, dass sich Arndt in der Erforschung der Kirchen- und Theologiegeschichte sowie der Erbauungsliteratur einer auffälligen Beliebtheit erfreut.[28] Die Erarbeitung einer historisch-kritischen Gesamtausgabe bleibt ein dringendes Desiderat, wie schon Schneider[29] mit Verweis auf Wallmann feststellte. Mittlerweile liegt jedoch bereits eine kritische Edition der Urausgabe von Buch 1 der *BWC* vor sowie ein Reprint der frühesten Auflage aller vier Bücher (1610). Aber auch die von Lukas Osiander an Arndt formulierte Kritik und die Vareniussche Entgegnung sind bislang, so weit ich sehe, nur beiläufig zum Gegenstand der Erforschung gemacht worden.[30] In vorliegendem Beitrag kann diesbezüglich nur ein erster Schritt unternommen werden. Anhand ausgewählter Aspekte—nämlich der Thematiken der geistlichen Geburt Christi im Menschen, der mystischen Union des Glaubenden mit Gott und der Verhältnisbestimmung von äußerem und innerem Wort—soll der Frage nachgegangen werden, in welcher Weise, mit welchem Interesse und mit welchem Hauptaugenmerk Osiander und Varenius Arndts *BWC* gelesen und interpretiert haben.

Osiander ist es darum zu tun, nachzuweisen, dass Arndt im Zuge seiner Rezeption der Theologie Johann Taulers, Thomas von Kempens u.a. verglichen mit der Lutherschen Theologie derart viel fremdartiges Gut importiert hat, dass seine *BWC* nicht nur nicht den Anspruch erheben können, mit der Heiligen Schrift und den Bekenntnisschriften übereinzustimmen, sondern zudem als Produkt spiritualistisch-enthusiastischer Irrlehre bezeichnet werden müssen. „Wann du meinest/ man lehre dich das wahre (vnnd also vnmangelhaffte) Christenthumb/ so hast du ein Daulerthumb",[31] was in besonderem Maße—so Osiander (27)—für das dritte Buch gilt. In der Tat, und hierin liegt das unbestreitbare Verdienst des Tübinger Theologen, gelingt es Osiander, eine Fülle von Passagen und Textkomplexen namhaft zu machen, in denen

---

[27] Ebd. 30ff u. Brecht 1993: 184.

[28] Ausgewählte jüngere Forschungsliteratur zu Arndt ist im Anhang zu diesem Beitrag genannt.

[29] Schneider 1983: 205.

[30] So von Wallmann 1980 (24, 28–31), Brecht 1993 (147f, 184) und 2002 (27, 138) sowie in der älteren Forschungsliteratur von Koepp (110–118).

[31] Osiander 22.

Arndt sich von der lutherischen Theologie nicht nur graduell, sondern prinzipiell unterscheidet. Hierzu gehören, um nur ein Beispiel zu nennen, die in den *BWC* häufig wiederkehrenden Aussagen, die einen deutlichen Dualismus von Fleisch und Geist, von Äußerlichem und Innerlichem implizieren.

Nach Ansicht Luthers und der lutherisch-orthodoxen Theologie vermittelt sich der den Glauben stiftende Geist nicht anders als durch äußerliche Medien, mithin durch das verbum externum, d.h. durch das geschriebene und gepredigte Wort Gottes sowie durch die Sakramente. Die Matrix für dieses nicht zuletzt in hermeneutischer Hinsicht grundlegende Theologumenon liegt in der im Sinne Luthers interpretierten altkirchlichen Christologie, wonach sich Gott im Menschen Jesus offenbart, der vere Deus und vere homo zugleich ist. Ebenso bindet sich die dritte trinitarische Person, sich gleichsam in ihnen inkarnierend, an die äußeren media salutis. Bei Arndt dagegen finden sich mehrfach Stellen, an denen er das äußere Wort und den Geistempfang schroff gegeneinandersetzt, ja den biblischen Buchstaben zugunsten der Rede von einer direkten, unvermittelten Geistbegabung im Inneren des Menschen abwertet. Zu Recht weist Osiander (103f) auf einen Teiltext in *BWC* 3: Kap. 1, 6 hin, wo es heißt: „Denn für Gott gilt nichts eusserliches, sondern das innerliche/ nicht was in Buchstaben bestehet/ sondern was aus dem Geist gehet/ vnnd im Geist bestehet. Darumb ist [...] ein grosser vnterscheidt/ vnter einem Weltgelerten vnd Gottesgelerten [...] der Gelerte lernet von aussen auß dem Buchstaben/ der Heilige lernet aus GOtt: Jnwendig/ aus dem H. Geist". Es kann kaum bezweifelt werden, dass diese Formulierung ebenso gut bei Valentin Weigel stehen könnte[32] und sehr ähnliches in der Tat bei diesem Autor in *Der gülden Griff* (Kap. 16, 86ff) zu lesen ist. Gerade dort, wo Osiander Arndts Verhältnisbestimmung von Fleisch und Geist, Welt und Nachfolge Christi, Äußerlichem und Innerlichem thematisiert, gelingt es ihm, die heterodoxen Einflüsse bei Arndt und dessen Affinität zu spiritualistischen Autoren wie Weigel und Schwenckfeld namhaft zu machen.

Was Osiander dagegen nicht oder kaum gelten lässt, ist der unübersehbare Umstand, dass sich in den *BWC* sehr wohl auch solche Abschnitte finden lassen, in denen Arndt durchaus in einer mit Luther

[32] Vgl. zu Arndts theosophisch tingiertem Spiritualismus Geyer 1: 325–338. Bezüglich Arndts „Lehre von Wort und Geist" von einer spiritualistischen „Akzentverschiebung" zu sprechen, „die seiner eigenen reformatorischen Intention nicht entspricht" (Braw 1985: 157), halte ich für problematisch.

vergleichbaren Weise davon spricht, dass sich Gott um willen der Stiftung des Glaubens, und das heißt des Heils, äußerlicher Medien bedient. Dies macht Varenius in seiner Entgegnung (1: 131) zu Recht deutlich, wenngleich die Tendenz des Varenius erkennbar ist, Arndt orthodoxer erscheinen zu lassen als er in Wahrheit ist, während Osiander umgekehrt ein Interesse daran hat, Arndt von seinen spiritualistischen Aussagen her zu interpretieren und alles andere als nebensächliches Beiwerk erscheinen zu lassen. Weder Osiander noch Varenius—so viel wird hieran schon deutlich—gelingt es mithin, die ungeheure Spannung, die in den *BWC* zwischen den orthodoxen und den heterodoxen Teiltexten herrscht, auszuhalten, ohne sogleich bestrebt zu sein, diese in die eine oder in die andere Richtung hinein aufzulösen. Beiden fällt zudem nicht auf, dass Arndts Definition des Verhältnisses von äußerem und innerem Wort nicht stets dieselbe geblieben ist. Vergleicht man Arndts diesbezügliche Ausführungen in den *BWC* mit seinen späteren Schriften, so wird sichtbar, dass beim späteren Arndt neben der Akzentuierung des inneren Wortes eine kirchlich-lutherische Linie verstärkt hervortritt, der zufolge die Begegnung des Menschen mit dem verbum externum die unabdingbare Voraussetzung für den Glauben darstellt.[33]

Dass man es sowohl bei Osiander als auch bei Varenius mit zwei sehr unterschiedlichen, jedoch gleichermaßen höchst tendenziösen Arndt-Interpretationen zu tun hat, verrät auch die Art und Weise, wie die beiden Arndts Rede von der geistlichen Geburt Christi im Menschen beurteilen. Arndt spricht in Buch 1, Kap. 6, 56 davon, dass sich die leibliche Geburt Christi im Menschen in geistlicher Weise ereignen muss, indem er sagt: „Denn wie Christus ist durch den heiligen Geist im Glauben von Maria leiblich empfangen vnd geboren/ also mus er in mir geistlich empfangen vnd geborn werden/ Er muß in mir geistlich wachsen vnnd zunemen". Osiander nun hält Arndt vor, er allegorisiere unerlaubterweise den Artikel von der Menschwerdung Christi, verkürze damit die soteriologische Bedeutung der Inkarnation und verdecke den wahren Sinn des Evangeliums, das davon spricht, dass die Glaubenden in Christus neu geboren werden. Die Aussage aber, dass Christus im Herzen des Glaubenden geboren wird, sei unbiblisch. „Dann nirgend gesagt wird/ daß CHRistus in vns newgeboren/ sondern daß wir/ wir/ wir/ in Christo/ ein Newe Geburt vnnd newe Creatur müssen werden" (*Bedencken* 92). Osiander hält es damit bereits für eine ausgemachte

[33] Hamm 59f.

Sache, dass die inkriminierte Lehre Arndts—wie er meint—nicht schriftgemäß ist und auf Tauler sowie auf Weigel zurückgeht.

Varenius nun ist bestrebt, die Vorwürfe Osianders zu entkräften, und bietet hierzu eine bemerkenswert facettenreiche theologische Gelehrsamkeit auf. Varenius bringt nicht nur Belege dafür bei, dass auch Martin Luther von der geistlichen Geburt Christi im Inneren des Menschen spricht (*Rettunge* 1: 330f), vielmehr gelingt ihm zudem der Nachweis, dass auch solche lutherische Theologen wiederholt auf diesen Topos zu sprechen kommen, deren Rechtgläubigkeit beileibe nicht in Frage steht und auch von Osiander kaum in Zweifel gezogen worden sein dürfte. Neben Johann Gerhard bezieht sich Varenius auf David Chytraeus, Johannes Winckelmann, Aegidius Hunnius, Johannes Tarnow u.a. und reiht Zitat an Zitat. Am ausführlichsten kommt Johann Gerhard als Gewährsmann Arndts zu Wort, insbesondere dessen Weihnachtspredigt über Gal 4, 19 ('Meine lieben Kinder/ welche ich abermal mit engsten gebere/ bis das Christus in euch eine gestalt gewinne'). Den Osianderschen Vorwurf, Arndt allegorisiere den Artikel von der Menschwerdung in unerlaubter Weise, entkräftet Varenius nicht nur dadurch, dass er aufweist, dass auch der orthodoxe Gerhard ähnliches zu bieten hat, sondern auch mit Hilfe des ebenfalls Gerhard entnommenen Hinweises darauf, dass der Topos von der geistlichen Geburt Christi im Herzen insofern schriftgemäß ist, als Paulus in Eph 3, 17 von der inhabitatio Christi in corde spricht, was einen wie immer gearteten Einzug des Gottessohnes in das Herz doch voraussetze. Darüber hinaus spreche Paulus in Gal 2, 20 und 4, 19 davon, dass der Sohn Gottes im Menschen lebt bzw. geistlich Gestalt gewinnt (329, 336): „Jst/ lebet/ wohnet vnd wircket nun CHristus/ in einer gläubigen Seele/ so muß er auch in derselbigen seyn empfangen/ vnd Geistlich geboren/ denn was in einem Menschen nicht wird empfangen/ das kan auch in jhm nicht leben" (336). Zudem—so Varenius (322, 338)—könne die geistliche Interpretation der Geburtsgeschichte schon insofern nicht verfehlt sein, als sie in Analogie steht zu der Art und Weise, in der Paulus selbst in Röm 6 den Tod und die Auferstehung Christi geistlich deutet und auf die Taufe bezieht, womit Varenius einem von Johann Gerhard zu Beginn seiner Predigt geäußerten Gedanken[34] folgt. Völlig korrekt ist

[34] Gerhard, *Postilla*, Jena 1613, 3 Teile und Appendix (HAB Wolfenbüttel 419–420 Theol.), hier: 1: 159f: „GLeicher weise/ wie der Apostel Paulus Rom. 6. vnd Coloss. 3. lehret/ daß wir mit Christo geistlicher weise aufferstehen/ vnd gen Himmel fahren müssen/ soll vns anders Christi Aufferstehung vnd Himmelfarth zu gute kommen. Wir

ebenfalls Varenius' Hinweis darauf, dass wie Arndt auch Gerhard die leibliche Geburt Christi durch Maria und dessen geistliche conceptio im Herzen der Glaubenden parallelisiert und sich hierbei (ähnlich wie Luther,[35] worauf Varenius freilich nicht Bezug nimmt) auf Augustins berühmte Wendung stützt, der zufolge Maria zuallererst zu preisen ist, weil sie Christus geistlich in ihrem Inneren gebar, wohingegen die leibliche Empfängnis und Geburt Folgeerscheinungen dieser conceptio spiritualis sind.[36]

Varenius konkludiert in seiner *Rettunge*: Zwar handele es sich bei Arndts Rede von der geistlichen Geburt Christi um einen Topos, der sich auch bei Tauler findet,[37] doch sei dies nicht zu bemängeln, da diese Lehre nicht nur schriftgemäß sei, sondern sich auch bei lutherisch-orthodoxen Theologen finde. Außerdem macht Varenius zu Recht geltend (1: 353), dass zwischen dem, was Arndt und Weigel über die geistliche Geburt sagen, recht tiefgreifende Unterschiede bestehen.

Die Ehrenrettung Arndts durch Varenius (zumindest an dieser Stelle) scheint geglückt, das Gewicht der beigebrachten orthodoxen Autoritäten erdrückend und die Argumentation Osianders entkräftet. Indes lohnt es sich, einen Moment innezuhalten, um nicht voreilig aus der Deckungsgleichheit eines Topos bei Arndt und Gerhard vorschnelle Schlüsse zu ziehen.

---

sind/ spricht er/ mit Christo begraben durch die Tauffe in den Tod/ auff daß/ gleich wie Christus ist aufferwecket von den Todten/ durch die Herrligkeit des Vaters/ also sollen auch wir in einem newen Leben wandeln. So wir aber sampt jhm gepflantzet werden zu gleichem Tode/ so werden wir auch der Aufferstehung gleich seyn. Vnd abermal: Seyd jhr nun mit Christo aufferstanden/ so suchet was droben ist/ da Christus ist/ sitzend zu der Rechten Gottes/ Trachtet nach dem/ das droben ist/ nicht nach dem/ das auff Erden ist. Also können wir gleichsfals sagen/ daß Christus in vnsern Hertzen geistlicher weise müsse gebohren werden/ sollen wir anders derselben heilwertigen Frucht/ die er in seiner Geburt vns mitbracht/ theilhafftig werden".

[35] Vgl. Martin Luther, *Werke*, Weimar 1883ff (fortan zit. *WA*), hier: *WA* 7: 189, 4–7. 11–14 („Sermon von der Geburt Christi" [...] 1520): „Aber dorinne stehet das recht wunder, das die junckfraw Maria glaubt, das dise dinge solten in yhr geschehenn, das ist so groß, das wir uns daruber nicht genug mögen vorwundern [...] Darumb sagt auch Augustinus, das sie vil genadenreicher unnd seliger gewest sey in dem, das sie Christum im hertzen (durch den glaubenn) hat entpfangen, dan im fleysch, das sie ehr sein mutter ist worden im hertzen dan im fleisch".

[36] Augustin, „De sancta virginitate", *Opera*, Pars III, Wien 1900 (= Corpus Scriptorum Ecclesiasticorum Latinorum 41), cap. 3, 237, 12f. 17–19: „beatior ergo Maria percipiendo fidem Christi quam concipiendo carnem Christi. [...] sic et materna propinquitas nihil Mariae profuisset, nisi felicius Christum corde quam carne gestasset". Vgl. hierzu Varenius 1: 339 und Gerhard, *Postilla* 1: 160f.

[37] Meinhold 191.

Darum noch einmal zurück zu Arndt. Es ist auffällig, dass Arndt aus der oben zitierten Parallelisierung von leiblicher und geistlicher Geburt Christi die Folgerung zieht, dass der Mensch, in dem Christus spiritualiter geboren wird, diesem im Sinne der imitatio nachzufolgen hat. Arndt bietet eine ganze Liste von Dingen, die ein Christ zu tun hat, damit die Geburt Christi in seinem Inneren zu ihrem Ziel kommen kann: „Jch muß mit jhm [scil. Christus] vnd in jhm im Exilio vnd Elende seyn/ ich muß mit jhm in Demut vnnd Verschmehung der Welt/ in gedult vnd Sanfftmut/ in der Liebe wandeln/ Jch mus mit jhm meinen Feinden vergeben/ barmhertzig seyn/ die Feinde lieben/ den Willen des Vaters thun/ ich mus mit jhm vom Sathan versuchet werden/ vnd auch vberwinden/ ich muß mit jm vmb der Warheit willen/ die in mir ist/ verspottet/ verachtet/ verhönet/ angefeindet werden/ vnd so es sein sol/ auch den Tod vmb seinet willen leiden".[38] Symptomatisch für die *BWC* dürfte sein, dass hiermit die Bedeutung der sowohl leiblichen als auch geistlichen Geburt Christi pro me in den Hintergrund tritt zugunsten der Aufforderung zur imitatio Christi. Nicht von ungefähr wird so zumindest der Verdacht virulent, es könnte in Arndts Sicht der Dinge die imitatio Christi sein, die die geistliche Geburt Christi zu ihrer Vollendung führt bzw. überhaupt erst ermöglicht.

Dieser Verdacht wird durch eine Vielzahl von Teiltexten erhärtet, von denen eine Stelle, auf die auch Osiander eingeht, herausgegriffen sei. In Buch 3, Kap. 2 kommt Arndt ausführlich auf die Gelassenheit zu sprechen, die nicht nur die Voraussetzung dafür sei, dass der Mensch—Taulersch—in sein Inneres einkehren kann. Vielmehr geht Arndt so weit, dass er (auch hierin mit Tauler durchaus einer Meinung) die Gelassenheit und die Distanzierung des Menschen von allem Irdisch-Weltlichen und Äußerem als die Bedingung der Möglichkeit des Wirkens Gottes im Menschen apostrophiert: „Also ist des Menschen Lauff der aller Edelste vnd vollkomneste/ wenn er wiederkehret in seinen Vrsprung/ welcher ist GOtt. Das kan aber nicht geschehen denn wenn ein Mensch in sich selbst gehet mit allen seinen Kräfften/ vnd seinem Verstand/ Willen vnnd Gedechtniß erlediget von der Welt/ vnnd von allen Fleischlichen Dingen/ vnd seine Seele mit allen jhren Begierden zu GOTT wendet durch den Heiligen Geist/ ruhet vnd feyret von der Welt durch einen stillen Sabbath: Als denn fehet GOTT an in ihm zu wircken/ denn auff diesen Hertzen Sabbath wartet GOTT"

[38] Arndt, *BWC* 1: 1610, Kap. 6, 56.

(3: Kap. 3, 16f.). (Ganz ähnlich spricht im übrigen auch Valentin Weigel im *Gebetbuch* von der „gelassenhaitt“ als einem „Sabbath“.)[39] Die hier angesichts des Lutherschen Verständnisses der Rechtfertigung sola gratia unschwer zu entdeckende Problematik besteht darin, dass das Wirken Gottes von Arndt abhängig gemacht wird von vorgängigen Frömmigkeitsübungen der mortificatio carnis und abnegatio sui seitens des Menschen.[40]

In diesen Kontext nun zeichnet Arndt auch sein Verständnis der geistlichen Geburt Christi ein. So wie Maria Christus als Jungfrau empfing, so kann nur derjenige Mensch Christus spiritualiter empfangen, der in geistlicher Hinsicht jungfräulich ist, sich also durch die Übung von Gelassenheit bereits zuvor von der Welt abgekehrt und in seinen Seelengrund versenkt hat: „Gleich wie die Jungfrawe Maria eine reine vnbefleckete Jungfraw war (vnd bleibt in Ewigkeit) als sie CHristum Leiblich empfing/ also vnsere Seele/ soll sein wie eine reine vnbefleckte Jungfraw/ das ist/ sie sol nicht mit der Welt Liebe beflecket sein/ so wird sie Christum Geistlich empfangen [. . .] ist sie aber mit der Welt vermehlet/ wie kan sie mit GOTT vermehlet werden?“[41] In dieser Hinsicht ist Osianders Kritik recht zu geben, wenngleich dieser bisweilen nicht davor zurückschreckt, zusätzliche Belegstellen in den *BWC* durch entstellende Auslassungen von Satzteilen zu fingieren. So zitiert Osiander Arndt mit folgenden Worten und macht seinen Leser glauben, Arndt spreche davon, dass der Mensch sich durch Übung von

[39] Weigel 118, 10f.—Geyer 2: 218f weist darauf hin, dass dieser Topos „im nachreformatorischen Spiritualismus“ sehr verbreitet war, sich „jedoch nicht im Werk Taulers findet“.

[40] Geyer 2: 219 spricht meines Erachtens zu Recht angesichts der oben zitierten Passage aus den BWC von einem „zumindest indirekten Synergismus“. Eine vergleichbare Stelle, an der allerdings die durch die Arndtsche ‘wenn-dann’-Formulierung aufkommende gesetzliche Färbung fehlt, findet sich indes auch bei Gerhard, *Postilla* I, 244: „Wie die Eltern des Kindleins JEsu vermeyneten/ bey den Freunden vnd Geferten dasselbe zu finden/ es war aber alles vergeblich/ im Tempel musten sie jhn suchen: Also gedencken wir offt mit auslauffen Christum zu ergreiffen/ aber er wil sich am liebsten im Tempel des Hertzens suchen vnnd finden lassen/ Es muß ein starck Einkehren geschehen/ spricht Taulerus, von allen jrrdischen dingen/ das Hertz muß in stille Ruhe gesetzt werden/ so wirstu Christum mit seinem Trost vnd innerlichem alloquio im Tempel deines Hertzens finden/ Jch wohne bey denen/ spricht er Esa. 57. so zuschlagenes vnd demütiges Geistes sind/ auff daß ich erquicke den Geist der Gedemütigten/ vnd das Hertz der Zurschlagenen. Eine solche Seele/ welche demütig/ in stiller Ruhe/ von allem Jrrdischen abgewendet vnd entfreyet/ göttliches Trosts erwartet/ die wird Christum in jhr selber finden/ vnnd die Süssigkeit seiner Güter vnnd seines Trosts am ehesten empfinden“.

[41] Arndt, *BWC* 3: Kap. 3, 21.

humilitas selbst in die Vereinigung mit Gott schwingen kann: „Es ist nun kein besserer Weg/ dadurch man zu Gott/ vnnd IN GOTT gezogen werde/ dann gründtliche Demuth deß Hertzens/ vnd Geistliche wesentliche Armuth deß Geists".[42] Tatsächlich aber ist bei Arndt zu lesen: „Aber in warheit ist nehest dem wahren Lebendigen Glauben [...] kein besser vnd leichter Weg darzu/ denn die ware gründliche Demuth" 3: Kap. 5, 43).

Gravierender aber ist die Tatsache, dass Osiander im Zuge des Gefechtseifers Formulierungen in die Feder fließen, die derart überzogen sind, dass sie selbst als häretisch einzustufen sind. Osiander hält es nicht für erstrebenswert, sich in den Seelengrund einzusenken, da auch nach der regeneratio „in dem Abgrund der Seelen vnd Hertzens [...] die Grundsuppen der Erbsünd" verbleibe. Ist diese Aussage durchaus noch mit Luthers dialektischer Bestimmung des Glaubenden als 'simul iustus et peccator' kompatibel, so gilt dies nicht für die Wendung, die Osiander an das soeben Zitierte anschließt und die den Eindruck erweckt, als sei im Inneren des Gerechtfertigten ausschließlich Böses zu finden. Osiander schreibt, „daß/ wann ein Christ sein Seel vnnd Hertz/ recht durchsihet/ vnnd auff den Grund schawet/ so findet er nicht allein nichts Gutes/ sondern lauter abschewlichen Jammer vnd Grewel vor Gott. Wie kan dann da ein Mensch durch das jnnerliche Einkehren/ in sein jnnerstes der Seelen vnnd Hertzens/ ein Liecht vnd Ruhe vor GOtt haben: da nicht[s] dann Sünden/ Jammer vnd Elend zu finden?"[43] Dies widerstreitet nicht nur der Auffassung Luthers, der zufolge mit dem Glauben Christus in das Herz des Menschen einzieht (worauf Varenius im übrigen deutlich hinweist),[44] sondern ist zudem mit dem biblischen Zeugnis in keiner Weise deckungsgleich. Varenius jedenfalls bereitet es ein diebisches Vergnügen, dem orthodoxen Arndt-Kritiker Osiander nachzuweisen, dass er zu weit gegangen ist, den Boden der Heiligen Schrift verlassen hat und nun selbst den Vorwurf, Irrlehren zu verbreiten, auf sich zieht, indem er die Tatsache, dass dem

[42] Osiander 337.

[43] Osiander 185.

[44] Varenius 1: 342 bezieht sich auf Luther, *WA* 10/I, 1, 48, 14–20 (Kirchenpostille 1522): „drumb kanstu yhn [scil. Christus] auch mitt keynem andern, denn mit dem hertzen auffnehmen, das thustu wenn du auffthuist unnd sprichst mit hertzen, ya ich glewb es sey alßo. Sihe alßo geht er durchs Euangeli zu den oren eyn ynn deyn hertz und wonet alda durch deynen glawben, da bistu denn reyn und gerecht, nit durch deyn thun, ßondernn durch den gast, den du ym hertzen durch den glawbenn hast empfangen". Vgl. auch *WA* 10/I, 1, 161, 11–13.

Christenmenschen aufgrund der Taufe seine Sündhaftigkeit nicht mehr zugerechnet wird, diese vielmehr 'zugedeckt' ist, konterkariert.[45]

Ein verglichen mit Arndt gänzlich anderes Bild indes zeigt sich, wenn man die bereits genannte Predigt Gerhards näher betrachtet, auf die sich Varenius mehrfach stützt, um Arndts Rechtgläubigkeit nachzuweisen. Ähnlich wie Arndt kommt auch Gerhard im Zusammenhang mit der conceptio Christi spiritualis auf Tauler zu sprechen und zitiert ihn mit den Worten: „da sollen wir verleugnen alle Eigenschafft/ Willens vnd Begerens vnd Wirckens/ denn soll bleiben ein bloß lauter Meinen Gottes/ allein innen zu seyn/ vnd jhm stat zu geben/ daß er seines Wercks in dir bekommen möge/ vnd von dir an dem vngehindert bleibe. Jtem/ wie Christus vmb Mitternacht gebohren ward/ da es alles still war/ Also soltu schweigen/ so mag diß Wort in dir gesprochen vnd gehöret werden".[46] Gerhard nun—und hierin liegt eine wesentliche Differenz zu Arndt—übernimmt das, was Tauler über die Selbstverleugnung, die Gelassenheit und das innere Stillwerden sagt, nicht unkommentiert. Vielmehr konzediert Gerhard, dass Taulers Redeweise nicht nur missverständlich, sondern „den Vngeübten etwas vnuerständlich" (1: 167) ist, weswegen er seine Hörer bzw. Leser ausführlich in sieben Punkten dazu anleitet, die geistliche Geburt Christi recht zu verstehen.

Hierbei hebt Gerhard anders als Arndt unmissverständlich, ja programmatisch hervor, dass Christus im Menschen allein durch den Glauben geistlich geboren wird, so wie Maria Christus sola fide leiblich empfangen hat. Im Unterschied zu Arndt kommt bei Gerhard an keiner Stelle der Verdacht auf, als sei eine wie immer geartete spirituelle Jungfräulichkeit des Menschen die durch ihn erst zu erbringende Vorbedingung für diese geistliche Geburt. Andererseits sind—und hierin ist Gerhards Argumentation derjenigen Osianders weit überlegen—das

---

[45] Varenius 1: 441–444, bes. 443f: „So heissets nach Pauli Außsag/ daß nichts verdamlichs sey/ an denen die in CHRJSTO JESV sind/ zun Römern am 8. cap. So gar ist die Sünde in den getaufften vnnd wiedergebornen Christen vergeben vnnd zugedecket/ daß sie denselbigen nicht verdammen kan. Ja nicht allein daß/ sondern es sind auch noch darzu die getaufften Christen zu Kindern GOttes/ zu Erben der ewigen Seligkeit/ vnd zu Gnaden Tempeln vnd Wohnungen Gottes gereiniget worden/ wie hernacher weiter sol bewiesen werden. Vnd also sage ich recht/ daß in einem getaufften Christen nicht nur die Sünde vnd tieffe Verderbung stecket/ sondern die Sünde ist in der Tauffe/ in dem roten Blut Christi ersäufft/ zugedecket/ vnd das Hertz dermassen gereiniget/ daß auch Gott selbst in eines getaufften Christen Hertzen wohnen wil/ vnd kräfftiglich wircken/ 1. Cor. 6. v. 11. 19. Tit. 3. v. 5. 6."

[46] Gerhard, *Postilla* 1: 166f.

Wiedergeborenwerden des Menschen in Christus und die Geburt Christi im Menschen keine sich ausschließenden Gegensätze. Vielmehr—und einzig hierin wird man eine angemessene Interpretation des diesbezüglichen neutestamentlichen Zeugnisses erblicken dürfen, die weder Gefahr läuft, die Arndtschen noch auch die Osiandrischen Einseitigkeiten zu wiederholen—koinzidieren die regeneratio hominis in Christo und die conceptio Christi spiritualis in homine als gewissermaßen wechselseitige Geburtsprozesse: Indem der Mensch in Christus wiedergeboren wird, wird dieser zugleich im Menschen geistlich geboren. „[...] daher sagt man recht/ daß wir durch die Widergeburt aus Christo/ vnd Christus in vns gebohren werde".[47]

So wie also die fides nicht ein Werk des Menschen, sondern ein Werk Gottes im Menschen ist, so ist auch die Geburt Christi in corde nichts, was der Mensch tut, sondern Wirkung des Glaubens schlechthin und nichts anderes. Nicht zufällig verweist Gerhard in diesem Zusammenhang auf 1 Petr 1, 23 ('als die da widerumb geborn sind/ Nicht aus vergenglichem/ sondern aus vnuergenglichem Samen/ nemlich/ aus dem lebendigen wort Gottes/ das da ewiglich bleibet') und Jak 1, 18 ('Er hat vns gezeuget nach seinem willen/ durch das wort der Warheit/ Auff das wir weren Erstlinge seiner Creaturen'),[48] um die Alleinwirksamkeit des verbum Dei zu unterstreichen und jeglichen Verdacht aus der Welt zu räumen, als habe das Tun des Menschen, etwa die Übung von Gelassenheit oder das Still- und Leerwerden, irgendeine vorbereitende oder gar konstitutive Bedeutung bezüglich des in Rede stehenden Geburtsprozesses—ein Verdacht, der sich bei Arndt, aber auch bei Tauler immer wieder aufdrängt.

Doch noch eine weitere Differenz zwischen Arndt und Gerhard darf nicht unerwähnt bleiben: Nach Arndt muss sich die geistliche Geburt Christi notwendig in der imitatio Christi und im Erleiden von leiblichen bzw. seelischen Unbilden fortsetzen. Anders jedoch

[47] Gerhard, *Postilla* 1: 168.

[48] *Postilla* 1: 168: „Gleich wie der Engel zu Maria der werthen Mutter des HErrn gesendet wird/ welcher jhr bringet verbum aedificatorium vitae, Er verkündiget jhr/ daß Christus in jhr soll empfangen vnd gebohren werden/ Luc. 1. Also brauchet gleichsfals Gott der HERR zu dieser geistlichen Geburt sein göttliches Wort/ welches ist der v[n]uergängliche Same/ daraus wir geistlicher weise gezeuget werden. 1. Petr. 1. Jhr seyd wieder gebohren/ nicht aus vergänglichem/ sondern aus vnuergänglichem Samen/ nemlich aus dem lebendigen Wort Gottes/ das da ewiglich bleibet. Jacob. 1. Er hat vns gezeuget nach seinem Willen durch das Wort der Warheit/ auff daß wir Erstlinge weren seiner Creatur".

Gerhard, dem zufolge Gott „manchmal" „Verfolgung/ Drückung vnd Widerwertigkeit" sendet, um die geistliche Geburt zu befördern. Zeigen sich bei Arndt auf Schritt und Tritt—manchmal deutlicher, zuweilen weniger klar—Affinitäten zu der mystisch-spiritualistischen Ansicht, der zufolge der Mensch erst durch nicht nur willig angenommenes, sondern auch durch selbstauferlegtes Leiden christusförmig wird, so ist dies bei Gerhard anders. Während Gerhard zu denen spricht, die sich in der tentatio befinden, und ihnen eine Perspektive eröffnet, ihre bereits vorfindliche Situation als eine von Gott initiierte Beförderung der innerlichen Geburt Christi zu begreifen, behauptet Arndt, dass das Kreuz erlitten sein muss, damit diese überhaupt erst in Gang kommen kann.

Gerhard zitiert in seiner Weihnachtspredigt über die geistliche Geburt Christi zwei Strophen aus Luthers „Vom Himmel hoch" [...]: „Ach mein hertzliebes JEsulein/ Mach dir ein fein sanfft Bettelein/ Zu ruhen in meins Hertzen Schrein/ Daß ich nimmer vergesse dein. Davon ich allzeit frölich sey/ Zu springen/ singen jmmer frey/ Das rechte Susaninne schon/ Von Hertzengrund den süssen Thon".[49] Auch nach Luther und Gerhard gibt es mithin eine Vorbereitung der inhabitatio Christi und der aus ihr folgenden mystischen Union. Sie aber zu leisten, ist einzig und allein die Aufgabe des künftigen Einwohners, nicht jedoch des Bewohnten. Auch in diesem Punkt unterscheidet sich der Ansatz Arndts von demjenigen Gerhards, insofern ersterer verstärkt das Augenmerk darauf richtet, welche praeparationes der Mensch zu treffen hat, um dem Sohn Gottes die Einwohnung zu ermöglichen. Zudem ist nach Vom Himmel hoch [...] nicht das Ruhen und Stillwerden des Menschen das Ziel und schon gar nicht als Voraussetzung für die dann erst erfolgende Geburt Christi in der Seele. Vielmehr ist umgekehrt das Ruhen des Gottessohnes im „Bettelein" des Herzens Voraussetzung schlechthin—nicht für das Ruhen oder sonstigerlei Quietismus, sondern für die geistliche Freude, die den Menschen neumacht und sich emotional im Springen und Singen artikuliert.

Mit einem Wort: Gerhard führt in seiner Weihnachtspredigt vor, wie eine lutherisch-orthodoxe Adaptation Taulerschen Gedankengutes vorgenommen werden kann, nämlich durch dessen klare Einbettung in die Botschaft von der iustificatio sola fide, mithin auf dem Wege der reformationstheologischen Transformation der vorgegebenen

[49] Gerhard, *Postilla* 1: 171; vgl. Wackernagel 3: 23.

mittelalterlich-mystischen Traditionslinien. Genau hieran indes lässt es Arndt—wenn nicht durchgehend, so doch immer wieder—mangeln. Dies zu konzedieren jedoch ist Varenius nicht bereit, sondern behauptet, Arndt habe Tauler ausschließlich in schriftgemäß-rechtgläubiger Weise adaptiert.[50] Wenn Varenius sagt, Arndt habe seine *BWC* einzig und allein auf die Heilige Schrift gegründet und Tauler nur dann herangezogen, wenn dieser als Zeuge der biblischen Wahrheit aufzutreten geeignet sei, so beschreibt er hiermit die Verfahrensweise Gerhards, nicht aber diejenige Arndts, in dessen *BWC* sehr viel Taulersches 'Sondergut' vorkommt, das gleichsam unbearbeitet weitergegeben wird.

Gerhard selbst hat (neben anderen Autoren) nicht nur in seinen Predigten, sondern vor allem in den *Meditationes Sacrae*, aber auch in der *Schola Pietatis*, mit der er bekanntlich die *BWC* ersetzen wollte,[51] einen Weg gewiesen, wie die mittelalterlich-mystische Frömmigkeitstradition innerhalb des Luthertums heimisch werden kann, ohne mit den reformatorisch-theologischen Grundlinien zu kollidieren, und im Gegenteil geeignet ist, diese zu vertiefen und zu bereichern. Dass diese Bemühungen um eine produktive Aneignung und Interpretation der mittelalterlichen Mystik in der ersten Hälfte des 17. Jahrhunderts jedoch nicht einfach lutherisches Gemeingut waren, zeigt sich exemplarisch anhand von Osianders *Bedencken*, wenngleich es zutreffend ist, dass Osiander hiermit eine „schon damals völlig isolierte Extremposition"[52] einnahm. Denn Osiander richtet sich keineswegs nur gegen die Art und Weise der Rezeption mittelalterlicher Autoren speziell bei Arndt. Vielmehr hält er die Beerbung der vorreformatorischen pietas im Kontext der

---

50 Varenius, 1: 65f, 69.

51 Vgl. Johann Gerhard an Aegidius Hunnius d. J.... (2.2.1625), in: Philipp Julius Rehtmeyer, *Historia Ecclesiastica Inclytae Urbis Brunswigae*, Teil 5, Braunschweig 1720 (UB Leipzig Hist. Sax. inf. 246: 4/5), 237f, hier: 238. Es handelt sich hierbei um denselben Text, der auch abgedruckt ist in: Johann Andreas Gleich Hg, *TRIFOLIVM ARNDTIANVM SEV B. IOANNIS ARNDTI TRES EPISTOLAE HACTENVS INEDITAE, DE LIBRIS VERVM CHRISTIANISMVM CONCERNENTIBVS, AD D. PETRVM PISCATOREM, THEOLOGVM QVONDAM IENENSEM, SVB AVSPICIVM SVPERIORIS SECVLI SCRIPTAE, NVNC VERO PRIMVM IN LVCEM PVBLICAM EMISSAE, QVIBVS ANNEXA EST PECVLIARIS EPISTOLA D. IOANNIS GERHARDI [...] DE EODEM LIBELLO*. Wittenberg [1714] (BSB München 4 Diss. 2483 [21]). Eine zweite Auflage hiervon erschien 1726. Der Text findet sich ebenfalls in: Johann Ernst Gerhard, *EPISTOLA AD AMICVM DE OBTRECTATIONIBVS, quibus B. IOHANNES ARNDIVS cum libris suis de vero Christianismo expositus fuit, & deriuata hinc IN B. IOHANNEM GERHARDVM calumnia, Qua simul maculam famae suae aspersam abstergit*, Frankfurt a. M./Leipzig 1720 (UB Greifswald 536/Disp. theol. 25, 14) ([1]1705 [UB Greifswald 536/Disp. theol. 25, 13]), 10f.

52 Mahlmann 162.

Erbauungsliteratur generell für unsinnig. Man täte besser daran—so Osiander (432, 436)—die Heilige Schrift selbst zu lesen und im Rahmen der praxis pietatis den Katechismus zu traktieren, Gebete zu sprechen und Kirchengesänge zu singen. Auch wäre Arndt Osiander zufolge (41) besser beraten gewesen, sein *Wahres Christentum* aus Luthers Schriften zu kompilieren.

Der hauptsächliche Grund, weswegen Osiander der eklektischen Rezeption mittelalterlicher Autoren abhold ist, ist jedoch darin zu sehen, dass er zwar bereit ist, zuzugeben, dass es Tauler, Thomas von Kempen u.a. „gut gemeinet" (38) haben, ihre Texte aber gleichwohl mit derart vielen Irrtümern durchsetzt seien, dass stets die Gefahr besteht, dass auch diese erneut virulent werden. „Vnd kan ich nicht sehen/ wann man solche verdächtige/ vnd in der Lehr seuchtige vnd verfürhische [sic!] Leut/ canonisieren zu heiligen machen/ vnd gut machen will/ warzu es dienen solte Als wann ein Falscher Würtzkrämer neben gantzen Pfunden vnnd Centnern böser vnnd verdorbener Gewürtzen/ ettliche wenige Loth guter Würtz auch hette/ vnnd man dieselbige herfür ziehen/ den Leuten rhümen vnnd loben will/ biß man dem falschen Würtzkrämer [...] einen Namen vnnd Zulauff machen/ vnnd wann man jhme nuh alles gutes anfahet zutrawen/ man alsdan auch seine falsche Wahr annemmen muß" (40). Das Arndtsche Prinzip, das auch Varenius verficht, 1 Thess 5, 21 folgend alles zu prüfen und das Gute zu behalten, will Osiander folglich gar nicht einleuchten. Darum rät Osiander von der Lektüre der *BWC* ab, weil den meisten Lesern die theologische Urteilskraft abgehe, die Arndtschen Texte im Sinne von 1 Thess 5, 21 kritisch zu lesen.[53] Und mehr noch: Osiander appelliert nicht nur an die Theologen, alles daran zu setzen, den Arndtianismus zurückzudrängen, vielmehr ruft er auch die weltliche Obrigkeit dazu auf, das Ihre zu tun, um das drohende Verderben der Kirche zu verhindern. Wenn nicht explizit, so doch implizit und unüberhörbar verlangt Osiander von den politischen Instanzen, die Verbreitung der Schriften Arndts zu verhindern. Nachdruck verleiht er dieser Forderung, indem er die Schreckgepenste des Bauernkrieges und der Münsteraner Täufermonarchie an die Wand malt und behauptet,

[53] „Ja/ sagen ettliche ferner/ spricht doch S. Paulus: Prüfet alles/ vnd das gute behaltet: 1. Thessal. 5. Darumb kan ich Arndten Christenthumb wol lesen/ vnnd was gut darinnen ist/ für mich herausser klauben. Antwort. Wer die Gnad hat/ die Geister zu vnterscheiden/ der möchte solches wol thun können. Weil aber solche nicht jederman hat/ so ist es sicherer/ solcher vnreinen Bücher müssig zugehen" (430).

dass diese aus eben dem enthusiastischen Geiste entstanden seien, den auch Arndts Schriften atmeten. „Wie nicht weniger auch Christliche Obrigkeiten wol auffzusehen haben/ daß nicht ein solches Vnheil in der Kirchen Gottes einreisse. Darbey auch die Obrigkeiten wol zu Gemüth zu führen/ was in den vorigen nahend Hundert Jahren/ als eben auch dieser Geist die Ohren gerecket/ durch jhne/ vnter dem Schein Gött[l]icher jnnerlicher Erleuchtungen vnnd Offenbahrungen/ für Jammer/ Noth vnnd Zerrüttung erfolget/ daß Müntzer dadurch den leidigen Bawrenkrieg erwecket/ die Münsterische Westphälische Wiedertäuffer aber/ eben vnter solchem Schein/ sich gerottieret/ Auffruhr/ vnd Mord angefangen/ jhnen auch fürgenommen/ die gantze Welt zu reformieren/ vnnd alle Könige/ Fürsten vnnd Obrigkeiten zu tödten vnd zuvertilgen." (444f).

Mit der Generalkritik Osianders an der im zeitgenössischen Luthertum sehr breiten Wiederaufnahme und Verarbeitung mittelalterlicher Gebets- und Frömmigkeits-Traditionen will es jedoch nicht so recht zusammenpassen, dass Osiander (436) seinen Lesern rät, statt auf Arndts *BWC* auf Johann Habermanns (1516–1590) Gebetbüchlein zurückzugreifen. Denn, wie Traugott Koch zeigt, werden auch in ihm vielerlei mittelalterliche Texte verarbeitet, allen voran pseud-augustinische und pseud-bernhardinische, wenngleich diese eine konsequentere lutherisch geprägte Bearbeitung erfahren als dies bei Arndt der Fall ist.

Osiander hält Arndt weiterhin vor, er verstehe die unio mystica weigelianisch und vertrete zudem die Lehre von der wesentlichen Vereinigung des Christen mit Gott. Hierbei bezieht sich Osiander (333) auf Buch 2, Kap. 6, 75 der *BWC*, wo es heißt: „Dann gleich wie Göttliche vnd Menschliche Natur in Christo Persönlich vereiniget ist. Also müssen wir alle mit Christo durch den Glauben aus Gnaden vereiniget werden/ auff das die tieffe verderbung vnser sündlichen Natur verbessert werde/ als mit dem höchsten ewigen Gute". Es wird nicht übertrieben sein, zu behaupten, dass Osiander diese Stelle absichtlich missversteht, um seine Pauschalkritik voranzutreiben. Denn eigentlich hätte ihm auffallen müssen, dass Arndt—wie andere lutherische Theologen, die von der mystischen Lebensgemeinschaft mit Christus handeln—die unio mystica in der in Rede stehenden Wendung als eine solche bezeichnet, die durch den Glauben bzw. die göttliche Gnade zustandekommt und nicht anders. Zudem unterstellt Osiander Arndt, er setze die unio hypostatica von göttlicher und menschlicher Natur in Christus und die unio mystica gleich, bezeichne also auch letztere als eine unio essentialis. Das Arndtsche „gleichwie" versteht Osiander mithin im Sinne einer Identifikation. Varenius dagegen, der das von Arndt Intendierte meines Erachtens an dieser Stelle besser trifft, meint, dass Arndt hier wie Gerhard und andere lutherische Theologen von einer Analogie zwischen der Vereinigung der beiden Naturen in Christus und der als unio spiritualis verstandenen mystischen Union spricht (1: 422, 429). Varenius hat insofern Recht, als Arndt in der von Osiander kritisierten Wendung keineswegs eine wesentliche Vereinigung im Blick hat. An Arndt zu kritisieren bliebe allenfalls, dass er es—wie häufiger—an der notwendigen Trennschärfe der theologischen Begrifflichkeit mangeln lässt, indem er zwar von einer Analogie der beiden uniones spricht, nicht aber auch das sie Unterscheidende klipp und klar benennt, was leicht durch die Näherbestimmung der unio mystica als einer geistlichen Vereinigung hätte geschehen können.

Osiander zieht eine weitere Formulierung Arndts heran, um den (in die Irre gehenden) Verdacht, Arndt lehre eine unio essentialis von Gott und Glaubendem, zu erhärten. Hier jedoch spricht Arndt lediglich davon, dass Gott „wesentlich die Liebe selbsten" ist, dem die Menschen durch die Liebe „ehnlich vnnd gleichförmig" werden sollen (1: Kap. 29, 296). Osianders Behauptung, Arndt lehre eine Verschmelzung des Menschen mit dem göttlichen Wesen, hat an dem beanstandeten Text nicht den geringsten Anhalt, sondern entpuppt sich als Unterstellung. Zu Recht weist Varenius darauf hin, dass „Arndt nirgend das Wörtlein Wesentlich [scil. gebraucht]/ da er von der Vereinigung mit GOtt redet" (1: 410). Andererseits jedoch ist nicht von der Hand zu weisen, dass Arndt, worauf Osiander (339) wiederum zu Recht aufmerksam macht, ein Tauler-Zitat bringt, das besagt, dass der Mensch in der unio mystica mit Gott verschmolzen wird: „Durch ein solch innerlich gebet wird man denn gefüret zu dem vbernatürlichen gebet/ welches geschicht wie Taulerus sagt/ durch ware vereinigung mit Gott durch den Glauben/ da vnser erschaffner geist verschmeltzt vnd versenckt wird in den vnerschaffnen geist Gottes".[54] Osiander ist darin Recht zu geben, dass Arndt in den *BWC* im Anschluss an Tauler verschmelzungsmystische Topoi aufgenommen hat. Hierin unterscheidet sich Arndts Sicht der Dinge etwa von derjenigen Gerhards oder Philipp Nicolais, die die unio mystica in den Kategorien des seligen Tausches und Wechsels interpretieren, wodurch gewährleistet ist, dass in der mystischen Vereinigung von Mensch und Gott das Gegenüber im Sinne

[54] Arndt, *BWC* 2: Kap. 20, 233.

eines Prozesses lebendiger Wechselbeziehung gewahrt bleibt.[55] Zwar findet sich das Theologumenon des commercium auch in den *BWC*, jedoch erhebt Arndt es hier noch nicht zum Deuteprinzip, weswegen er verschmelzungsmystische Gedanken nicht entsprechend revidiert, sondern stehen lässt, wodurch eine deutliche Spannung bezüglich der Definition der unio wirksam wird. Auffällig indes ist, dass Arndt in späterer Zeit (vor allem in seinem Traktat *De unione credentium cum Christo Jesu*) sehr viel intensiver von der Gedankenfigur des commercium Gebrauch gemacht und zudem stärker betont hat, dass die unio mystica allein durch das den Glauben stiftende Wort zuwegegebracht wird: „Per VERBUM fit unio Dei & hominum"[56]. In diesem Zusammenhang nimmt es nicht wunder, dass Varenius um willen des Nachweises der Orthodoxie Arndts recht häufig Zitate nicht aus den *BWC*, sondern aus *De unione* herbeizieht,[57] ohne freilich darauf hinzuweisen, dass Arndt sich hier sehr viel weniger missverständlich artikuliert als dies in den *BWC* noch der Fall gewesen ist und zudem die Taulersche Diktion vom Seelengrund stark in den Hintergrund tritt.[58]

Osiander fasst seine Vorwürfe gegen Arndt zusammen, indem er sagt: „Heißt dann das nicht Wigelianische vnnd Enthusiastische grobe Schwärmerey/ neben dem Jrrthumb von der wesentlichen Göttlichen

---

[55] Dazu Steiger 1997: 94–123.

[56] Johann Arndt, *DE UNIONE CREDENTIUM, CVM CHRISTO JESV, Capite Ecclesiae*, in: Ders., *Lehr= vnd TrostBüchlein/ Vom Glauben vnd heiligen Leben/ Zum wahren Christenthumb gehörig* [...] *Mit beygefügtem Büchlein/ von der Vereinigung der Gläubigen mit Christo Jesu ihrem Häupt* [...], Lüneburg 1620 (Forschungsbibliothek Gotha Theol. 8° 547/1 [3]), 10. Es handelt sich um die Überschrift zu Kap. 3. Vgl. 12f: „Sic conjunguntur mentes hominum cum DEO, quae per peccatum à DEO avulsae sunt, ut sedem suam figat iterum altißimus in ipsorum animis. Vehiculum DEI verbum est, Ezech. 1. Zach. 6." Vgl. auch Johann Arndt, *POSTILLA, Oder AVslegung der Sontages und aller Festen Evangelien/ durchs gantze Jahr* [...], Lüneburg 1656, 3 Teile (HAB Wolfenbüttel Th 2° 4), hier: 3: 845–849, wo Arndt „Von der Vereinigung des Menschen mit GOtt" handelt. Zu den „Merckzeichen der gnädigen Einwohnung Gottes und Vereinigung mit GOtt" führt Arndt aus: „DAs erste Kennzeichen ist GOTtes Wort und der Glaube an das Wort". Es „ist das Wort GOTtes nichts anders als ein Band/ welches GOtt und Menschen zusammen hält und bindet. Darumb hat GOtt mit dem Wort der Gnaden und Verheissung/ ja durch das ewige wesentliche Wort/ welches ist Christus/ mit dem Menschen einen ewigen Bund gemacht/ und derselbe Bund vereiniget Gott und Menschen/ durch die Verheissung" (S. 848). Vgl. Braw 1985: 46: „Arndt hat also die Mystiker in derselben Weise gelesen, wie er hofft, dass man sein W[ahres] Chr[istentum] entgegennehmen wird: mit dem rechtfertigenden Glauben als Voraussetzung [...] Er hat die Mystiker mit seinem evangelischen Glauben als Fundament gelesen, ein Fundament, das ihnen in seinem reformatorischen Gepräge unbekannt war". Vgl. auch Braw 1987 sowie Repo.

[57] Varenius 1: 118, 464 u.ö.

[58] Braw 1985: 215, 228.

Gerechtigkeit vnnd Seligkeit deß Menschen einführen/ so kan ich doch nicht sehen/ wie mans jmmermehr solte gröber machen können" (336). Hier drängt sich die Vermutung auf, dass Osiander darum ein Interesse daran hat, die Lehre von der unio essentialis des Menschen mit Gott in den *BWC* aufzuspüren, um deren Verfasser nicht nur des Weigelianismus, Schwenckfeldianismus etc., sondern auch der Ketzerei seines Großvaters Andreas Osiander zu bezichtigen. (Dass Lukas Osiander nicht die Freiheit hatte, auch diese 'Irrlehre' [wie sonst bei ihm üblich] mit dem Namen ihres Hauptvertreters in Verbindung zu bringen, spricht für sich.) Andreas Osiander jedenfalls hatte gelehrt, dass die göttliche Natur Christi die Gerechtigkeit der Sünder sei. Die sündigen Menschen werden laut Andreas Osiander nicht dadurch von Gott angenommen, dass ihnen die fremde Gerechtigkeit Christi angerechnet wird. Vielmehr vollzieht sich die iustificatio dergestalt, dass den peccatores die natura divina Christi und mit ihr die iustitia essentialis eingegossen wird. Kerngedanke des sog. effektiven Rechtfertigungsverständnisses Andreas Osianders ist mithin die Rede von der Vergottung (Theosis) des Menschen.[59] Indes: So viele Stellen Lukas Osiander auch aus den *BWC* anführt, es mag ihm nicht gelingen, Arndt stichhaltig als Anhänger seines Großvaters zu überführen.

Varenius weist gegen Osiander zu Recht und quellenkundig darauf hin, dass die geistliche Geburt Christi auch bei lutherisch-orthodoxen Theologen eine wichtige Rolle spielt. Somit gelingt es Varenius, klarzustellen, dass Osianders Kritik an Arndt zu pauschal ist und ihr die notwendige Differenzierung fehlt. Zugleich aber zeigt sich, dass auch Varenius' Ehrenrettung Arndts häufig die nötige Differenzierung fehlt, wenn man die Quellen näher betrachtet, aus denen Varenius die Orthodoxie Arndts erweisen will. Varenius leitet aus der vorschnellen Konstatierung der Identität gewisser Topoi bei Arndt und orthodoxen Theologen eine vermeintlich unerschütterliche lutherische Rechtgläubigkeit Arndts ab, die bei näherer komparatistischer Analyse der von Varenius für inhaltlich mit Arndt deckungsgleich erklärten Quellen orthodoxer Provenienz nicht Stich hält. Anders formuliert: Varenius weist in treffsicherer Weise einen Weg, wie die *BWC* 'in meliorem partem' zu rezipieren wären, indem er Arndt aus dem Blickwinkel prominenter orthodoxer Theologen liest und vielfältiges Material zu einem rechtgläubigen Arndtianismus beisteuert, der sich

[59] Zu Andreas Osiander vgl. Seebaß 1995: 511f, 1967: 273 und Hauke.

zudem in Philipp Jakob Speners Hochschätzung Arndts fortsetzt.[60] Bekanntermaßen hat Spener in eine von ihm besorgte Ausgabe der *BWC*, die 1674 erschien, viele Exzerpte aus Varenius' Arndt-Apologie aufgenommen,[61] was u.a. in der ersten dänischen Edition der *BWC* Schule gemacht hat.[62]

Das ungelöste Problem aber, das Varenius indes nicht reflektiert, besteht darin, dass sich das von ihm herbeigezogene orthodoxe Material, in dessen Sinne er die *BWC* verstanden wissen will, bei Arndt selbst eben häufig nicht findet. Letztlich trifft somit die beiden Kontrahenten Osiander und Varenius derselbe Vorwurf, nämlich derjenige der Einseitigkeit, ohne dass bei ihnen eine Empfänglichkeit für die Tatsache zu entdecken wäre, dass hierin bei aller sonstigen Differenz eine Gemeinsamkeit bestehen könnte.

Im Streit zwischen Osiander und Varenius trug letzterer den Sieg davon, denn das Stuttgarter Konsistorium bemängelte, dass Osiander sein *Bedencken* nicht zur Zensur eingereicht hatte. Der Replik, die Osiander gegen Varenius abgefasst hatte und publizieren wollte, verweigerte das Konsistorium die Druckerlaubnis.[63] Bei aller Unzulänglichkeit des von Osiander und Varenius Geleisteten offenbaren deren Schriften jedoch zumindest zweierlei: Nach der Lektüre von Osianders *Bedencken* wird niemand mehr bestreiten können, dass Arndts *BWC* einer konsequent spiritualistischen Interpretation nicht nur offenstehen, sondern diese streckenweise auch nahelegen. Wer Varenius gelesen hat jedoch, wird die Kunst, Arndt eine interpretatio orthodoxa angedeihen zu lassen, mit all ihren Schwierigkeiten gründlich erlernt haben. Dass sowohl Osiander als auch Varenius auf ihre, d.h. unterschiedliche Weise recht haben, belegt die weitere Wirkungsgeschichte des Arndtschen Oeuvres im 17., 18. und 19. Jahrhundert und findet eine Bestätigung noch in der heutigen Arndt-Forschung.

Es bleibt zu prüfen, ob die übrigen Publikationen zu Arndts *BWC* aus der ersten Hälfte des 17. Jahrhunderts (darunter Johann Saubert, Melchior Breler) ein höheres Maß an Differenzierung aufweisen als die Schriften Osianders und Varenius'. Eine Sichtung der bei Breler gedruckt vorliegenden Gutachten, die von den Theologischen Fakultäten

---

60 Zu diesem Urteil kommt auch Wallmann 1980: 29f.

61 Wallmann 1980: 30. Zur Ausgabe der *BWC* von 1674 vgl. Blaufuß sowie Brecht 1979: 123ff.

62 Dazu Bach-Nielsen 15.

63 Brecht 1993: 148.

in Königsberg und Jena abgefasst worden sind, lassen durchaus unterschiedliche Grade an Differenzierungsvermögen erkennen. Das Königsberger Gutachten nimmt Arndt in Schutz, verweist auf dessen von ihm selbst behauptete Rechtgläubigkeit und meint, die Probleme durch den Hinweis darauf aus der Welt schaffen zu können, dass Arndt nicht für die justificandi, sondern für die bereits Gerechtfertigten geschrieben[64] und die Beförderung von deren praxis pietatis im Blick gehabt habe, wie Breler meinte.[65] Der Duktus des Königsberger Gutachtens und dessen Wirkabsicht sind der Arndt-Apologie Varenius' durchaus vergleichbar.

Die Theologische Fakultät der Universität Jena hatte sich im Jahre 1623 aufgrund einer Anfrage Herzog Christians von Braunschweig-Lüneburg zu der Frage zu äußern, ob es ratsam sei, die lateinische Übersetzung der *BWC* (gemeint ist diejenige von Melchior Breler)[66] zu publizieren. Im Vergleich mit dem Königsberger verrät das Jenenser Gutachten, das im wesentlichen auf Johann Gerhard zurückgeht,[67] ein differenzierteres Urteilsvermögen. Die Problematik der bei Arndt nicht selten zu findenden mangelnden begrifflichen Trennschärfe wird zunächst sehr deutlich benannt, sodann aber mit folgender Verallgemeinerung entschärft: „Multi etiam ex orthodoxis rectius senserunt, quam locuti sunt".[68] Dies deckt sich im übrigen mit der Einschätzung, die Gerhard in seinem Brief an Aegidius Hunnius d. J. vom 2. Februar 1625 vertritt. Auch hier gibt er zu, dass in Arndts *BWC* „ambiguas & quidem ejusmodi phrases complures occurrere, quae in Weigelianum sensum trahi facile possunt".[69] Da Arndt aber kein Heuchler sei, müssten seine Beteuerungen, er wolle sein Werk im Rahmen der libri symbolici interpretiert wissen, ernstgenommen werden. „Interim hoc apud me certum, quod rectius sentiat, quam loquatur, quod finem & scopum hunc respiciat unicum, ut ad pietatis

[64] So Saubert nach Sommer 1999: 250.

[65] Breler 93–101.

[66] *REV. D. IOHANNIS ARND* [sic!] *De Vero Christianismo LIBRI QUATUOR Nunc primum Latine editi [...]*, Lüneburg [1625] (Forschungsbibliothek Gotha Theol. 8° 659/9). Die lateinische Version der *BWC* erschien demnach unbeschadet der Verweigerung der Approbation durch die Theologische Fakultät Jena.

[67] Vgl. Johann Gerhard 1625 an Aegidius Hunnius d. J. (Rehtmeyer 237), wie oben Anm. 51.

[68] Breler 104.

[69] Rehtmeyer 237.

studium torpescentes hominum animos excitet" (238). Nicht nur Salomo Glassius ist Gerhard in diesem Urteil gefolgt.[70]

Arndt ist nach dem Urteil der Jenenser Zensur keinerlei „Schwärmerey" zu bescheinigen. Im Gegenteil sei seine Übereinstimmung mit der Lehre der *Confessio Augustana invariata* und der *Formula Concordiae* offensichtlich. Trotz der klar erkennbaren Sympathie für Arndt und des Bestrebens, die *BWC* in meliorem partem zu entziffern, gibt die Jenenser Fakultät jedoch zu, dass es bei Arndt, vor allem im dritten Buch des *Wahren Christentums*, zahlreiche Texte gibt, die eine spiritualistische Interpretation nahelegen, obgleich dies nicht der Intention des Autors Arndt entspreche. „Warzu er zum dritten/ eine solche Art im reden gebraucht/ die nicht allzeit auff den gemeinen Strassen bleibet/ sondern sich nach den Phrasibus Tauleri alter deutschen Theologiae Thomae de Kempis lencket vnd accommodiret/ vnd demnach offtmals/ bevorab im dritten Buch zimlich hart vnnd seltzam lautet/ vnd daher wider des Autoris beständige Meynung von Vngelehrigen vnd Leichtfertigen in Mißverstand; Ja als wenn sie gantz Weigelianisch vnd Schwenckfeldisch were angezogen vnd außgeleget worden".[71] Darum empfiehlt die Fakultät, mit dem Druck der lateinischen Übersetzung der *BWC* so lange zu warten, bis ein hierfür geeigneter Theologe die zur Verhinderung eines spiritualistischen Missverständnisses der *BWC* nötige Redaktion des Textes durchgeführt und ein entsprechendes „Tractätlein" abgefasst haben wird.[72] Die Fakultät also verweigerte die von Herzog Christian erwünschte approbatio.[73] Man mag dieser Einschätzung der Dinge zustimmen oder nicht. Klar aber ist, dass die Jenenser Fakultät anders als Osiander und Varenius fähig war, diejenige Ambivalenz zu benennen, die die *BWC* prägt und die man auch durch noch so detaillierte historisch-theologische Erforschung dieses Werkes niemals restlos wird auflösen können.

---

[70] Albrecht-Birkner 40.

[71] Breler 103.

[72] Breler 103f.

[73] Gerhard an Hunnius (Rehtmeyer 237, wie Anm. 51): „[...] unde etiam approbationem scripti recusauimus". Vgl. hierzu Geyer 2: 267f.

Literaturverzeichnis

Albrecht-Birkner, Veronika. „Zur Rezeption Johann Arndts in Sachsen-Gotha (1641/42) und in den Auseinandersetzungen um den Pietismus der 1690er Jahre". *Pietismus und Neuzeit* 26 (2000): 29–49.

Werner Anetsberger. *Tröstende Lehre: Die Theologie Johann Arndts in seinen Predigtwerken*. München 2001.

Arndt, Johann. [*BWC:*] *Vier Bücher Von wahrem Christenthumb. Die erste Gesamtausgabe (1610)*, Hg. Johann Anselm Steiger (= Philipp Jakob Spener. *Schriften*. Sonderreihe, Bd. V.1–3 = Johann Arndt-Archiv, Bd. II.1–3). Hildesheim u.a. 2007.

——. *Von wahrem Christenthumb. Die Urausgabe des ersten Buches (1605)*. Kritisch hg. und mit Bemerkungen versehen v. Johann Anselm Steiger (= Philipp Jakob Spener. *Schriften*. Sonderreihe, Bd. IV = Johann Arndt-Archiv, Bd. I). Hildesheim u.a. 2005.

Axmacher, Elke. „Das Spiegelbild Gottes. Johann Arndts theologische Anthropologie". *Belehrter Glaube, Festschrift Johannes Wirsching*, Hg. E. Axmacher u. Klaus Schwarzwäller. Frankfurt a. M. 1994, 11–43.

——. *Johann Arndt und Paul Gerhardt: Studien zur Theologie, Frömmigkeit und geistlichen Dichtung des 17. Jahrhunderts* (Mainzer hymnologische Studien 3). Tübingen 2001.

Bach-Nielsen, Carsten. „Spuren von Arndt in dänische[r] Missionsgeschichte, Hymnologie und Emblematik". *Johann Arndt: Rezeption und Reaktion im Nordisch-Baltischen Raum*, Hg. Arndt Jarlert. (Bibliotheca Historico-ecclesiastica Lundensis 41) Lund 1999, 15–25.

Bibliographia Gerhardina (1601–2002). Verzeichnis der Druckschriften Johann Gerhards (1582–1637) sowie ihrer Neuausgaben, Bearbeitungen und Übersetzungen, bearb. von Johann Anselm Steiger unter Mitwirkung von Peter Fiers (Doctrina et Pietas I, 9). Stuttgart 2003.

Blaufuß, Dietrich. „Arndts 'Wahres Christentum': Die Ausgabe 1673/74 unter Beteiligung von Philipp Jacob Spener". *Wolfenbütteler Barock-Nachrichten* 5 (1978): 216f und 6 (1979): 378.

Braw, Christian. *Bücher im Staube: Die Theologie Johann Arndts in ihrem Verhältnis zur Mystik* (Studies in Medieval and Reformation Thought 39). Leiden 1985.

——. „Das Gebet bei Johann Arndt". *Pietismus und Neuzeit* 13 (1987): 9–24.

Brecht, Martin. „Philipp Jakob Spener und das Wahre Christentum". *Pietismus und Neuzeit* 4 (1979): 119–154.

——. „Das Aufkommen der neuen Frömmigkeitsbewegung in Deutschland". *Geschichte des Pietismus*, Hg. M. Brecht. Bd. 1. Göttingen 1993, 113–203.

——. „Luthertum mit Johann Arndt zwischen Calvinisten und Katholiken: Die Chronik Sweder Scheles von Weleveld/Welbergen (1569–1639)". *Rezeption und Reform. Festschrift für Hans Schneider zu seinem 60. Geburtstag*. Hg. Wolfgang Breul-Kunkel und Lothar Vogel (Quellen und Studien zur hessischen Kirchengeschichte 5). Darmstadt 2001, 137–155.

——. *J. V. Andreae und Herzog August zu Braunschweig-Lüneburg. Ihr Briefwechsel und ihr Umfeld* (Clavis Pansophiae 8). Stuttgart 2002.

Breler, Melchior. *Warhafftiger Glaubwirdiger vnd gründlicher Bericht von den vier Büchern vom Wahren Christenthumb Herrn Johannis Arndten auß den gefundenen brieflichen Vrkunden zusammen getragen*. Lüneburg 1625.

Egard, Paul. *Ehrenrettung Johannis Arndten/ Das ist/ Christliche vnd in Gottes Wort wolgegründete Erinnerung/ was von D. Lucae Osiandri, Theologiae Professoris zu Tübingen Vrtheil vnd Censur, uber Johan Arndten wahres Christenthumb/ sey zu halten. Allen Frommen vnd Gottfürchtigen zu nothwendiger Nachrichtung/ aus Liebe der Warheit beschrieben vnnd ans Liecht gegeben* [...]. Lüneburg 1624.

Fritz, Eberhard. „Bäuerlicher Bücherbesitz in Bernloch 1603 und 1749". *Blätter für Württembergische Kirchengeschichte* 89 (1989): 27–47.

Gerhard, Johann. *Meditationes Sacrae (1606/7). Lateinisch-deutsch*. Kritisch hg., kommentiert und mit einem Nachwort versehen von Johann Anselm Steiger, 2 Bde. (Doctrina et Pietas I, 3). Stuttgart 2000.

——. *Postilla: Das ist/ Erklärung der Sontäglichen vnd fürnehmesten Fest=Euangelien/ vber das gantze Jahr* [...]. 3 Teile und Appendix. Jena 1613.

Geyer, Hermann. *Verborgene Weisheit. Johann Arndts 'Vier Bücher vom Wahren Christentum' als Programm einer spiritualistisch-hermetischen Theologie* (Arbeiten zur Kirchengeschichte 80/I–III). 3 Bde. Berlin 2001.

Greschat, Martin. „Die Funktion des Emblems in Johann Arnds 'Wahrem Christentum'". *Zeitschrift für Religions- und Geistesgeschichte* 20 (1968): 154–174.

Halverscheid, Heinrich. *Lumen Spiritus prius quam Scriptura intellecta: Hermann Rahtmanns Kritik am lutherischen Schriftprinzip*. Diss. Marburg 1971.

Hamm, Berndt. „Johann Arndts Wortverständnis: Ein Beitrag zu den Anfängen des Pietismus". *Pietismus und Neuzeit* 8 (1982): 43–73.

Hauke, Rainer. *Gott-Haben—um Gottes Willen: Andreas Osianders Theosisgedanke und die Diskussion um die Grundlagen der evangelisch verstandenen Rechtfertigung. Versuch einer Neubewertung eines umstrittenen Gedankens* (Kontexte. Neue Beiträge zur Historischen und Systematischen Theologie 30). Frankfurt a.M. 1999.

Ingen, Ferdinand van. „Die Wiederaufnahme der Devotio Moderna bei Johann Arndt und Philipp von Zesen". *Religion und Religiosität im Zeitalter des Barock*, Hg. Dieter Breuer u.a. (Wolfenbütteler Arbeiten zur Barockforschung 25). Wiesbaden 1995, 467–475.

Jarlert, Anders, Hg. *Johann Arndt—Rezeption und Reaktion im Nordisch-Baltischen Raum* (Bibliotheca Historico-ecclesiastica Lundensis 41). Lund 1999.

Jöcher, Christian Gottlieb. *Allgemeines Gelehrten=LEXICON* [...]. 4 Teile. Leipzig 1750f (Reprint Hildesheim 1960f).

Koch, Ernst. „Dorfpfarrer als Leser. Beobachtungen an Visitationsakten des 18. Jahrhunderts im Herzogtum Sachsen-Gotha". *Pietismus und Neuzeit* 21 (1995): 274–298.

Koch, Traugott. *Johann Habermanns 'Betbüchlein' im Zusammenhang seiner Theologie* (Beiträge zur Historischen Theologie 117). Tübingen 2001.

Koepp, Wilhelm. *Johann Arndt: Eine Untersuchung über die Mystik im Luthertum* (Neue Studien zur Geschichte der Theologie und Kirche 13). Berlin 1912.

Krummacher, Hans-Henrik. „Andreas Gryphius und Johann Arndt: Zum Verständnis der 'Sonn- und Feiertags-Sonette'". *Formenwandel* (Festschrift Paul Böckmann), Hg. Walter Müller-Seidel. Hamburg 1964, 116–137.

——. *Der junge Gryphius und die Tradition: Studien zu den Perikopensonetten und Passionsliedern*. München 1976.

Inge Mager: „Gottes Wort schmecken und ins Leben verwandeln: Johann Arndts Schriftverständnis". *Jahrbuch für finnisch-deutsche Literaturbeziehungen* 24 (1992): 149–158.

——. „Spiritualität und Rationalität: Johann Arndt und Georg Calixt in Norddeutschland im 17. Jahrhundert". *Jahrbuch der Gesellschaft für Niedersächsische Kirchengeschichte* 90 (1992): 31–41.

Mahlmann, Theodor. „Die Stellung der unio cum Christo in der lutherischen Theologie des 17. Jahrhunderts". *Unio. Gott und Mensch in der nachreformatorischen Theologie. Referate des Symposiums der Finnischen Theologischen Literaturgesellschaft in Helsinki 15.–16. November 1994*, Hg. Matti Repo und Rainer Vinke (Veröffentlichungen der Finnischen Theologischen Literaturgesellschaft 200 [Schriften der Luther-Agricola-Gesellschaft 35]). Helsinki 1996, 72–199.

Meinhold, Peter. „Die Marienverehrung in der Deutschen Mystik". *Saeculum* 27 (1976): 180–196.

Milchsack, Gustav. „Johann Arndts 'Vier Bücher vom wahren Christentum' 1605". *Braunschweigisches Magazin* 29 (1923): 15.

Müller-Mees, Elke. *Die Rolle der Emblematik im Erbauungsbuch aufgezeigt an Johann Arndts '4 Büchern vom wahren Christenthum'*. Diss. phil. Köln. Düsseldorf 1974.

Nagy, Márta. „'Pietas Hungarica': Die ungarischen Übersetzungen des Wahren Christentums und des Paradiesgärtleins von Johann Arndt". *Halle und Osteuropa. Zur europäischen Ausstrahlung des hallischen Pietismus*, Hg. Johannes Wallmann u. Udo Sträter. Tübingen 1998, 275–282.

Osiander, Lukas d. J. *Theologisches Bedencken/ Vnd Christliche Treuhertzige Erinnerung/ welcher Gestalt Johann Arndten genandtes Wahres Christenthumb/ nach Anleitung deß H. Worts Gottes/ vnd der reinen Evangelischen Lehr vnd Bekandtnüssen/ anzusehen und zuachten seye* [...].Tübingen 1624.

Pleijel, Hilding. „Die Bedeutung Johann Arndts für das schwedische Frömmigkeitsleben". *Der Pietismus in Gestalten und Wirkungen. Festschrift Martin Schmidt*, Hg. Heinrich Bornkamm. (Arbeiten zur Geschichte des Pietismus 14). Bielefeld 1975, 383–394.

Peil, Dietmar. „Zur Illustrationsgeschichte von Johann Arndts 'Vom wahren Christentum'. Mit einer Bibliographie". *Archiv für Geschichte des Buchwesens* 18 (1977): 963–1066.

——. *Zur 'angewandten Emblematik' in protestantischen Erbauungsbüchern: Dilherr—Arndt—Francisci—Scriver* (Beihefte zum Euphorion 11). Heidelberg 1978.

Schilling, Michael. *Imagines Mundi. Metaphorische Darstellungen der Welt in der Emblematik* (Mikrokosmos. Beiträge zur Literaturwissenschaft und Bedeutungsforschung 4). Frankfurt a. M. 1979.

Repo, Matti. „Die christologische Begründung der Unio in der Theologie Johann Arndts". *Unio: Gott und Mensch in der nachreformatorischen Theologie*, Hg. Matti Repo u. Rainer Vinke. (Referate [...] 1994). Helsinki 1996, 72–199.

Rost, Georg. *Amica ac fraterna Admonitio Super Controversijs DE VERO DN. JOANNIS ARNDTEN generalis in Ducatu Luneburgico Superintendentis, p. m. Christianismo, inter D. D. Lucam Osiandrum & M. Henricum Varenium, Dn. Paulum Eggardum, aliosque Theologos & politicos ortis* [...]. Rostock 1626.

——. *EXAMEN BREVIS CONSIDERATIONIS VARENIANAE APOLOGETICUM, Seu Justa ac modesta defensio amicae ac fraternae Admonitionis super controversijs de vero ARNDI p. m. Christianismo ortis, IN QVA ARNDIANI CHRIstianismi orthodoxia approbatur, heterodoxia fraternè emendatur, & quaestiones ac controversiae motae solidè ac nervosè discutiuntur, & à Φλυαρίαις, convitijs & corruptelis M. HENRICI VARENII vindicantur.* Rostock 1628.

Seebaß, Gottfried. *Das reformatorische Werk des Andreas Osiander* (Einzelarbeiten aus der Kirchengeschichte Bayerns 44). Nürnberg 1967.

——. „Osiander, Andreas". *Theologische Realenzyklopädie* 25 (1995): 507–15.

Schneider, Hans. „Johann Arndt und die makarianischen Homilien". *Makarios-Symposion über das Böse*, Hg. Werner Strothmann (Göttinger Orientforschung, Reihe 1: Syriaca 24). Wiesbaden 1983, 186–222.

——. „Johann Arndt und Martin Chemnitz: Zur Quellenkritik von Arndts 'Ikonographia'". *Der zweite Martin der Lutherischen Kirche* (Festschrift zum 400. Todestag von Martin Chemnitz), Hg. Wolfgang A. Jünke. Braunschweig 1986, 201–223.

——. „Johann Arndts Studienzeit". *Jahrbuch der Gesellschaft für Niedersächsische Kirchengeschichte* 89 (Festschrift Hans-Walter Krumwiede) 1991: 133–175.

——. „Johann Arndt als Lutheraner?". *Die lutherische Konfessionalisierung in Deutschland*, Hg. Hans-Christoph Rublack (Schriften des Vereins für Reformationsgeschichte 197). Gütersloh 1992, 274–298.

——. „Johann Arndts 'verschollene' Frühschriften". *Pietismus und Neuzeit* 21 (Festschrift Johannes Wallmann) 1995: 29–68.

——. „Johann Arndt als Paracelsist". *Neue Beiträge zur Paracelsus-Forschung*, Hg. Peter Dilg u. Hartmut Rudolph. Stuttgart 1995, 89–110.

Sommer, Wolfgang. „Johann Arndts Wirken in Celle als Generalsuperintendent des Fürstentums Lüneburg und Herzog Christian d. Ä.. von Braunschweig-Lüneburg". *Celler Chronik* 2 (1985): 7–38.

——. *Gottesfurcht und Fürstenherrschaft. Studien zum Obrigkeitsverständnis Johann Arndts und lutherischer Hofprediger zur Zeit der altprotestantischen Orthodoxie* (Forschungen zur Kirchen- und Dogmengeschichte 41). Göttingen 1988.

——. *Politik, Theologie und Frömmigkeit im Luthertum der Frühen Neuzeit: Ausgewählte Aufsätze.* (Forschungen zur Kirchen- und Dogmengeschichte 74). Göttingen 1999: „Johann Sauberts Eintreten für Johann Arndt im Dienst einer Erneuerung der Frömmigkeit", 239–262.

——. „Gespräche zwischen Arndt und Spener im Reich der Toten. Ein Beitrag zur

Wirkungsgeschichte Johann Arndts im sächsischen Pietismus". *Rezeption und Reform* (Festschrift Hans Schneider), Hg. Wolfgang Breul-Kunkel u. Lothar Vogel (Quellen und Studien zur hessischen Kirchengeschichte 5). Darmstadt 2001, 267–293.

Steiger, Johann Anselm. *Johann Gerhard (1582–1637): Studien zu Theologie und Frömmigkeit des Kirchenvaters der lutherischen Orthodoxie* (Doctrina et Pietas I, 1). Stuttgart-Bad Cannstatt 1997.

——. „'Das Wort sie sollen lassen stahn [...]': Die Auseinandersetzung Johann Gerhards und der lutherischen Orthodoxie mit dem Danziger Pfarrer Hermann Rahtmann und deren abendmahlstheologische und christologische Implikate". *Zeitschrift für Theologie und Kirche* 95 (1998): 338–365.

——. „Rahtmann, Hermann". *Religion in Geschichte und Gegenwart.* 4. Aufl., Bd. 7 (2004), 24.

Sträter, Udo. „Egard, Paul". *Religion in Geschichte und Gegenwart.* 4. Aufl., Bd. 2 (1999), 1065.

Tiililä, Osmo. „Die ältesten Übersetzungen der deutschen pietistischen Literatur in Finnland". *Theologische Literaturzeitung* 84 (1959): 330–334.

Tschizewskij, Dmitrij. „Das 'Wahre Christentum' Arndts in Rußland". *Evangelium und Osten* 8 (1935): 41–47.

Varenius, Heinrich. *Christliche/ Schrifftmässige/ wolgegründete Rettunge der Vier Bücher vom wahren Christenthumb/ des seligen/ vmb die Kirche Christi wolverdienten recht Lutherischen Evangelischen Theologi H. Johannis Arndten* [...]. 2 Teile. Lüneburg 1624 [u. Nachdruck 1689].

——. *Eine Antrits=Predigte Von Dem Ampte aller getrewen Lehrer vnnd fleissigen Zuhörer/ voll schöner Lehr vnd Erinnerunge/ Aus Dem ordentlichen Evangelio Dominica Sexagesimae, Jn der Kirchen zu S. Marien in Vltzen gehalten* [...]. Lüneburg 1628.

——. *BREVIS CONSIDERATIO ADMONITIONIS GEORGII ROSTII SUPER CONTROVERSIIS DE VERO B. ARNDI CHRISTIANISMO ORTIS. In qua B. ARNDI innocentia ejusdemque Christianismi orthodoxa veritas asseritur, & a Φλυαρίαις & corruptelis G. Rostij vindicatur, quaestionesque motae dextrè & solidè explicantur* [...]. Lüneburg 1626.

Wackernagel, Philipp. *Das deutsche Kirchenlied von der ältesten Zeit bis zu Anfang des XVII. Jahrhunderts.* Bd. 3. Leipzig 1870.

Walch, Johann Georg. *Historische und Theologische Einleitung in die Religions-Streitigkeiten der Evangelisch-Lutherischen Kirche.* Faksimile-Neudruck der Ausgabe Jena 1733–1739 mit einem Nachwort von Dietrich Blaufuß. Bd. 3/1. Stuttgart 1985.

Wallmann, Johannes. „Herzog August zu Braunschweig und Lüneburg als Gestalt der Kirchengeschichte unter besonderer Berücksichtigung seines Verhältnisses zu Johann Arndt". *Pietismus und Neuzeit* 6 (1980): 9–32.

——. „Johann Arndt und die protestantische Frömmigkeit. Zur Rezeption der mittelalterlichen Mystik im Luthertum". *Frömmigkeit in der frühen Neuzeit. Studien zur religiösen Literatur des 17. Jahrhunderts in Deutschland*, Hg. Dieter Breuer (Chloe 2). Amsterdam 1984, 50–74.

——. „Johann Arndt und die protestantische Frömmigkeit". *Jahrbuch der Hessischen Kirchengeschichtlichen Vereinigung* 35 (1984): 371–379.

——. „Osiander, Lukas (II.)". *Religion in Geschichte und Gegenwart.* 4. Aufl., Bd. 6 (2003), 721.

Weber, Edmund. *Johann Arndts Vier Bücher vom wahren Christentum als Beitrag zur protestantischen Irenik des 17. Jahrhunderts: Eine quellenkritische Untersuchung* (Studia irenica 2). Hildesheim [3]1978 (Marburg [1]1969).

Weigel, Valentin. *Gebetbuch (Büchlein vom Gebet). Vom Gebet. Vom Beten und Nichtbeten.* Hg. Horst Pfefferl (*Sämtliche Schriften.* Neue Edition. Bd. 4). Stuttgart 1999.

——. *Der güldene Griff. Kontroverse um den 'Güldenen Griff': Vom judicio im Menschen.* Hg. Horst Pfefferl (*Sämtliche Schriften.* Neue Edition. Bd. 8). Stuttgart 1997.

Wimmel, Herbert. *Sprachliche Verständigung als Voraussetzung des 'Wahren Christentums'. Untersuchungen zur Funktion der Sprache im Erbauungsbuch Johann Arndts* (Kasseler Arbeiten zur Sprache und Literatur 10). Frankfurt a. M. 1981.

Zoltán, Csepregi. „Az Igaz keresztyénség hamis impresszuma. Adalékok a dunántúli Arndt-recepcióhoz". *Kulturelle Bestrebungen in der frühen Neuzeit. Festschrift für Bálint Keserü.* (Materialien zur Geschichte der Geistesströmungen des 16.–18. Jahrhunderts in Ungarn). Szeged 1997, 51–63.

DER BILDUNGSWEG DES JUNGEN BERNHARD VARENIUS

Klaus Lehmann

Abstract

The Formative Years of Bernhard Varenius

Varenius attended school at a time when the traditional scholarly disciplines were entering a new phase and still had to establish legitimate positions. Learning was strictly governed by statutes and regulations. Varenius was growing up when the world was no longer explained on the bases of texts, especially the Bible; gradually observation, experience and experiment, in addition to mathematical models, were beginning to dominate.

At Uelzen, Helmstedt and Hamburg Varenius was taught by men who were widely read in different fields of learning, among them polymaths like Conring or Jungius; and through them he became aware of the changing views of the time. Established subjects of teaching were splitting apart and new disciplines being developed. Thus the early seventeenth century marks the beginning of geography as a subject in its own right, composed of previously independent fields of enquiry, and Varenius, keenly observing his teachers' endeavours, eventually contributed to this development. In fact, we can see from Jungius' correspondence with his students that Varenius was one of the best pupils he ever had during his teaching career of over twenty-eight years.

Although he was trained in comprehending what classical authors had written, Varenius soon progressed to systematising knowledge himself. New fields of scholarship opened up with reproducing knowledge in taxonomic surveys. In 1997 Peter Burke described the process of developing programmes in order to reform classifications of knowledge. This is exactly the task Varenius set himself. Jungius had provided him with the necessary skills in logical reasoning and in organising his material in slip boxes. Without filing his notes in such a way (Jungius' practice must have been his model), it would have been impossible for him to have arranged his huge number of excerpts in such a short time. In *Geographia Generalis* Varenius made use of this technique, but whether Jungius' work on his own geography project influenced Varenius directly is a question beyond the scope of this study. Varenius certainly used his sources critically in designing a classification of geography. Göllnitz was consulted, and Varenius had been familiar with Clüver's book as a standard textbook since his studies at Helmstedt University. But in contrast to Clüver, Varenius did not place the earth at the centre of the universe.

In Helmstedt Aristotle was still taught in the humanist manner, which was not the case in Hamburg—where Jungius (not least through his own *Logica*) encouraged Varenius to question Aristotle radically.

Thus we can say that Varenius' attraction to geography owes something to the example of his teachers' scientific interests. Jungius' lectures in geography may have led him to study the earth and the universe. Jungius' notes, however, much as their title "Geographia generalis" might suggest, do not allow us to presume that Varenius could have consulted them for his own work, as they are merely unstructured notes Jungius used for his lectures. Only the terms *geographia generalis* and *geographia specialis* turned out to be useful for Varenius.

Wir sprechen heute gerne von einer Wissens- oder Informationsgesellschaft, die von Fachleuten und ihren wissenschaftlichen Methoden beherrscht wird. Es ist reizvoll, sich derer zu erinnern, die um 1600 begannen, Strukturen und Systematisierungen zu entwickeln, die den Weg zu einer Wissensökonomie bahnten. Um solch einen ‚Systematisierer' handelt es sich bei Bernhard Varenius in der Mitte des 17. Jahrhunderts. Dies soll in diesem Beitrag gezeigt werden, indem die bisher bekannt gewordenen biographischen Daten seiner Kindheit und Jugend aktualisiert werden und sein Werdegang im Kontext der Bildungslandschaft seiner Zeit dargestellt wird. Um Varenius und sein Werk aus seiner Zeit heraus zu verstehen, ist gerade auch der frühe Lebensweg wichtig, so dass die Forschungs- und Lehrinteressen seiner Lehrer im Vordergrund dieser Studie stehen. Ihre Denkanstöße, Einflüsse, Arbeitsweisen und -haltungen weckten seine Neigungen und den Wissensdurst und nahmen direkt oder indirekt Einfluss auf seine Schriften. Sie lassen uns besser verstehen, wie ein Doktor der Medizin seinen Forschungsdrang inmitten der geistigen Auseinandersetzungen seiner Zeit auf die Mathematik und schließlich auf geographische Grundfragen richten konnte.

Varenius' bekanntere Schriften, die Kompilation zu Japan (1649) und die *Geographia Generalis* (1650), sein Hauptwerk, lagen zwar in mehreren Auflagen, zum Teil nachträglich bearbeitet, in verschiedenen Ländern vor, nur zu seiner Person wusste man wenig. Fast 200 Jahre nach dem Tode Bernhard Varenius' bezog sich Alexander von Humboldt—einer der letzten großen Universalgelehrten—in seinem Spätwerk *Kosmos* (1845) auf Varenius.[1] Er spricht „von dem großen

[1] Siehe auch den Beitrag von Sandra Rebok in diesem Band.

Geographen Bernhard Varenius",[2] weiß aber über seine Herkunft nicht mehr, als dass „dieser scharfsinnige Geograph ein Deutscher und zwar ein Lüneburger" war.[3] Lange galt er als ein Sohn der Stadt Uelzen. Im Gymnasium, das er in Hamburg bei Joachim Jungius besuchte, wird sein Geburtsort als „Ulzensius Luneburgicus" angegeben. Erst mit der Veröffentlichung des Jungius-Briefwechsels 1850 durch Guhrauer und eines weiteren Jungius-Schülers konnte der Zusammenhang zwischen den beiden Brüdern August und Bernhard Varenius und somit der Geburtsort bestimmt werden. 1880 trug Breusing[4] die wenigen Lebensdaten des Bernhard Varenius zusammen und benannte Hitzacker (Elbe) eindeutig als seinen Geburtsort.

Zuletzt hat sich Rainer Kastrop 1982 im Zusammenhang mit der Bedeutung des Varenius knapp mit seiner Biographie beschäftigt. Doch der Blick war eher auf die Vorläufer gerichtet, die Varenius bei der Erstellung seiner *Geographia Generalis* herangezogen hat bzw. haben könnte. Neue Beiträge zur Jungius-Forschung in den 1980er und 1990er Jahren (genannt sei hier besonders Christoph Meinel), die Kastrop noch nicht berücksichtigen konnte, regen dazu an, neue Akzente zur Person und zum Verständnis des Varenius zu setzen.

Persönliche Lebenszeichen aus der Kinder- und Jugendzeit des Bernhard Varenius fehlen leider gänzlich, und dürre Fakten können nur indirekt erschlossen werden. So ist es nicht verwunderlich, dass bislang Varenius' Zeit in Hitzacker, Uelzen und Helmstedt in seiner Biographie unberücksichtigt geblieben ist. Insbesondere fehlten Zeugnisse, wo er sich zwischen 1635, dem Tode seines Vaters, und dem Besuch des Johanneums bei Joachim Jungius ab 1640 aufhielt. Durch den Nachweis seiner Immatrikulation in Helmstedt, den Rohrbach schon erbracht, aber nicht veröffentlicht hatte, ist nunmehr festzustellen, dass weitere Lehrer und nicht nur Jungius Einfluss auf den Bildungsweg des jungen Varenius genommen haben. Ergänzendes Quellenmaterial sind hier die Immatrikulationseintragungen in Königsberg und Leiden (Niederlande) sowie neun Briefe von Varenius an seinen Lehrer Joachim Jungius in dem Zeitraum 1643–1648 sowie zwei Briefe von Jungius.

---

[2] Alexander von Humboldt, *Kosmos: Entwurf einer physischen Weltbeschreibung*, Bd. 1, Stuttgart 1845, 60.

[3] *Kosmos*, 75 Anm. 7.

[4] A. Breusing, „Lebensnachrichten von Bernhard Varenius," *Petermanns Geographische Mitteilungen* 26 (1880): 136ff.

Bei den schulischen Auseinandersetzungen und bei der Frage, was Varenius an Basiswissen mit auf den Weg bekommen hat, sind neben den Lehrerpersönlichkeiten auch deren didaktisch-methodisches Vorgehen und die Literatur, die ihm zur Verfügung stand, von großer Bedeutung. Hier bietet sich besonders an, die im Nachlass von Joachim Jungius befindliche „Geographia Generalis" heranzuziehen. Es handelt sich dabei um ein von Jungius verfasstes Zettelwerk, das wegen seiner eigenwilligen Schrift aber nicht ausgewertet worden ist und nun unten näher beschrieben werden kann.

Elternhaus und Jugend in Hitzacker

Bernhard Varenius wurde 1622 in Hitzacker (Elbe) als Sohn des Hofpredigers Heinrich Varen(ius) am Hofe Herzog Augusts d. J. geboren. Sein Geburtsjahr ergibt sich aus einem Eintrag im *Album studiosurum* der Universität Leiden: 1645, Maji. 6 Bernhardus Varenius Luneburgensis aet. 23 Stud. med. habitat apud Hermannum Bogemaker.[5] Der Vater stammte aus Herford. Heinrich Varenius promovierte 1617 zum Magister der Theologie und erhielt sogleich die Hofpredigerstelle, ohne erst wie üblich Hauslehrer oder Schulmeister werden zu müssen. Zur Persönlichkeit von Varenius' Mutter, die aus einer angesehenen Theologenfamilie aus Rostock (Enkelin des David Chytraeus) stammte, ist uns wenig bekannt. Sie starb 1623, im selben Jahr wie des Herzogs erste Frau, vielleicht bei der Geburt einer Schwester Bernhards im Kindbett.[6] Es ist möglich und entspricht auch der Zeit, dass der Vater sehr schnell wieder geheiratet hat. Einen weiteren Bruder nennt demnach die Helmstedter Matrikel vom 7. Mai 1640: Georgius Henricus Varenius Ultzensis.[7]

---

[5] Breusing 137.

[6] In Schomerus' Leichenpredigt auf August Varenius wird eine „Jungfr. Schwester" am Totenbett von August erwähnt.

[7] Es ist zu vermuten, dass dieser identisch ist mit Georg Varenius, der von 1656–1666 als Pfarrer in Wustrow im dortigen Kirchenregistraturbuch nachweisbar ist. Dort heißt es, er sei „ein wackerer, hochgelehrter aber auch liederlicher Mann gewesen, der sich besser zum Welt- und Kriegsmann als zum Geistlichen geschicket." Wustrow, Kirchenregistraturbuch, S. 451. Er wurde aus Wustrow vertrieben, angeblich wegen Ehebruchs, und man beschuldigte ihn, das Pfarramt und drei Bürgerhäuser angezündet zu haben. Später soll er nach Hamburg gegangen sein und ist verschollen. Vgl. *Wendland Lexikon*, Hg. Wolfgang Jürries, Bd. 2, sub „Varenius IV" (Alfred Kelletat), im Druck.

Der Vater Heinrich Varenius verfasste eine Verteidigungsschrift für Johann Arndt,[8] den Wegbereiter der Erbauungstheologie. Sein rationaler Widersacher, der aber an gemeinsamer Front gegen die Orthodoxie stand, war Calixt, der Helmstedter Professor, den wir noch kennen lernen werden. Damit nahm Varenius Partei in einer heftigen Auseinandersetzung, die zur Folge hatte, dass er seine Pläne, Professor zu werden, nicht mehr verwirklichen konnte. Da sowohl Herzog Christian zu Celle wie auch Herzog August in Hitzacker Arndt und Varenius unterstützten, entschädigte man Heinrich Varenius nach dem Tode Arndts mit der Präpositur in Uelzen Ende 1627 (als Propst und Superintendent), was einer erheblichen Aufwertung gleichkam. Die Familie(?) zog mit den Söhnen nach Uelzen, und dort starb Heinrich Varenius dann 1635.[9]

Bis zu seinem vierten Lebensjahr lebte Bernhard also in dem kleinen Inselstädtchen Hitzacker fern der Wirren des Dreißigjährigen Krieges. Das Elternhaus stand direkt bei der Kirche und ist heute das Gemeindehaus.[10] Seit 1617 gab es in Hitzacker eine Lateinschule, an der Heinrich Varenius vielleicht unterrichtete; sie war zeitweise auch eine Prinzenschule, solange bis Herzog August 1634 Hitzacker verließ.[11] Die Stadtinsel Hitzacker, eingeschlossen von der Jeetzel, die in Hitzacker in die Elbe fließt, hatte zu dieser Zeit 49 Hausstellen. Zu Beginn des Krieges hatte Hitzacker nur ungefähr 300 Einwohner, deren Zahl sich mit dem Hof und den durch die Residenz bedingten Dienstleistungen auf etwa 350 erhöht hat.[12] Das Herzogtum umfasste das Amt Hitzacker,

[8] Zum Theologenstreit siehe Steigers Studie in diesem Band.

[9] Vgl. Nachlass C. Rohrbach, unveröffentlichtes Manuskript, Eutiner Landesbibliothek LB 10/1058).

[10] Einen Hinweis auf Heinrich Varenius gibt eine Verkaufsakte im Archiv des Museums im Alten Zollhauses: „Verkauf eines Bürgergutes in Hitzacker durch Antonius Kahle, Ratsverwandter in ‚Borchtorff', an Schwiprecht Malchow und Regelung betreffend die darauf haftende Schuld gegenüber dem Pastor in Hitzacker Mag. Heinrich Varenius (‚Varen', ‚Farenius') 1624" in *Findbuch zum Archiv der Stadt Hitzacker*, Alte Abteilung, Hitzacker 1998, 29 Nr. 255.

[11] 1665 wurde der einzige Lehrer der Lateinschule in Hitzacker wegen seiner Lateinschwäche entlassen; so: Johannes Tütken, *Höhere und mittlere Schulen des Herzogtums Braunschweig-Wolfenbüttel, der Herrschaft Dannenberg und der Grafschaft Blankenburg im Spiegel der Visitationsprotokolle des Generalschulinspektors Christoph Schrader (1650–1666)*, Wiesbaden 1997, 16, 536f.

[12] Gerd Heinrich, „Nova-Ithaka: Fürstliches Landleben und soziale Wirklichkeit im Herzogtum Dannenberg-Hitzacker zwischen 1605 und 1635," *Fruchtblätter: Freundesgabe für Alfred Kelletat*, Hg. Harald Hartung et al., Berlin 1977, 262.—Gegen Ende des Kriegs hatte sich die Bevölkerungszahl halbiert, und rund die Hälfte der Hausstellen war unbesetzt (Heinrich 275).

der Herzog war also wenig begütert; er hatte sich wohl nur durch die Apanage seiner ersten Gattin den Bau eines Schlosses in der Stadt leisten können.

Der schulische Weg

Varenius' Bildungsweg wird maßgeblich von bedeutenden Lehrern begleitet und beeinflusst: durch Magister Georg Krukius, seinen Lehrer an der Lateinschule in Uelzen, später durch die Professoren Calixt, Conring, Schrader und Hornejus von der Helmstedter Academia Juliae der Universität, und schließlich dem Rektor Joachim Jungius und dem Lehrer Johann Adolf Tassius, die beide am Akademischen Gymnasium Johanneum in Hamburg lehrten.

Die Lateinschule in Uelzen war eine Einrichtung der Stadt. „Lateinschule" war eine Sammelbezeichnung für höhere und mittlere Schulen, und der Name verweist bereits auf ihr dominantes Merkmal.[13] Uelzen sowie Salzwedel und Lüneburg hatten eine Lateinschule 1. Ordnung, zu der vornehmlich auch Schüler aus dem Gebiet Lüchow-Dannenberg hinüberwechselten.[14] In Uelzen diente die Lateinschule, deren Chronik Steinmeyer erforscht hat,[15] nicht nur dem Unterricht derer, die studieren wollten, sondern war zugleich die allgemeine Stadtschule, in der die Knaben für einen bürgerlichen Beruf vorbereitet wurden. Die Schüler waren auf drei Klassen, Prima, Sekunda, Tertia—zu verstehen als locker getrennte Schulabteilungen—in einem oder zwei Schulzimmern verteilt. Rector, Conrector und Cantor unterrichteten sie. Innerhalb der Abteilungen wurden die besten und zuverlässigsten Schüler zusammengefasst. Sie unterstützten die Lehrer beim Unterricht, indem sie

[13] Tütken 1997 setzt sich mit der Schulordnung Herzog Augusts von 1651 und den Visitationsberichten auseinander und erstellt so ein Bild der Schule in der Mitte des 17. Jahrhunderts. Er nimmt z. T. auch Bezug auf die Schulsituation vor der Zeit der Visitation sowie auf die Ausbildungssituation der Lehrer in Helmstedt. Zur Lateinschule führt er aus (1997: 14), dass, während in den katholischen Territorien sich die Bezeichnung „Gymnasium" schon durchgesetzt habe, die Schulverwaltung im Herzogtum Braunschweig-Wolfenbüttel diese Typenbezeichnung erst nach der westfälischen Zeit verwandte.

[14] Der Generalschulinspektor Schrader unterschied nur in seltenen Fällen zwischen 1. und 2. Ordnung, sondern verwandte den Begriff Schule (Tütken 1997: 14). Im Zusammenhang mit Uelzen wird von 1. Ordnung gesprochen (193ff).

[15] Rudolf Steinmeyer, *Die Chronik der Herzog-Ernst-Schule, der früheren Lateinschule in Uelzen*, Uelzen 1964, 17–31.

den Schülern aufgegebene Fragen abhörten und selbst dabei von ihren Lehrern überprüft wurden.

Das 1589 erlassene Schulgesetz und die Schulordnung geben Auskunft, womit Varenius in seiner ersten Schulzeit konfrontiert wurde: Glaubenslehre, Sprachen und vorbereitendem Unterricht in den philosophischen Wissenschaften (artes). Hinzu kamen Musik und Gesangstunden, dafür kein Unterricht im Deutschen, in Geschichte und Erdkunde, in Mathematik und Naturwissenschaft, wohl aber ein Computum astronomiae, in dem die astronomischen Elemente mit der kirchlichen Festrechnung vereinigt wurden. Der Katechismus des Chytraeus (Bernhards Urgroßvater) war Arbeitsvorlage, die Sonntagspredigt, zuvor in der Schule repetiert, gehörte mit zum Unterricht. Der Latein- und Griechisch-Unterricht (Latein wurde bevorzugt) sollte dazu befähigen, ein gewandtes Latein zu sprechen, Reden zu halten, Texte und Gedichte zu verfassen. Cicero war als Lektüre besonders beliebt. „Die Schüler der ersten Klasse sollen beim Unterricht und jeder Zusammenkunft nicht das geringste Wort in der Muttersprache reden, wenn sie die Strafen ihrer Lehrer vermeiden wollen."[16] Zum Unterricht gehörten ebenso Theateraufführungen eines Terenz, Plautus oder Seneca.

Der griechische Unterricht blieb weit hinter seinem Anspruch zurück. Öffentliche Reden und Deklamationen, sorgsam nach der vorgegebenen Form ausgearbeitet, wurden zu verschiedenen Gelegenheiten gefordert. Der vorbereitende Unterricht in den philosophischen Wissenschaften beschränkte sich auf die Dialektik eines Melanchthon.

Schulbeginn war im Sommer um 6 Uhr, im Winter eine Stunde später. Mit Pausen zur Vertiefung der Arbeiten wurde außer am Sonntag meistens bis 15 Uhr unterrichtet. Sonntags mussten sich die Schüler um 5 und um 8 Uhr in einer feierlichen Prozession von der Schule zum Gottesdienst begeben, wobei sie Chorgesänge anstimmten. Das Schuljahr endete mit einer öffentlichen Prüfung aller Klassen. Dem Propst der Stadt, also Bernhard Varenius' Vater, oblagen die Inspektion der Schule und die Leitung der Examen, deren Lektionen er sogar noch bestimmte.[17] Der Schulalltag war durch eine Vielzahl von Paragraphen geregelt. Pünktlichkeit und Fleiß waren Pflicht.

---

[16] Steinmeyer 21.

[17] Loc. cit.

> Während des Lehrervortrages sollen die Schüler aufmerksam zuhören. Sie sollen Bücher, Papier und Tinte bereithalten und was nützlich ist, notieren. Die gehörten Lektionen sollen sie, ohne dass jemand ihnen zuflüstert, bei geschlossenem Buch mit klarer und deutlicher Stimme nachsagen. Dabei ist für die Primaner die lateinische Sprache die Umgangssprache. Zu Haus sollen die Schüler die gehörten Lektionen wiederholen und jeden Müßiggang vermeiden.[18]

Zu den Lehrern des Varenius in Uelzen muss Georg Krukius (Krugh/Kruck) gehört haben. Krukius (1598–1659),[19] aus Eisenach gebürtig, studierte an der frühneuzeitlichen Reformuniversität Jena, die nach Melanchthons humanistisch-reformatorischem Schulkonzept begründet war, mit ihren vier Fakultäten: Philosophie, Theologie, Recht und Medizin. Jena gehörte bis zur Zeit des Dreißigjährigen Krieges zu den vier großen deutschen Universitäten neben Wittenberg, Frankfurt-Oder und Helmstedt.[20] Danach ging er nach Wien. Mit erst 28 Jahren wurde er Rektor der Lateinschule zu Uelzen. Der allgemeine Lehrkanon in Uelzen verrät nicht, welche Position Georg Krukius hier einnahm. Stand er den Gedanken Melanchthons nahe, wo man versuchte zwischen der mathematischen Theorie des Copernicus und dessen Weltbild und den Texten der Offenbarung eine Trennung mittels rationalen Beweises zu vollziehen? Oder stand er der Mathematik als Erfahrungswissenschaft—wie die Schüler Melanchthons—nahe? Aus den Akten ist zu entnehmen, dass der Rat mit seinem Amt wohl nicht zufrieden war. Wie fast alle Lehrer und auch Rektoren war auch Krukius gezwungen, seinen Unterhalt durch Nebenerwerb aufzubessern. Man lastete ihm aber an, durch seine Nebentätigkeit (als Arzt?)[21] nicht rechtzeitig zum Unterricht erschienen zu sein, und überhaupt mangele es in der Schule an Disziplin. Als man ihn zwingen wollte, einen Revers zu unterschreiben und er sich weigerte, erhielt er 1632, so Steinmeyer, seine Demission. Also wird er Varenius in der Zeit von 1628–1632 unterrichtet haben. Krukius wandte sich wieder akademischen Studien zu, und zwar der Medizin wie später sein Schüler Varenius. Das Studium der Medizin im 17. Jahrhundert hatte eine intensive Beschäftigung mit den Naturwissenschaften zur Folge. Krukius

[18] Steinmeyer 23.

[19] Zu Krukius siehe *Scandinavian Biographical Archive* 1990, microfiche 182, sowie Steinmeyer.

[20] Vgl. Peter Baumgart, „Die Anfänge der Universität Helmstedt im Spiegel ihrer Matrikel, 1576–1600", *Braunschweigisches Jahrbuch* 50 (1969): 7.

[21] Vgl. Steinmeyer 31.

erwarb den Dr. med. in Rostock 1635, ging 1638 nach Dänemark und wurde 1641 als Dr. med. et phys. Professor an der Ritterakademie in Sorø (sie war eng verbunden mit dem dänischen König Christian IV.), wo er bis zu seinem Tod lebte.

Die Umstände in Uelzen während Varenius' Kindheit waren geprägt durch die Wirren des Dreißigjährigen Krieges. Als Dreizehnjähriger erlebte er 1635, im Todesjahr seines Vaters, wie die schwedischen Truppen plündernd und brandschatzend in das Fürstentum Lüneburg einfielen. Schon durch Tillys Truppen 1629, nun durch die Schweden wurden Einquartierungen in die Stadt gelegt. Die allgemeine Notlage verschärfte sich noch mit dem Ausbruch der Pest.

An der Universität in Helmstedt (1636–1640)

Helmstedt hatte eine stadteigene Lateinschule, doch nicht dort, sondern auf der Academia Juliae Carolinae finden wir den dreizehnjährigen Bernhard und seinen fünfzehnjährigen Bruder August wieder.[22] Sein älterer Bruder August (1620–1684)[23] studierte danach in Königsberg und Rostock und wurde mit 23 Jahren bereits Professor für hebräische Sprache, promovierte noch zum Doktor der Theologie und wurde anschließend auf den theologischen Lehrstuhl in Rostock berufen. In wieweit sich der ältere Bruder um Bernhard kümmerte, ist unklar. Ihre Lebenswege kreuzten sich jedenfalls in Helmstedt, Hamburg und Königsberg.[24]

Auch Helmstedt, der neue Studienort der Brüder Varenius, litt unter den Kriegsfolgen, vor allem den Besatzungen. Mehrfach hatte die Universität zuvor schon wegen Pestepidemien geschlossen werden müssen. 1636 traten neue Fälle von Pest auf, Land und Leuten drohte „Verderbnis". Erst 1637 besserte sich die Lage. Gegen Ende des Jahres 1637 hielten sich 250 Studenten in Helmstedt auf, wenngleich man 1639 noch daran dachte, „der gefährlichen Läufte halber" die Akademie an einen anderen Ort zu verlegen.[25]

---

[22] *Album Academiae Helmstadiensis*, 105.—Semester 1634/36. August Varenius, Hitzackerensis: 17. Sept. 1634 (333, Nr. 99); Bernhard Varenius, Hitzackerensis 10. Jan 1636, (334, Nr. 187)

[23] Zu August Varenius siehe oben Schuchards Beitrag über die Familie.

[24] S. Günther, *Varenius* (Klassiker der Naturwissenschaften.4), Leipzig 1905. Nachdruck Amsterdam 1970, 135 Anm. 66.

[25] Uwe Alschner, *Universitätsbesuch in Helmstedt 1576–1810*, Braunschweig 1998, 67f.

Die Universität Helmstedt gehörte in der Mitte des 17. Jahrhunderts zu den bedeutendsten Universitäten in Deutschland, bis sie gegenüber der ebenfalls von den Welfen 1737 gegründeten Georgia Augusta in Göttingen das Nachsehen hatte.[26] (Im Zuge der französischen Besatzung unter Napoleon wurde 1810 die Universität Helmstedt geschlossen, und später fand sich kein Interesse mehr sie wieder zu öffnen.) Helmstedt ging der Ruf voraus, die Hochburg der protestantischen ‚Spielart' des Aristoteles zu sein.[27] Gegründet wurde die Universität im reformatorisch-humanistischen Geiste eines Melanchthon; zwischen 1614–1656 setzte sich nach heftigen Richtungskämpfen mit den orthodoxen Lutheranern ein moderner Späthumanismus durch mit den klassischen vier Fakultäten;[28] so spielte die Bildungsstätte mit ihrem späthumanistischen Profil eine entscheidende Rolle als Landesuniversität.

Die meisten Studenten waren in der Philosophischen oder Artisten-Fakultät, wie die Eingangsfakultät für alle genannt wurde, eingeschrieben. Der Übergang von den Lateinschulen zur Universität war kaum problematisch, das Abitur noch eine unbekannte Größe. Der Leitgedanke der vorherrschenden Didaktik war die Imitatio, die Nachahmung der kaum erreichbaren Vorbilder der Vergangenheit. Die Grundausbildung orientierte sich an der „humanistischen These": Disputation und Deklamation waren Paradestücke der oft hervorragenden Schulung. „Lehrstoff waren die christliche Elementarlehre, die lateinische (Schwerpunkt Cicero) und griechische Sprachlehre, Dialektik und Rhetorik, sodann Arithemetik und ein wenig Astronomie und Weltgeschichte. Auf der zweiten Stufe folgten Ethik, Geometrie, Astronomie und Physik."[29]

Ein Blick auf die Unterrichtsmethode der Lehrer des Varenius in dieser Zeit zeigt, welch zentrale Rolle das Gedächtnis spielte, das „als Schatzhaus des Sach- und Weltwissens" im Mittelpunkt des Lernprozesses stand. Der Lehrer präsentierte lesend das Regelwissen der neuen praecepta und erläuterte sie durch exempla. Lesen ist in Schraders Protokollen einem Synonym für Lehren gleich, da die

[26] Tütken 1997: 110f.
[27] Tütken 1997: 91.
[28] Tütken 1993: 57.
[29] Carl Haase, „Bildung und Wissenschaft von der Reformation bis 1803," *Geschichte Niedersachsens*, Hg. Hans Patze, 3. Bd., Teil 2, Hildesheim 1983, 321.

Informationseingabe weitgehend im Stil einer Vorlesung (lectio) erfolgte.[30]

Der klassische Bildungsgegenstand und die Lehrkompetenz wurden so zur ultima ratio erhoben als notwendige Vorbedingung für eine ausreichende Voraussetzung des Lehrens. Willens- und Moralbildung wurden als kognitive Informationen vermittelt: nach dem Schema Text—Lehre bzw. Studium—Verstand—Wille. Tütken wirft hier die Frage auf, ob Schrader diese Unterscheidung psychischer Instanzen im Zusammenhang mit moralischem Lernen schon vertraut war.[31] Auf die Fragwürdigkeit didaktisch-methodischer und pädagogischer Prämissen hatten schon Zeitgenossen (Comenius) hingewiesen.[32]

Die Grenzen der in Helmstedt gepflegten humanistischen Weltsicht entsprachen auch der generellen Einstellung vieler Humanisten zur Mathematik, „die durch Distanz und geringe Kompetenz gekennzeichnet war. Sie machte auf der Wissenschaftsebene diese Disziplin—insbesondere in ihrer anspruchsvollen Kombination von Astronomie/Mathematik—weitgehend zu einer außerakademischen Wissenschaft, [...]"[33] Es ging ihr weniger um die praktische Anwendung, wie sie in speziellen Rechenschulen gepflegt wurde, als um „den kognitiven Aspekt formaler Denkschulung". [34]

Die Statuten der Helmstedter Universität verpflichteten die Professoren „ihre wissenschaftlichen Leistungen zu Gottes Ruhm und zum Nutzen der Kirche" zu erbringen.[35] Dies ermöglichte ihnen bei geschickter Handhabung auch einen Interpretationsspielraum, ansonsten hatten Lehre und Forschung den anerkannten Autoritäten zu folgen. Beide Brüder Varenius begegneten in diesen Kriegszeiten in Helmstedt einerseits wohl dem Quadrovirum der Blütezeit der Universität: Calixt, Conring, Schrader und Hornejus, und erlebten andererseits wahrscheinlich eher indirekt die ersten Veränderungen nach dem Tod des Herzogs Friedrich Ulrich (1634) mit, dessen Nachfolge Herzog August der Jüngere aus Hitzacker antrat. Im Hinblick auf diese hoch angesehenen Lehrer, wobei sicher weitere zu nennen sind, die die Brüder prägten, unterschied sich die Ausbildung für Bernhard

---

30 Tütken 1997: 159.
31 Tütken 1997: 238f.
32 Tütken 1993: 94f.
33 Tütken 1997: 298.
34 Tütken 1997: 299.
35 Tütken 1997: 90.

Varenius wesentlich von Schulen auf dem Lande, auch wenn die überwiegende Zahl der Lehrer in Helmstedt ausgebildet wurde. Denn die Kontroverse um die Ansichten des Ramus (1515–1572), der für eine Vereinfachung gegenüber dem komplizierten System des Aristoteles eintrat, worauf die Lehrer gern zurückgriffen, machte die Academia eher zum Bollwerk des Aristoteles als zu einem der Ramisten, die die scholastische Handhabung der Logik des Aristoteles wegen seiner oft „weltfremden Spitzfindigkeiten" kritisierten.[36] Ramus gehörte übrigens zu den Systematisierern wie Keckermann, Alstedt und Kircher, die „die Nachwelt weniger ernst genommen hat, als Bacon, Descartes, Locke und Leibniz".[37] Varenius wurde von diesem Umfeld sicher nachhaltig berührt.

Georg Calixt (1586–1656) war gewiss einer der schillerndsten Professoren der Helmstedter Theologie. Das von ihm entwickelte Konzept einer humanistischen Erziehung auf der Grundlage der alten Sprachen gewann großen Einfluss. Er, der eine Vorliebe für die Sprache der Römer hatte und Deutsch nur im Notfall nutzte, stritt sich mit seinen orthodoxen Kollegen, war aber mehr Außenseiter, als dass er sich gegen das strenge Luthertum gestellt hätte.[38] Calixt stimmte der fortschreitende Krieg nachdenklich, und er versuchte eine Annäherung der Konfessionen zu propagieren. Er hoffte, die Einheit der Kirche ließe sich wiederherstellen, wenn die verschiedenen Konfessionen auf fundamentale Glaubenslehren verzichteten. Das musste natürlich heftige Kontroversen auslösen. Eng verband ihn dies aber auch mit der Auffassung seines Kollegen Conring, dass die Geschichte—und zwar die Profan- wie die Kirchengeschichte—für die Bewältigung der Gegenwart von grundlegender Bedeutung sei, was sich so widerspruchsfrei mit dem konfessionellen Pazifismus und mit dem politischen Pazifismus des Landesherren in Einklang bringen ließ.[39] Vielleicht erfuhr Varenius hier ein Stück ‚Religionsöffnung', so wie bei seinem weiteren Lehrer Conring vielleicht seine Ambitionen für Geschichte, Genealogie und

---

[36] Tütken 1997: 66.

[37] Peter Burke, *Papier und Marktgeschrei: Die Geburt der Wissensgesellschaft*, Berlin 2000, 119.

[38] Vgl. Inge Mager, „Reformator. Theologie u. Reformationsverständnis an der Univ. Helmstedt im 16. u. 17. Jh.," *Jahrbuch der Gesellschaft für niedersächsische Kirchengeschichte* 74 (1976): 32.

[39] Jörg Jochen Berns, „Herzog August—Frömmigkeit und kirchliche Tradition," *Sammler, Fürst, Gelehrter: Herzog August zu Braunschweig Lüneburg 1579–1666* (Ausstellungskatalog Herzog August Bibliothek 27), Braunschweig 1979, 348.

Geographie geweckt wurden. Konrad Hornejus (1590–1649) war ein früherer Studienfreund von Calixt gewesen und wurde zu seinem Mitstreiter für ein im Geiste Melanchthons gemäßigtes Luthertum. 1628 erhielt er die zweite Professur für Theologie in Helmstedt.[40]

Seit 1632 unterrichtete Hermann Conring (1606–1681) als Professor der Naturphilosophie in Helmstedt; 1636, im Jahr seiner Promotion zum Doktor der Medizin, nahm er den Ruf zum Professor der Physiologie an, doch heute gilt er als der Begründer der Rechtsgeschichte. Er gehörte zu der Gruppe der Gelehrten seiner Zeit mit einem fast enzyklopädischen Wissen. Hier wird Varenius auf einen Menschen getroffen sein, der begann, „empirische Wege der methodischen und vorurteilskritischen Wahrheitsfindung“[41] einzuschlagen und der sich dafür einsetzte, mit Hilfe der Lehre des Aristoteles den Versuch zu unternehmen, den Staat zu säkularisieren und ihn von religiösen Grundlagen abzulösen. Dahinter verbargen sich erste Schritte unseres heutigen Verständnisses von Geschichte. Und so hörte Varenius hier einen Mann mit der Fächerkombination Medizin und Politik, der ganz nach dem Grundsatz dachte: Das Wohl des Volkes ist oberstes Gebot! Dabei hatte der Lehrende aber zu bedenken, dass es in Helmstedt galt, Forschungsfelder nur zu interpretieren in Bezug auf anerkannte Autoritäten und „sich aller Verwirrung stiftenden Neuerungen in der Sache und Methode zu enthalten“.[42] Ein unvorstellbarer Balanceakt!

Zur gleichen Zeit 1636 wurde ein weiterer maßgeblicher Lehrer in die Artistenfakultät aufgenommen: Christoph Schrader, der spätere Generalschulinspekteur des Herzogtums Braunschweig-Wolfenbüttel, ein enger Freund Conrings. Auch er muss ein sehr beliebter und vielseitiger Lehrer gewesen sein.[43] Schrader verstand sich nicht nur auf Theorie und Praxis der Beredsamkeit in der lateinischen Schrift und Rede. Er unterwies seine Schüler in dem Verständnis der römischen Schriftsteller. Aber auch die deutsche Sprache, das Hebräische, Geschichte und Geographie beherrschte er perfekt. Seine geographischen Privatvorlesungen begann er 1635 mit dem klassischen Autor Pomponius Mela.[44]

Tütken interessierte 1997 insbesondere die Wirkung der Fächer Geographie und Geschichte durch die Schulordnung von 1651, die ja

[40] Bautz in *BBKL*.
[41] Tütken 1997: 93.
[42] Tütken 1997: 92.
[43] Tütken 1997: 98.
[44] Tütken 1997: 100.

maßgeblich Schraders Handschrift trug. Von eigenen Lehrinteressen bewegt, habe Schrader seine Aufgabe darin gesehen, den Lateinschulen historische und geographische Betrachtungsformen zu verschaffen. Die Belege (da es noch keine Fächer in dem Sinne gab, wie wir sie heute kennen) sprächen eindeutig dafür, „daß in den untersuchten Schulen die Anwendung der Geographie und Geschichte auf die Alte Welt und ihre Nutzung als fallweise Interpretationshilfe biblischer und klassischer Autoren im Vordergrund"[45] stand, dass sie aber selbst 1653 noch weit von der Anerkennung als reguläre Schulfächer entfernt gewesen seien. So wird es auch Varenius kennen gelernt haben. Zu Schraders Lehrprogramm gehörte Clüver, der bis zu seinem Tod 1622 in Leiden gelehrt hatte. Es sollte Jahrzehnte dauern, bis Geschichte und Geographie unter neueren Zielsetzungen den Rang eigenständiger Fächer im Schulkanon erstreiten konnten. Noch 1666 prüfte Schrader, der in der Schulordnung die Autoren Cluverius und Sleidan eigens genannt hatte, in Helmstedt die alte Geographie, die „Geographia Vetus zu beßerem verstäntnis der Lateinischen auctorum".[46]

Hier sei die Bemerkung eingefügt, dass Helmstedter akademische Lehrer wie Conring, Schrader und Hornejus in Leiden, einer kalvinistischen Universitätsgründung von 1575, studiert hatten an eben der Universität, an der sich später Varenius 1645 immatrikulierte. Die Leidener Universität[47] war geschickt geleitet worden, man lockte regelrecht berühmte Gelehrte mit interessanten Angeboten an die Universität, Fachbereiche wurden aufgetan, die es für sich gesehen eigenständig anderswo noch gar nicht gab, so u.a. Botanik und Sprachforschung. Wurde andernorts strikt an den Traditionen der Lehre und Denkschulen festgehalten, öffnete man sich hier, was nicht zuletzt Varenius' Beobachtungen in seinen Briefen an Jungius bezeugen. Leiden war in seiner formalen Struktur nicht neu, doch zwei relativ junge Disziplinen der Geisteswissenschaften, Geschichte und Politik, erlangten hier schnell eine ungewöhnliche Bedeutung, mit dem Ergebnis, dass es zwischen 1613 und 1697 in Leiden allein 762 Studenten der Politik gab.[48] Wie ungewöhnlich die Bemühungen der Universität Leiden waren, zeigt beispielsweise 1616 die Einrichtung einer bezahlten Forschungsstelle für Geographie an der Universität

[45] Tütken 1997: 289.
[46] Tütken 1997: 287.
[47] Dazu siehe Vermij in diesem Band.
[48] Burke 52.

Leiden für Philipp Clüver/Cluverius (1580–1622). Diese war wahrscheinlich eine der ersten dieser Art.[49]

In Helmstedt wird Varenius auch erstmals mit dem weit verbreiteten Werk Clüvers konfrontiert worden sein, *Phillippi Cluverij Introductionis in universam geographiam, tam veterem quam novam, libri VI*, das postum zuerst 1624 erschienen war und dem eine Rede von Daniel Heinsius (1580–1655) auf den Verstorbenen beigegeben wurde.[50] Bei Clüvers Buch handelte es sich um ein Standardwerk, das vielfach bearbeitet wurde. Im deutschsprachigen Raum erschien es erstmals 1641 in Braunschweig bei Müller.[51]

Vier Jahre blieb Varenius in Helmstedt. Durch die Zielsetzungen der in Helmstedt praktizierten Didaktik ist Varenius von einem literarisch geprägten Humanismus Helmstedter Spielart beeinflusst worden, der sicher auch in Teilen im Widerspruch zu dem stand, was ihn bei Jungius—dem ‚eisernen Logiker'—erwartete. Varenius erlebte also ein Studium, dem es an naturwissenschaftlicher und mathematischer Innovationsfreudigkeit nicht mangelte. Dennoch galt vorrangig in Helmstedt die enge Bindung zwischen der Heiligen Schrift und Aristoteles, und die Physik hatte als Lehre der Natur nur „Magddienste für die Kirche zu verrichten".[52] So ist Varenius noch von einem mathematisch-naturwissenschaftlich ausgerichteten Bildungsweg entfernt, sein Blick dafür aber bereits geschärft. Mela, Strube, Heidmann, Clüver, historisch und länderkundlich ausgerichtete geographische Bücher mag er im Handgepäck mitgebracht haben. Seine Lehrer kannten die Standardwerke. Und Conring wusste bereits von Varenius' neuem Lehrer in Hamburg: ein Schüler von Jungius schrieb 1640 nach Hamburg, wie hoch Conring die *Logica Hamburgensis* von Jungius schätzte, während man in Wittenberg dagegen öffentlich zu Felde zog.[53] In diesem Geiste, mit Aristoteles wohl vertraut, zog Varenius nach Hamburg.

---

[49] Burke 122.

[50] Heinsius gehörte zu Conrings und Schraders Lehrern in Leiden mit ihren Fächern Poesie, Griechisch, Politik und Geschichte, ebenso Gerhard Johannes Vossius (1577–1649).

[51] Elf Jahre nach Varenius Tode erschien 1661 in Wolfenbüttel (Guelferbyti) von dem Lüneburger Lehrer Buno eine Überarbeitung des Clüver mit dem Hinweis, „wer mehr wissen wolle, wird solches bey dem gelehrten Varenius in seiner GG gründlich und weitläufig antreffen." Varenius hatte Eingang in das in deutschen Landen bis ins erste Drittel des 18. Jahrhunderts am häufigsten aufgelegte praktische Geographiebuch gefunden.

[52] Tütken 1997: 401.

[53] Robert C.B. Avé-Lallemant, *Das Leben des Dr. med. Joachim Jungius*, Breslau 1882, 96.

Varenius war achtzehn Jahre alt, als er Helmstedt verließ. Hat er nun zu lange studiert, womöglich ohne Abschluss? Ein Blick auf den Werdegang seiner Lehrer ermöglicht es zu vergleichen. Christoph Schrader (geb. 1601) besuchte ab 1610 mit neun Jahren die Lateinschule in Celle und ging mit sechzehn auf die Lateinschule nach Hannover. Mit zwanzig begann er in Helmstedt sein Studium. Mit vierundzwanzig Jahren setzte er es in Leiden fort. Er verließ 1632 Leiden, um zunächst eine Tätigkeit als Privaterzieher zu übernehmen. Dies war eine nicht seltene Station in der Karriere eines Professors. In Helmstedt bekam er 1635 mit vierunddreißig Jahren eine Professur für Rhetorik. Hermann Conring (geb. 1606) dagegen nahm wie Bernhard Varenius mit vierzehn Jahren sein Studium in Helmstedt auf, mit zwanzig ging er nach Leiden und nach sechs Studienjahren nahm er 1631 zunächst eine Stelle als Privatdozent an, bevor er 1632 mit sechsundzwanzig Jahren die Professur für Naturphilosophie, Physik und Rhetorik erhielt. Es war also nicht ungewöhnlich, dass Varenius seine schulische Ausbildung mit achtzehn Jahren in Hamburg fortsetzte.

Varenius bei Jungius in Hamburg (1640–1643)

Am 1. April 1640 wurde Varenius am Johanneum, dem Akademischen Gymnasium in Hamburg, immatrikuliert. Hamburg hatte in der ersten Hälfte des 17. Jahrhunderts durch eine geschickte Handels- und Neutralitätspolitik in seinen starken Mauern für Stabilität und Prosperität in den Wirren des Krieges gesorgt. In dem streng lutherischen Hamburg herrschte gleichzeitig Bekenntnisvielfalt.[54] Der Zustand des Schulwesens war im Vergleich zu den Territorialstaaten hinter den Erfordernissen der Zeit zurückgeblieben. Der gewerblichen und kaufmännischen Mittelschicht fehlte eine ihren Bedürfnissen entsprechende Bildungseinrichtung.

Hamburg besaß zu dieser Zeit nur eine einzige bedeutende Lateinschule, das Johanneum. Das Akademische Gymnasium, dem Johanneum angegliedert, offerierte ein breites Angebot freier Vorlesungen, zu denen Absolventen der achten Klasse, etwa im Alter von 16 bis 19 Jahren nur angenommen werden konnten, wenn sie die

[54] Christoph Meinel, *Die Bibliothek des Joachim Jungius: Ein Beitrag zur Historia litteraria der frühen Neuzeit* (Veröffentlichung der Jungius-Gesellschaft 67), Göttingen 1992, 49f.

„fundamenta linguarum et artium logicarum" beherrschten.[55] So war dieses Gymnasium in seiner Funktion in der Mitte zwischen Schule und Hochschule angesiedelt (eine Universität gab es in Hamburg nicht), blieb aber von inneren und äußeren Krisen nicht verschont. Die karge Bezahlung, die Konkurrenz privater Hauslehrer oder auswärtiger Schulen und die Verpflichtungen, mit denen die Lehrer dem Scholarchat der Hansestadt unterworfen waren, forderten 1629 regelrecht ein Reformprogramm heraus. Um dieses zu entwickeln, wurde Joachim Jungius als Rektor des Akademischen Gymnasiums berufen, denn er konnte eine vorzügliche akademische Karriere vorweisen.

Joachim Jungius (1587–1657) war im Jahr 1609 mit gerade 22 Jahren bereits Professor für Mathematik in Gießen. Nach einem Studium in Padua, der Hochburg der neuaristotelischen Naturphilosophie, promovierte er dort zum Doktor der Medizin. 1623 stand Jungius vor der Wahl, in Rostock die Professur für griechische Sprache, für Mathematik oder für Ethik zu übernehmen. Er wählte die Mathematik. Das zeigt, welch breit angelegtes Wissen dieser noch junge Mann vorweisen konnte. Wegen des Krieges ging er kurzzeitig als Professor der Medizin nach Helmstedt. 1626 finden wir ihn wieder als Professor der Mathematik in Rostock. Hier begann er auch seine ersten Entwürfe für eine Wissenschaftsreform zu skizzieren. Ausgangspunkt seines Forschungsdranges waren für ihn die konkreten Körper einer stofflichen Welt.[56] Die alten, schon bei Aristoteles zu findenden Fragen nach Materie und Form, nach Prinzipien und Ursachen, Substanz und Attributen waren ihm präsent, doch er suchte sie von ihrer empirisch-naturwissenschaftlichen Basis neu zu begründen. Meinel spricht von einem kritischen Verstand des Jungius, einer kompromisslosen Schärfe, von Sarkasmus und von Spaß an polemischer Pointe. Er muss ein präziser, kritischer Leser „mit einem Hang zur Pedanterie" gewesen sein, wie es sich bei den Eintragungen in seinen Büchern zeigt.[57]

Sein wissenschaftliches und methodologisches Selbstverständnis formulierte Jungius bereits 1629 präzise für das Akademische Gymnasium des Johannes bei seiner Antrittsrede als eine Art Reformprogramm: „Nach dem Vorbild von Arithmetik und Geometrie sollten sämtliche Wissenschaften in der Ordnung zunehmender Komplexität ihrer

[55] Meinel in: Joachim Jungius, *Praelectiones Physicae*, Historisch kritische Edition, Hg. Christoph Meinel, Göttingen 1982, 9.
[56] Meinel 1992: 48f.
[57] Meinel 1992: 74.

Gegenstände neu errichtet werden."[58] Bereits in seinem ersten selbständigen Werk, *Geometrica empirica* (Rostock 1629), stellte Jungius schon ein schlichtes Übungsbüchlein vor, das in der Geschichte der Mathematik bisher kaum Parallelen hatte: es galt, die Ergebnisse nicht mittels Euklidischer Axiomatik und geometrischer Beweise, sondern per experientiam, durch Versuch und Probieren, zu erbringen. Sein Unterrichtsziel war so angelegt und es galt für ihn als Kriterium für Wahrheit und Wissenschaft, dass die empirische Methode zum gleichen Ergebnis führen müsse wie der strenge Beweis.[59] So kristallisierten sich für ihn künftig drei Aufgaben heraus: das nach falschen Prinzipien errichtete System der Schulwissenschaft sei zu beseitigen, die Grundaussagen (Axiome) der Natur seien empirisch zu ermitteln mit dem Ziel, in einem dritten Schritt darauf „eine neue Philosophie, die neue Wissenschaft einzurichten".[60] Nur konsequent entstand 1630 sein Curriculum, die *Praelectiones Physicae*, das die Fächer Mathesis, Logica, Metaphysica, Physica, Philosophia practica (Ethik), Eloquentia, Historia, Studium humanitatis et linguarum umfasste.[61] 1638 lag Jungius' grundlegende Schrift, die *Logica Hamburgensis*, die zum wichtigen Lehrbuch für seine Schüler wurde, in drei Bänden komplett vor

In Vorlesungszyklen ging es um die Sichtung und Prüfung der herrschenden Lehren (Doxoskopie). Jungius' Ziel war es, die „Naturwissenschaft von der Metaphysik zu trennen" und ihr eine neue, in der Erfahrung gegründete Beweislage zu geben. Im Vordergrund stand die Frage nach dem stofflichen Aufbau der Welt. Seine Schüler sollten aber nicht empirisches Faktenwissen anhäufen, die große Bibliothek der Schule und die eigene von Jungius könnten dies vermuten lassen, sondern sich in eine kritische Auseinandersetzung mit Autorität und literarischer Überlieferung begeben. Jeder der Professoren las wöchentlich 4–8 Stunden publice, hinzu kamen Collegia privata. Mittwochs und samstags gab es mehrstündige Disputationen und jährlich bis zu acht öffentliche Exercitia zu einzelnen Fragen der Naturphilosophie.[62]

Die Anforderungen an die Schüler waren folglich streng, der Schüler musste eine Aufnahmeprüfung bestehen. Wie schwer es war, an die Schule zu kommen, zeigen die Bemühungen eines Vaters namens

---

58 Meinel 1992: 51.
59 Meinel 1992: 48.
60 Meinel 1992: 52.
61 Meinel in Jungius 1982: 10.
62 Loc. cit.

Drepper, der Jungius mehrfach darum bat, seinen Sohn am Gymnasium aufzunehmen und für einen „Disch", also Unterkunft bei einem Lehrer, zu sorgen. In seinem Brief vom 17. Juli 1642 erläuterte Jungius die Durchsetzung seiner Anforderungen: Der Sohn solle einen Praeceptor (Nachhilfelehrer) nehmen, etwa einen älteren Academicus (gelehrten Mitschüler), der in „lingua latina, rhetorica und Logica rudimentis" ihn „abrichte" und ihn so für „die Studiis Gymnasii tüchtig mache".[63]

Auf welche Weise Varenius der Eintritt in das Akademische Gymnasium 1640 ermöglicht wurde, ist nicht bekannt. Jedenfalls fand er bald das Wohlwollen von Jungius. Mittlerweile erlaubt uns die Forschungslage zu Jungius, dass wir uns ein Bild von der Schulzeit des Varenius in Hamburg machen können, dessen Studien unter Jungius' Führung für sein späteres Leben so entscheidend wurden. Die Persönlichkeit dieses herausragenden Lehrers forderte im Schulbetrieb die Schüler auf eigene Weise heraus. „Gleichgültig war er niemandem", er habe „nur Freunde und Feinde, im vollsten Maße" gehabt.[64] Seine Schüler hatten sicher nicht viel bei ihm zu lachen, Jungius verlangte viel mit eiserner Disziplin. Die enge Freundschaft mit ihm über ihre Schulzeit hinaus zeigt uns aber auch, wie sehr sich Jungius mit seinen Schülern verband; oft war freilich diese Freundschaft nicht ganz uneigennützig, denn ihn interessierten insbesondere die Studienvorgänge an anderen Universitäten. Das zeigen nicht zuletzt Varenius' Antwort auf die Frage nach der Situation in Königsberg und seine Beschreibung der Auseinandersetzung zwischen den jeweiligen Anhängern des Aristoteles und Descartes in Leiden.[65] Auch konnte Jungius, unter dessen Leitung an die vierzig Disputationen entstanden sind (darunter auch die Arbeit des Varenius), nicht zuletzt über seine Schüler den Absatz seiner Disputationen fördern. Und er benutzte sie als Agenten für seine Bibliothek, um an Bücher heranzukommen bzw. von Neuerscheinungen zu erfahren.[66] Aus den Briefen des Varenius an seinen früheren Lehrer und Förderer, die er schrieb, nachdem er Hamburg längst wieder verlassen hatte, sehen wir, dass Jungius Varenius vertraut mit „mein Bernhard" ansprach. Varenius unterschrieb mit „Dein Dir ergebenster

[63] Robert C.B. Avé-Lallemant, *Des Dr. Joachim Jungius aus Lübeck Briefwechsel mit seinen Schülern und Freunden [...], aus den Briefen der Hamburger Stadtbibliothek zusammengestellt*, Lübeck 1863, 124.
[64] Avé-Lallement 1863: 117.
[65] Avé-Lallement 1863: 310ff.
[66] Avé-Lallement 1863: 315.

B.V."[67] In jedem Brief drückte er die Ehrerbietung gegenüber seinem Lehrer aus und betonte, wie wichtig ihm der väterliche Rat in schwierigen Lebenssituationen sei, was nicht heißt, dass er sich in jedem Fall für diesen Rat auch entschieden hätte.

Varenius stellte sich bei seinen Studien in Hamburg und darüber hinaus der Forderung des Jungius, die Kritik habe im Vordergrund zu stehen. War es bisher traditionell um die Ermittlung, Bestätigung und Einübung festliegender Wahrheiten gegangen, galt es nun in erster Linie Fehler und Widersprüche in der herrschenden Lehre aufzudecken.

Ein treffendes Beispiel, wie Jungius den kritischen Geist der Schüler zu wecken verstand, findet sich auch in der Einleitung zur Disputation eines seiner Schüler, Nicolaus Ropers, von 1639:

> Uns hat der hochberühmte Herr Praeses nachdenklich darauf hingewiesen, dass alle seine Disputationen, die er bisher gegen bestimmte übernommene Lehrmeinungen gehalten hat, darauf zielen, dass endlich die trügerischen und verworrenen Texte der physikalischen Bücher des Aristoteles verlassen und so das große Buch der Natur selbst aufgeschlagen werde, das heißt, dass man sorgfältig und genau Beobachtungen anstelle, von den Beobachtungen zu Indikationen, von diesen zu Demonstrationen in der Naturlehre voranschreite, zumal es außerordentlich viele Gegenstände der Naturwissenschaft gibt, über die Aristoteles entweder wenig oder nichts schriftlich hinterlassen hat. Denn diejenigen, die ihre Lehren nur aus den Werken des Aristoteles beweisen, seien eher Ausleger von Philosophen als selber welche, und endlich sei eine solche Art des Philosophierens servil und eines freien und Christenmenschen unwürdig.[68]

Hier wurde die intellektuelle Standfestigkeit im Stile der argumentativen Methode der Schulwissenschaft geprobt, wobei Jungius aber forderte, eigene experimentelle Erfahrungen nicht zum methodischen Prinzip zu erheben.[69]

Zu den Arbeitszielen von Jungius

Jungius' Curriculum *Praelectiones Physicae* wurde zu seinem Unterrichtsmaterial und zugleich, wie Meinel in seiner Edition ausführt, zum Ausgangspunkt einer drei Jahrzehnte umspannenden Auseinandersetzung

---

[67] Avé-Lallement 1863: 327.
[68] Meinel 1984: 19.
[69] Meinel 1992: 53f.

mit der Naturlehre und Naturphilosophie seines Zeitalters, die „wie kein anderes zum Umsturz des alten naturwissenschaftlichen Weltbildes beigetragen hat" (9).[70] Es sei alles andere als eine systematische Darstellung der Naturlehre.[71] Im Zentrum stehe vielmehr die Auseinandersetzung mit der neuaristotelischen Naturphilosophie eines Zabarella, die von Padua ausgehend zusammen mit der neuscholastischen Aufwertung der Metaphysik (wie in Helmstedt gepflegt) und dem Aristotelismus Wittenberger Prägung in der Trennung der Providentia von der Naturlehre die philosophischen Diskussionen im protestantischen Deutschland bis zum Ausgang des 17. Jahrhunderts beherrschen sollten.[72] Jungius wagte also nicht den radikalen Bruch mit der alten Physica. Es war eher der Versuch, Empirie und Metaphysik einander anzunähern und der letzteren eine auf Erfahrung bezogene Beweislehre zu geben.

Wissenschaftliche Naturbeobachtung, so lehrte Jungius, habe vom konkreten Einzelding auszugehen, und dieses sei vorerst unbestimmt-ganzheitlich und ohne theoretisches Vorverständnis allein mit den Sinnesorganen zu erfassen, ähnlich wie es die Aufgabe des Arztes sei, den Körper zunächst von schädlichen Stoffen zu befreien, ehe dieser wieder kräftigende Nahrung erhalte. Die Aufgabe einer wissenschaftlichen Naturforschung bestehe nun darin, so Jungius, die unbestimmte Primärerfahrung in zunehmend schärferer Beobachtung und präziserer Beschreibung zu differenzieren. Es müsse eine unterste Stufe der Wahrnehmung geben, die von jeglichem theoretischen Vorverständnis frei sei, jedoch zunächst unbestimmt und verschwommen bleibe und eine „vorwissenschaftliche Erfahrung" abgebe.[73] Eine wissenschaftliche Naturforschung müsse folglich mit größter Exaktheit, durch immer schärfere Beobachtungen und präzisere Beschreibungen differenzieren und in wissenschaftliche Erfahrung (distincta experimenta) überführen. Wo Aristoteles noch behauptete, mathematische Genauigkeit lasse sich beispielsweise bei der Physica nicht fordern, erwiderte Jungius, dies sei eine Ausrede, wenn anstelle von Demonstrationen nur Wahrscheinlich-

[70] Interessant ist hier Meinels Bewertung. Er würdigt des Jungius Beitrag in dem Bewusstsein, dass er letztlich nicht Wortführer dieses Umsturzes wurde, noch seine Beiträge unmittelbar dazu beitrugen. Die „Geographie der Wissensgesellschaft seiner Zeit" (Peter Burke, 2000) hatte andere Zentren (vgl. Schlussbemerkung.)

[71] Hier finden sich Hinweise u .a. auf Strabo, Galilei, (Florenz 1632); Keckermann, Hanau 1616).

[72] Meinel 1992: 52.

[73] Meinel 1984: 22f.

keitsargumente angeführt würden. Denn dann sei die Physica notwendigerweise weniger genau als die Mathematik.[74]

Demnach sei einer wahrhaft wissenschaftlichen Naturlehre, die auf Demonstrationen beruhe, der höchste Wert beizumessen. Der Weg von der Einzelbeobachtung zur allgemeinen Aussage sei dabei streng induktiv zu bedenken. Die Fülle immer schärferer Beobachtung und die anschließende gedankliche Unterscheidung lieferten die distinktive (deutliche) Erfahrung, aus der sich Begriffe, Definitionen und Beweise gewinnen ließen. „Auf der Grundlage der induktiv-empirisch erschlossenen Allgemeinsätze kann Wissenschaft dann im zweiten Schritt zur Demonstration im engeren Sinne voranschreiten und allgemeine Schlußsätze formulieren".[75] Jungius beschreibt das so:

> Der Verstand schreitet niemals wahrhaft und sicher fort, wenn er nicht schrittweise und stufenweise vom Kleinen zum Großen, vom Leichten zum Schweren, vom Einfachen zum Zusammengesetzten und schließlich von zahlenmäßig begrenzten Prinzipien über zahlenmäßig begrenzte Mittelglieder zu zahlenmäßig begrenzten Endzielen entweder apodiktisch oder empirisch gelangt.[76]

Anders als Bacon sucht Jungius in dieser Methode nicht durch Ausschließung zu höheren Induktionsstufen zu gelangen. Wie sehr Jungius sich von seinem einstigen Lehrer Cremonini, Schüler und Nachfolger des Zabarella, entfernt, bemerkt Meyer im Vorwort (XV) zu *Logica Hamburgensis*. Für Cremonini war Erfahrung schon ein wesentlicher Bestandteil der Erkenntnis, Natur dagegen beruhe ganz im Sinne der aristotelischen Kosmologie noch nicht auf einem der korrekturfähigen Ergebnisse der Erfahrung, sondern auf einer a priori feststehenden „Vollkommenheit".[77]

Es sei deshalb völlig verkehrt, so erfuhr Varenius, zuerst vom Himmel und danach von den sublunaren Körpern zu sprechen, wie dies die Schulwissenschaft tue. Ja schlimmer noch sei, mit dem Menschen zu beginnen, der doch ein Mikrokosmos sei, ein mundi compendium, und von allen Naturkörpern am kompliziertesten zusammengesetzt. In einem ordo scientiarum wäre es seiner apodiktischen Lehrmeinung nach daher unzulässig, Prämissen in physikalische Demonstrationen aus der Kategorienlehre einzubeziehen, d.h. Annahmen über Substanz, Quantität, Qualität, Wirken und Leiden, Relation, Ort, Zeit, Lage und Haben.

---

74 Meinel 1984: 24f.
75 Meinel 1984: 27.
76 Loc. cit.
77 Meinel 1992: 52.

Letztes Ziel sollte eine scientia totalis, ein umfassend hierarchisch gegliedertes System bewiesener, durch wissenschaftliche Methoden erhaltener Schlusssätze sein, in didaktisch geordneter Form.[78]

In der *Logica Hamburgensis* von 1638 bringt Jungius dies in sein gedankliches System, mit dem Ziel, gemäß der Schulordnung, die logischen Formeln und Operationen in größtmöglicher Vollständigkeit anzuführen. Es kann hier nicht darum gehen, die *Logica* des Jungius eingehend zu würdigen. Einige Kerngedanken sollen aber kurz angedeutet werden, nicht zuletzt weil sie für das Verständnis des Bildungsweges des Varenius von Bedeutung sind. Bis heute gilt die Schrift als „das umfassende Kompendium der traditionellen Logik", deren Epoche sie zugleich abschließe. Seine Vorlagen nennt Jungius selber: Kirchmann und Dietericus. Zudem war alles versammelt, was auf diesem Wissensgebiet Rang und Namen hatte: 57 Autoren mit 116 Einzelveröffentlichungen.[79]

Jungius folgte seinem einstigen Lehrer darin, in der *Logica* eine Methodenlehre zu sehen, wobei diese für ihn aber mehr eine Kunst (ars) war, die „die Operationen unseres Verstandes leitet: den Begriff, die Aussage und den Schluß".[80] Im Gegensatz zu den Aristotelikern ging er jedoch nicht von der Vollkommenheit des Himmels aus, sondern sah in ihr die Möglichkeit, durch die Erfahrung korrigierend einzugreifen: „Der Verstand ist also nicht die Quelle der Erkenntnis, höchstens dient er dieser in der Funktion eines Instrumentes, das Allgemeine in der Verwirrung des Besonderen zu entdecken."[81] Burkhardt sieht in Jungius einen echten Atomisten, im Gegensatz zu Leibniz, der nur einen Begriffsatomismus vertrete und im Übrigen wie Aristoteles in der Physik Continua verwende.[82] Jungius sei in dieser Hinsicht somit auch ein Gegner des Aristoteles. Damit ist er bereits der Descartesschen Konzeption—die er erst 1650 kennen lernte—sehr nahe, dass die Begriffe die ersten Operationen unseres Verstandes seien und die Aussagen samt ihrer Wahrheit und Falschheit von diesen Begriffen abhingen. Das

---

[78] Meinel 1984: 28.

[79] Meinel 1992: 55.

[80] Hans Burkhardt, „Jungius, Leibniz und die Logica Nova," *Praktische Logik, Traditionen und Tendenzen*, Hg. Peter Klein (Veröffentlichung der Joachim Jungius Gesellschaft 61), Göttingen 1990, 62.

[81] Meyer in: Joachim Jungius, *Logica Hamburgensis*, Hg. Rudolf W. Meyer, Hamburg 1957, XV.

[82] Burkhardt 75.

heißt bei Jungius noch mit „kognitiven Denkakten möglichst geschickt umzugehen...", denn „die Begriffe repräsentieren die Gegenstände innerhalb und außerhalb der Welt, die Aussagen verknüpfen sie und führen zur Unterscheidung von wahr und falsch."[83]

Letztlich wurde aus alledem keine logica nova bei Jungius, sondern eine logica vetus, eine Sammlung des Lehrgutes der Schullogik des 16. und 17. Jahrhunderts (die dennoch deutlich seine Handschrift trägt). Das macht auch verständlich, wie Meinel erläutert,[84] weshalb sich nur die hypothetisch-deduktive Methode eines Galilei mit Experiment und mathematischer Abstraktion anstelle einer induktiven Erneuerung, ausgehend von den untersten Gegebenheiten der erfahrbaren Realität, durchsetzen konnte. Leibniz nannte dessen Auffassung einer Begriffsstruktur „ein Alphabet des menschlichen Denkens"[85] und sah die deutliche Anlehnung an Zabarella, den Lehrer von Jungius in Padua.

Zum Schluss dieser Ausführungen ein Beispiel aus dem Kapitel XI seiner Dialektik, wo er sich eingehend und systematisch mit dem Ganzen und seinen Teilen und deren weiteren Unterscheidungen und Präzisierungen beschäftigt. In dem Schema von Burkhardt[86] sieht das dann wie folgt aus.

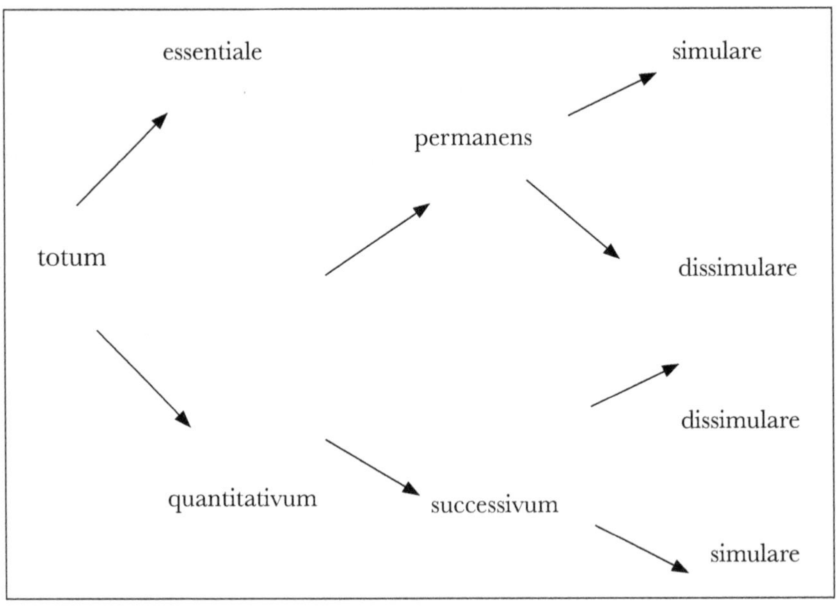

[83] Burkhardt 63.
[84] Meinel 1984: 43.
[85] Burkhardt 57.
[86] Burkhardt 70.

Aber, so hebt Meinel hervor (und wir täten gut daran, dies auch in der Bewertung des Varenius zu tun): Wir können „die Entstehung der neuzeitlichen Naturwissenschaft im Kontext ihrer Zeit nicht wirklich verstehen, wenn wir sie nur entlang dem schmalen Grat verfolgen wollten, der vom Fallgesetz und den Planetenbahnen bis hin zu Newtons Principia führt“.[87] Die historische Landschaft sei auch innerhalb der Grenzen, die der Aristotelismus vorgab, differenzierter und reicher an echten Alternativen, als uns dies aus der Perspektive der Gegenwart erscheinen mag.

Zu Jungius' Arbeitsweise und seinem Zettelwerk „Geographia Generalis“ im Blick auf Varenius

Um Jungius' Arbeitsweise besser zu verstehen, müssen wir einen Blick auf sein Handwerkszeug werfen, seinen Zettelkasten. Wo immer er war, scheint er Papier und Tinte bei sich gehabt zu haben. Kein Blatt blieb ungenutzt und waren es nur einseitig beschriebene, handschriftliche Aufzeichnungen von seinen Schülern.[88] Mit dem Zettelkasten verbanden sich auch zwei seiner Grundsätze: mittels der induktiven Forschungsmethode empirische Daten ohne weitere theoretische Vorannahmen zu sammeln, und zweitens, mit den Niederschriften auf den Zetteln Erkenntnisse so in ihre Bestandteile begrifflich zu zerlegen, dass sie sich hierarchisch wieder aufbauen lassen. Diese Kulturtechniken, Gewissenhaftigkeit, Pedanterie und Perfektionismus im Sammeln von empirischen Daten und Erkenntnisbausteinen dürften auch zum Rüstzeug unseres Varenius geworden sein. Die Führung eines Zettelkastens wurde von Jungius, so Meinel, bis zur Perfektion betrieben, wie eine Art Compendium Mundi.

Im Nachlass des Joachim Jungius befindet sich unter dem Titel „Geographia Generalis“ (im Folgenden GG) ein Manuskript, das vom

[87] Meinel 1984: 43.

[88] Zur Verzettelung des Wissens bei Jungius sei auf die einschlägigen Arbeiten von Meinel verwiesen.

Titel her Assoziationen zu Varenius hervorruft und die Frage aufwirft, ob Varenius bei seiner 1650 erschienenen *Geographia Generalis* von der Arbeit seines Lehrers, und sei es auch nur in indirekter Weise, profitieren konnte.

Bereits Kastrop verwies 1972 darauf und war der Auffassung, sie sei für Varenius nicht ohne Interesse gewesen: „Die Faszikel befindet sich unter der Signatur Pet. 37 in der Hamburger Universitätsbibliothek. Der Verfasser war aber trotz aller Bemühungen und Unterstützung von Prof. Schmidthüsen nicht in der Lage, diese Handschrift in ihrer Gesamtheit verständlich zu machen. Wir müssen uns daher mit gelegentlichen Bemerkungen begnügen."[89]

Den Entzifferungskünsten von Ulrich Staffhorst[90] und seiner Frau aus Karlsruhe ist es nun gelungen, Licht in dieses Dunkel zu bringen. Grundlage für die folgenden Ausführungen sind nicht nur das Original in der Hamburger Staatsbibliothek, sondern auch ein unveröffentlichter Manuskriptentwurf im Nachlass Prof. Karl Rohrbachs.[91]

Die GG des Jungius besteht aus einzelnen losen Zetteln, die in gefaltete Doppelbögen eingelegt wurden. Die Doppelbögen tragen Überschriften und kurze Erläuterungen. Die Rückseiten sind zum Teil beschrieben, einzelne Texte durchgestrichen. Die Aufzeichnungen scheinen aus dem Stegreif entstanden zu sein. Das unterschiedliche Schriftbild lässt dies erahnen. Es handelt sich mehr um ein Zettelwerk von Aufzeichnungen, aufgelesen und irgendwo abgelegt, gebündelt und zugeordnet, ohne dass die auf den Umschlägen gemachten Begrifflichkeiten durch den ergänzenden Text sich vollständig erklären ließen. Die Reihenfolge ist selbst nicht so gegliedert, dass sich vordergründig innere Zusammenhänge zu den *Praelectiones* oder zu *Logica* herstellen ließen. Es gibt vornehmlich lateinische Textstellen, aber auch kurze deutsche Passagen.

Die Thematik ist überwiegend mathematisch-topographischer Art (Lokalisierung auf Längen- und Breitengraden, Entfernungsangaben). Daneben findet sich Philologisches: lateinische (und griechische) Ausdrücke für Meer, Insel, Halbinsel, Kap, Sund usw.; dazu ein Sprachen-

[89] Rainer Kastrop, *Ideen über die Geographie und Ansatzpunkte für die moderne Geographie bei Varenius unter Berücksichtigung der Abhängigkeit des Varenius von den Vorstellungen seiner Zeit*, Diss. Saarbrücken 1972, 18. Anm. 65.

[90] Siehe seinen Beitrag in diesem Band.

[91] Eutiner Landesbibliothek 00/1048.

vergleich an Hand des Vaterunsers. Die Beispiele sind vielfach antiken Autoren (vor allem Strabon, Ptolemäus und Plinius, gelegentlich Thukydides, Herodot, Xenophon, Tacitus u.a.) entnommen, beziehen sich deshalb auch auf das antike Griechenland, Kleinasien, Italien. Daneben spricht Jungius u.a. von Atlas und Globus des Hondius, von Mercator, Jansonius, Copernicus, Hornius, Cluverius, Magirus. Jungius nennt hier Autoren, mit denen sich auch Varenius auseinandergesetzt hat.

Ein Autor wird in der GG des Jungius ausführlicher zitiert, Paulus Merula. Jungius wird ihm 1607 in Leiden begegnet sein. Kastrop geht davon aus, dass er auch Varenius bekannt gewesen sein dürfte. Er habe mit seiner *Cosmographia generalis* zum ersten Mal den Versuch unternommen, „zwei gegensätzliche Auffassungen der Geographie inhaltlich zu durchleuchten und methodisch zu klären." So sei Merula der erste, „der mit dem Begriffspaar ‚generalis' und ‚particularis' diese unterschiedlichen geographischen Richtungen auch begrifflich zu sondern sucht", wobei er aber nur alte Vorstellungen in ein neues begriffliches Gewand kleide.[92]

Von dem doch einigermaßen ausgearbeiteten Werk des Varenius sind die eher sporadischen Notizen des Jungius himmelweit unterschieden. Jungius gibt an keiner Stelle an, was er unter einer Geographia Generalis versteht. Am interessantesten sind noch die Konzepte über die Teilgebiete der Geographie (und Astronomie) bzw. Hilfswissenschaften (82f). Es wird nicht begründet, warum Themen wie Conventus (125), Prytaneum (17) und vor allem der Sprachenvergleich (107ff) zur GG gehören sollen. Er sagt nur, dass man bei der Unterscheidung der Völker die Sprachen betrachten muss (92). Kernbegriffe bei Varenius wie affectio scheinen bei Jungius zu fehlen.

Bei einer ersten Sichtung der 127 Seiten der GG lässt sich folgendes festhalten: So wie die Notizen vorliegen, ist es kaum vorstellbar, dass sie Jungius zur Grundlage eines kontinuierlichen Unterrichts (als Vorlesungsmanuskript) gedient haben. Es sind oftmals vereinzelte Beobachtungen, beinahe willkürlich zusammengestellt oder zitiert, eine Art Zettelkasten. Die Aufzeichnungen der „Geographia Generalis" des Joachim Jungius lassen nicht den Schluss zu, Varenius habe hier auf etwas Vorhandenes zurückgegriffen und es ausgearbeitet. Die GG des Jungius zeigt sich nach den von Staffhorst entzifferten und übersetzten

[92] Kastrop 1972: 35.

Textpassagen als ein Zettelwerk von Unterrichtsnotizen, jedoch ohne strukturierendes Element (vgl. dazu die am Ende dieses Beitrags zitierten Beispiele). Allein das Begriffspaar Geographia Generalis und Geographia Specialis dürfte Varenius angeregt haben.

Der Lehrer Johann Adolf Tassius

Eine weitere wichtige Persönlichkeit für den schulischen, aber auch späteren Lebensweg von Varenius war sein Lehrer und Professor der Mathematik Johann Adolphus Tassius (1585–1654). In den Briefen an Jungius wird Tassius mit einbezogen, und es muss einen eigenen Briefwechsel zwischen den beiden gegeben haben, der anscheinend nicht erhalten ist.

Tassius war, wie Günther ihn beschrieb, „ein fruchtbarer Schriftsteller und denkender Kopf".[93] Die meisten Veröffentlichungen erschienen erst nach seinem Tode. In drei von vier seiner geographisch orientierten Arbeiten beschäftigte er sich mit der mathematischen Seite der Erdkugel, die vierte (mir leider nicht zugänglich) sei „chronologischen Inhalts".[94] Interessant sind drei von ihm in Hamburg betreute Disputationen, dort erschienen in den Jahren 1635–36, bei denen es sich um eine Art fortgeschrittener Schülerarbeiten mit einem rein astronomisch-mathematischen Thema handelt.[95] Eine Arbeit aus dem Jahre 1635, die *Disputatio de rebus astronomicis et geographicis*, stammt von Henricus Krolovius aus Uelzen[96] und ist u.a. Heinrich Varenius, Bernhards Vater, in dessen Todesjahr gewidmet. Wenn Tassius, wie Günther (133) behauptet, einen „namhaften Einfluss auf seine Schüler ausgeübt" hat, dann haben diese Schriften es auf Varenius sicher nicht getan. Wären sie Unterrichtsmaterial für Varenius oder wären sie gar Quelle für seine *Geographia Generalis* gewesen, so hätte sich Varenius davon inhaltlich weit abgesetzt. Wie Staffhorst mir freundlicherweise in einem Kommentar zu diesen Disputationes mitteilte, wird dies an dem von Krolovius beschriebenen Weltbild in seiner *Disputatio* augenscheinlich

[93] Günther 132 Anm. 52.

[94] Günther 12.

[95] Tassius, *Disputatio de mundi axe et polis itemque terrae figura*, 1635; *Disputatio de rebus astronomicis et geographicis*, 1635; *Disputatio de constantia poli*, 1636.

[96] Dr. Egge vom Stadtarchiv Uelzen fand leider keinen weiteren Hinweis zu dem Autor.

demonstriert: Er sagt, dass die Erde unbeweglich in der Mitte liegt und sich das Weltall darum dreht. Dies geschehe mittels einer Achse durch den Mittelpunkt der Erde hindurch. An den jeweiligen Enden der Achse sei der Himmel. Interessant ist der Gebrauch der von den Schülern benutzten Termini. Zum Beispiel wird mit ‚tellus' und ‚Terrae figura' das gleiche ausgedrückt.

Mit seinem mathematischen Blick auf geographische Grundfragen dürfte Tassius jedenfalls Varenius sicher angeregt haben. Varenius' Auseinandersetzung mit den Zeitgenossen (u.a. Clüver, Gollnitz) zeigt aber, dass er in seiner *Geographia Generalis* eigenständig arbeitete und sich aus der Abhängigkeit von seinen Quellen befreit hatte.

Schlussbetrachtung

Varenius ging in einer Zeit zur Schule, in der die klassischen Wissenschaften im Umbruch waren und neu positioniert wurden. Uelzen, Helmstedt und Hamburg wurden seine Lernorte in einer Zeit sich verändernder Sichtweisen, als im nach wie vor strikt ausgerichteten Schulalltag das Lernen strengsten Statuten und Reglementierungen unterworfen war, die alles bis ins Kleinste festlegten. Unbeschwert konnte die Jugendzeit für Varenius als Waise ohne großen finanziellen Rückhalt ohnehin nicht sein. An seinen Schulorten erlebte er die wissenschaftlichen Kontroversen, und es ist deutlich erkennbar, dass die neuen Entwicklungen an ihm nicht vorbeigegangen sind. In dieser Zeit des geistigen Umbruches begann man, die Welt nicht mehr nur nach religiösen Vorstellungen zu erklären, sondern Beobachtung, Erfahrung und Experiment hielten neben der mathematischen Wahrscheinlichkeitsberechnung Einzug.

Varenius' Lehrer waren hoch geachtete Persönlichkeiten, großenteils Universalgelehrte mit einem breit angelegten Basis-Wissen wie Conring oder Jungius. Helmstedt präsentierte ihm einen Aristoteles humanistischer Spielart, den er in Hamburg mit Jungius aufs heftigste zu hinterfragen begann. Dass Varenius zu den bedeutenden Schülern des Jungius in dessen über 28jähriger Lehrertätigkeit gehört, geht nachweislich aus der Korrespondenz der beiden hervor. Indem die Gelehrten die klassische Fächereinteilung zu sprengen begannen, trugen sie mit dazu bei, dass neue Fachdisziplinen entstanden. So bildete sich aus mehreren Disziplinen die Geographie zu Beginn des 17. Jahrhunderts als Fach heraus und erlangte Bedeutung, zumal Handel und Seefahrt diese Entwicklung, zu der Varenius bald selbst so bedeutsam beitrug, geradezu herausforderten.

Die Art der Ausbildung, wie sie Varenius genossen hatte, konnte neben dem Lehrberuf an der Universität und der Schule auch zu einer selbständigen (etwa beratenden) Tätigkeit führen. Man muss leider sagen: hätte führen können, wenn es ihm gelungen wäre, den Lebensunterhalt in der Fremde zu sichern. An Aufgabenfeldern, die neu entstanden und die sich im Bereich von Reproduktion von Wissen ansiedelten, fehlte es nicht. Im akademischen Umfeld gab es nun Autoren, die (wie er) nicht im universitären Bereich ihr Brot verdienten. Eine wichtige Rolle übernahmen Einzelpersonen, Schriftsteller oder Drucker selbst. Wie Burke gezeigt hat, bildeten sich Studierstubengelehrte, die nicht mehr selbst hinaus gingen und aus eigener Anschauung schöpften; sie begannen Materialien zu sammeln, zu sichten, einzuteilen, zu systematisieren und wurden zu „Wissensmanagern". Burke[97] verweist auf Athanasius Kirchers und Jean-Baptiste Du Haldes Arbeiten über China oder die des Varenius über Japan, leichtfertig als Kompilation abgetan. Deren Ergebnisse förderten eine Informationsgesellschaft der frühen Neuzeit.

Wissenstransfer wurde dort möglich, wo sich Wissen zusammenballte. An solchen Orten arbeiteten Schriftsteller und Gelehrte zusammen, Kircher in Rom, De Laet über das Mongolenreich in Leiden, wo er den Zugang zu einer der bedeutendsten Sammlungen orientalischer Bücher und Manuskripte fand, und Varenius in Amsterdam und Leiden, Weltstädten in der Mitte des 17. Jahrhunderts. Frühere Zentren des religiösen Wissens in der Reformationszeit (Wittenberg unter Luther, Genf unter Calvin) spielten in der Zeit des Varenius keine Rolle mehr; die Informationszentren unter geänderten Wissensbedürfnissen hatten sich längst verschoben.

Ein Jungius unternahm in Hamburg den Versuch, der Logik einen neuen Impuls zu geben, ohne von Descartes zu wissen, der in Paris einen Schritt weiter ging. Englands Büchermarkt war im Vergleich mit dem Kontinent zu Beginn des 17. Jahrhunderts „im hohen Maße provinziell".[98] Man besorgte sich die Literatur von dort, wo die Zentren waren, bearbeitete sie und begann, Programme zur Reform der Klassifikation

---

[97] Burke 95.
[98] Burke 193.

des Wissens zu entwickeln.[99] Eine Art ‚Geographie des Wissens' könnte dies beschreiben.

In diese Welt wuchs Varenius hinein. Auf jeden Fall hatte er sein Rüstzeug bei Jungius erworben. Er konnte logische Operationen vornehmen und Zettelkästen führen. Denn wie anders hätte er innerhalb so kurzer Zeit mit Hilfe vorliegender Materialien (Zettelwerken seiner Schulzeit?) eine Systematik für das sich entwickelnde Fach Geographie ausführen und dabei kritisch sichten können, womit er sich auseinanderzusetzen hatte. Dazu gehörten unter vielen anderen Autoritäten Göllnitz und Clüver (dessen Bild der Erde als Mittelpunkt der Welt er nicht akzeptierte).[100] Bei seiner später entworfenen *Geographia Generalis* hat Varenius an der Methodik von Jungius partizipiert und das Lerngebäude seines Lehrers für seine Ziele eigenständig genutzt. In der Synthetisierung und Systematisierung seines Wissens belässt er es nicht bei der Sammeltätigkeit, hier ist er bei diesem Thema über Jungius weit hinausgegangen. Durch die „Geographia Generalis" des Jungius mag Varenius aber angeregt worden sein, eine solche überhaupt zu verfassen. Damit gelang ihm in der Verarbeitung des Wissens seiner Zeit ein gewichtiger Schritt auf dem Weg zu einer modernen Geographie.

* * *

Drei Seiten als Exempla aus dem Manuskript von Jungius'
„Geographia generalis"[101]
Transkription und Übersetzung von Ulrich Staffhorst

p. 12: GENERAL[IS] / GEOGRAPH[IAE] / PHILOLOGICA / Oceanus / mare, lacus, flumen [,] paludes / continens, insula, paeninsula sive / chersonnesus / portus, statio navium / sinus / litus, promontorium
PHILOLOGISCHES DER ALLGEMEINEN GEOGRAPHIE : Weltmeer, Meer, See, Fluss, Sümpfe, Festland, Insel, Halbinsel oder Chersonesus, Hafen, Reede, Bucht, Gestade, Vorgebirge

p. 27: Straus fl[uvius] qui / in Salam Fran-/cicum influit, pro-/lixior longe est, / h[oc] e[st] longiori meatu / a fontibus suis quam / ipse Sala. / Circa confluentes / eorum Neustat, / in q[uo] olim palatium / Caroli Magni. / Ita q[ui]dam fluvii e / Meliboco (Haarz) / in Leinam illaben-/tes.

---

[99] Burke 117–121.

[100] J. Partsch, *Philipp Clüver, der Begründer der historischen Länderkunde*, Wien 1891, 203.

[101] Staats- u. Universitätsbibliothek Hamburg: NL Jungius Pe 37.

Der Fluss Streu, der in die Fränkische Saale mündet, ist weit stattlicher, d.h. von seinen Quellen her hat er einen längeren Lauf als die Saale selbst. In der Nähe ihres Zusammenflusses liegt Neustadt, in dem sich einst eine Pfalz Karls des Großen befand.
Dies gilt auch für gewisse Flüsse aus dem Melibocus (Harz), die in die Leine münden.[102]

p. 105: GEOGRAPHIA. / Jod. Hond. in glob. anno 1601 edito terrestri / in locorum longitudine [...] mirifice [...] / orbis hydrographici peritis [...] / descriptae sunt, prout nautici in suis / navigationibus directionem duorum locorum ab uno loco ad / alterum inveniunt, [...] quo indifferenter[?] utuntur, quamvis in uno / non aeque atque in alio chalybea illa acus ponatur, / et a polo septentrione magis vel minus divertatur, / pro usu loci in quo directoria fabricata sunt, unde ne-/cessario longitudo locorum distorta est. Multi hos erro-/res frustra conati sunt emendare per polaris stellae, / alii per Lunae cursum, alii certius per eclipses. Verum / quis in tanta locorum multitudine eclipses observabit? At / cum jam tandem per variationem aut deviationem / acus nauticae, ut vocant, locorum longitudo inveniatur, / operae pretium me facturum putavi, si in hoc globo re-/giones omnes, saltem quarum longitudo jam cognita est, / suis quasi veris longitudinis gradibus delinearem. / Longitudinem incepimus non a Fortunatis, sed ab iis, quos / Azores vocant, quod acus nautica ibi recta in septentrionem [vergit].

Auf dem im Jahr 1601 herausgegebenen Erdglobus des Jod. Hond. sind im Längengrad der Örtlichkeiten in wunderbarer Weise für die der hydrographischen Welt Kundigen [...] beschrieben, so wie die Seeleute bei ihren Schifffahrten die Richtung zweier Orte von einem Ort zum anderen finden [...], woran sie sich unterschiedslos halten, wie sehr auch bei dem einen nicht in gleicher Weise wie bei dem andern jene stählerne Nadel die Stellung einnimmt und vom Nordpol mehr oder weniger abschweift je nach Gebrauch des Ortes, an dem die Routenvorschriften hergestellt wurden, weshalb notwendigerweise der Längengrad der Örtlichkeiten verzerrt ist. Viele haben vergeblich versucht, diese Irrtümer zu verbessern durch die Bahn des Polarsterns, andere durch die des Mondes, andere sicherer durch Sonnen- (bzw. Mond-)finsternisse. Doch wer wird bei einer so großen Menge von Örtlichkeiten die Finsternisse beobachten? Aber weil sich nun endlich durch das Ausschlagen bzw. die Abweichung der sogenannten nautischen Nadel [des Kompasses] der Längenkreis der Örtlichkeiten finden lässt, hielt ich es für der Mühe wert, wenn ich auf diesem Globus alle Gegenden—wenigstens die, deren Meridian schon bekannt ist,—sozusagen an den ihnen eigenen wahren Längengraden einzeichnete. Die Länge ließen wir nicht bei den Inseln der Seligen [den Kanarischen Inseln], sondern bei denen, die man Azoren nennt, beginnen [d.h. den Nullmeridian legten wir bei den Azoren fest], weil die nautische Nadel dort geradewegs nach Norden [gerichtet ist].

[102] Das Problem, das Jungius hier wohl anspricht, ist die Tatsache, dass in den genannten Fällen der beim Zusammenfluss kleinere Fluss mit seinem Namen siegt.

THE ROAD TO AUTHORSHIP AND PUBLICATIONS: PROJECTS, PATRONAGE AND THE ELZEVIERS

Margret Schuchard

Abstract

Der Weg zu Publikationen: Projekte, Patronage und die Elzeviers
Am Universitätsort Leiden und in Amsterdam suchte Varenius geistige Herausforderungen, angezogen von dem ausgezeichneten Ruf der Universität und der weltoffenen und durchaus toleranten Atmosphäre einer Handelsmacht. Er wurde freundlich gefördert durch seine Bekanntschaft mit Gerhard Johannes Vossius, der ihm (vergeblich) am Athenaeum Illustre in Amsterdam die Mathematik-Professur als Nachfolger von John Pell verschaffen wollte. Da alle Versuche fehlschlugen, durch Patronage seinen Lebensunterhalt und die Fortführung seiner Studien zu sichern, musste sich Varenius kümmerlich durchschlagen. Entweder starben seine Gönner zur Unzeit (der Bürgermeister von Amsterdam Albert Conraetsz. Burgh, Vossius) oder die Bitte um Unterstützung durch Buchwidmungen (an Königin Christina von Schweden, an die Stadtväter von Hamburg und Amsterdam) blieb ungehört. Seine geplanten wissenschaftlichen Projekte konnte Varenius deshalb nicht wie erhofft zum Abschluss bringen. Dennoch gelang es ihm, sein Medizin-Studium abzuschließen und im Hause Elzevier zu publizieren, sowohl in Leiden als auch in Amsterdam. Dass sein Japan-Buch als letzter, ursprünglich nicht geplanter Band in der Reihe der ‚Republiken' erscheinen konnte, ist sicher seiner Initiative zu verdanken und der Bereitschaft von Louis Elzevier, den Band statt seiner Vettern in Leiden zu übernehmen. In dessen Publikationsprogramm, fortschrittlicher als das der Leidener Verwandten, passte dann auch die *Geographia Generalis*, so dass Varenius das ‚richtige' Verlagshaus gefunden hatte.

The Dutch Republic offered opportunities to an aspiring scholar, and Varenius hoped to make the most of them. He was drawn to its highly reputed academic institutions and became no less fascinated by the worldwide connections of a seafaring nation, and its enterprising spirit. Elsewhere he could not have achieved what he managed to bring about as a scholar transmitting knowledge, even though in the end, his struggle against poverty prevented him from attaining his more far-reaching goals.

Bernhard, a polymath like Chytraeus, remained open to different areas of knowledge, but he lived in a changed temporal world from his great-grandfather, and by his time universalism was beginning to be replaced by specialisation. In the course of his studies Varenius slowly realised the need to concentrate his efforts. He intensively pursued his work in private studies when what the university had to offer did not satisfy him. Even Leiden, so highly reputed in the academic world, had its serious shortcomings in its teaching practice, as Varenius complained when he translated his frustrations into (on occasion) entertaining reports to Jungius. Would Paris be a better place to continue, was a question that persistently haunted him. Meanwhile, he intensified his work on self-defined projects. Not until later did he realise how absolutely essential it was for him to lose no time and acquire a degree in order to secure his material welfare. He had been in Leiden for just over a year when the Uelzen fire made him as poor as a church mouse and left him desperately in need of employment. He found it in the autumn of 1646 as a tutor in the house of a wealthy citizen in Amsterdam in the "Blue House" in the Heerengracht, a well-known landmark in the street (because of the blue stone in the façade). Who Varenius' employer was of the several families who lived in the house is not known,[1] but whoever he was, he would not let him go when Jungius offered Varenius an alternative in Hamburg. So he stayed in Amsterdam, gave up his employment as a tutor after a year, and muddled through by giving lessons in mathematics. He attended lectures at the Athenaeum Illustre in Amsterdam where Vossius, who had lived in Leiden until 1632, created an aura of academic study and drew large crowds of listeners from the Athenaeum and the citizens of Amsterdam. Vossius and Barlaeus were its leading figures, and Vossius, intimately familiar with Varenius' two places of study, Leiden and Amsterdam, became his highly esteemed and most valuable adviser.

Like his brother, Bernhard was no less aware of the vital importance of personal connections. They were even more essential in a foreign metropolis to which he had come as a stranger, and he made an effort to find some kind of patronage, but to no avail. Tragically the patron he had found, the mayor of Amsterdam and Vossius' old friend by the name of Albert Coenraetsz. Burgh, left for Moscow in 1647 with a group of delegates to negotiate a trade agreement with the tsar, but

---

[1] Dr. J.F. Ostermeier, personal communication.

he died on Christmas Eve 1647 in Novgorod. This wealthy merchant and skilled politician was also a very learned man and a trustee of the Athenaeum; he had become a doctor of medicine at Leiden, and he was liberal in his Protestantism and had the courage of his convictions in adversity. Burgh knew everyone, was widely travelled and had been to Denmark and Russia before. In short, Varenius could not have found a better patron. It was, however, not to be.

An important contact and mentor was undoubtedly Gerhard Johannes Vossius, who had come to Holland from Germany as a student and then stayed on.[2] In 1632 he became the professor of history and its first rector magnificus at the newly founded Athenaeum Illustre in Amsterdam after he had been forced to leave Leiden University in a period of heated religious conflicts, because as an Arminian he was no longer tolerated there. Varenius mentioned to Jungius that Vossius kindly gave him much good advice, and although Vossius was most generous in recommending his friends and students for positions where he could, he was unable to place Varenius at the Athenaeum as John Pell's successor as professor of mathematics.[3] Varenius had met Pell when he visited him soon after his arrival in Holland, which Pell mentioned in a letter to Tassius on 30 April 1645.[4] In the course of his deliberations about where he should direct his efforts, Varenius decided to make his mark as a mathematician, eventually with the idea in mind of qualifying himself for Pell's post. It had been vacant since Pell's departure for Breda in 1646. When Blaise Pascal's essay on conic sections was published in 1640, it made a huge impact, and the study of conic sections was still a 'hot' subject with Frans van Schooten and others in Leiden. Varenius devoted himself to this problem, but his study is lost, because no printer would print it at his own expense, and Varenius had no money for its production. His paper would have involved extra cost for the complicated setting up of type and the engravings with geometrical drawings. Useless as it is to speculate on what might have been, his pioneering contribution might, as has been

[2] See C.S.M. Rademaker, *Life and Work of Gerardus Joannes Vossius (1577–1649)*, Assen 1981.

[3] F.S. de Vrieze, „Academic Relations between Sweden and Holland", *Leiden University in the Seventeenth Century: An Exchange of Learning*, ed. Th. Lunsingh Scheurleer et al., Leiden 1975, 345ff; Christian Callmer, *Königin Christina, ihre Bibliothek und ihre Handschriften*, Stockholm 1977, 45–57.

[4] Pell to Tassius (Rohrbach in typescript, on Varenius' *De febri in genere*, p. 7f, Rohrbach papers in Eutiner Landesbibliothek, inventory no. LB 001052.

suggested, very well have anticipated what others were credited with afterwards. In any case, the professorial post for mathematics at the Amsterdam Athenaeum was cut in 1648 for financial reasons, because even in Pell's time there had been too few serious students of mathematics. Furthermore, Varenius was a Lutheran applicant, and Amsterdam proved less tolerant than expected.[5]

Varenius returned to his studies of medicine and finished them. His medical disputation on fever, delivered on 22 June 1649, was printed under the title *De Febri in Genere*. (In Carl Rohrbach's translation into German it was published in the Eutin Catalogue in 2001). Bonaventura and Abraham Elzevier, the Academic Typographers of Leiden University, printed it as they did most of these short disputations, although the students were not required to have them printed there.[6] When Varenius was not able to establish a reputation as a mathematician, he realised that it was absolutely essential for an aspiring young scientist to find a publisher, and the right one. By the earlier part of 1649, Varenius' rather hectic activities finally led to success when he managed to establish a working relationship with Lodewijk Elzevier of Amsterdam, who published his book on Japan.

Only three weeks before his medical disputation and dated on 1 June 1649 in Amsterdam, Varenius dedicated his *Descriptio Regni Japoniae* to the mayor and senators of the city of Hamburg (perhaps to recommend himself to them in case he should ever return, because for a long time he remained uncertain where he should go). On both the engraved title of the *Descriptio* as well as the printed title-page of the following *Tractatus in quo agitur de Iapone religione* "Med. D." is already added to Varenius' name. In his "Praefatio ad lectorem" he explains that he was encouraged and urged by learned men to complete this study and present it to the public. This obvious commonplace nevertheless reflects the truth, because he did look for such encouragement, and his preface implies that he himself must have taken the initiative in approaching Louis Elzevier, not vice versa.

The fact that he had found an esteemed publisher who subsequently printed his other works and the *Geographia Generalis* (1650) did not, however, improve his financial status. Louis Elzevier was not in the habit of

[5] On this issue see Vermij below.

[6] J. Bos in *Boekverkopers van Europa: Het 17de-eeuwse Nederlandse uitgevershuis Elzevier*, ed. B.P.M. Dongelmans et al., Zutphen 2000, 240.

paying his authors liberally, and in the seventeenth century publishers and authors would agree on a flat fee. If a book sold well, the publisher reaped the profit, not the author. He depended on patronage for an additional income. Without the backing of a person of authority such as Vossius, who was well known at the Swedish court, Varenius would surely not have dared to dedicate his *Tractatus in quo agitur de Iapone religione* to Queen Christina of Sweden on 1 August 1649. By that time, however, Vossius was dead. His son Isaac, head of the library in Amsterdam, was first invited by Christina in 1648, and in the spring of 1649 he arrived at her court in Stockholm, three weeks before his father's death in April. Varenius is sure to have focused his hopes on Isaac Vossius. Everyone knew of the learned young queen's patronage, and young authors from all over Europe hopefully dedicated their works of art and science to her, often elaborately and beautifully bound.[7] She was also known to be a passionate collector of books; by way of an agent she bought books from the Elzeviers in November of 1648.[8] Through Isaac Vossius, Hermann Conring later received an invitation from her to leave Helmstedt University, Varenius' old alma mater, which he left for Stockholm. Descartes was also urged to travel to Sweden. But for Varenius none of his dreams came true, and to date no evidence has surfaced that the queen remunerated him in any way for the dedication to her. In any case, his small booklet could not rival the splendour of what was usually presented to Christina.

The story of why the *Descriptio* was not published in Leiden, like all the others in the series, is not quite clear. It looks as if the Leiden Elzeviers, Abraham and Bonaventura, had decided on ending the series of the 'republics' before Varenius presented his Japan project. Luckily for Varenius, Louis stepped in and pragmatically took it on as an additional number, the 33rd and last volume. This procedure appears plausible when we consider the circumstances and Varenius' statement, which turned out to be true, that his *Descriptio Regni Japanorum* would be the last of the 'republics'. Perhaps Bonaventura and Abraham had exhausted their list of countries to be described or had become a little weary after more than twenty years and did not want to make the effort to modernise the concept in the last years of their lives; they died in

[7] Sten G. Lindberg, „Queen Christina Bindings", *Queen Christina of Sweden: Documents and Studies*, ed. Magnus von Platen, Stockholm 1966, 205ff.

[8] Lindberg 199.

1652. A suitable successor who could write in Latin and was willing to do the work may not have been in sight.

Abraham and Bonaventura Elzevier had developed the profile of the series from 1625 on. In 1626 the two Elzeviers had obtained a privilege for ten years from the States General which prohibited the reprint of fifteen treatises "van alle regieringen van verscheide coninckrycken ende republijcquen".[9] In 1628 their list of *Respublicae diversorum auctorum* catalogued all the books already published in the series and those intended to be published in future years. This core list of 'republics' contained descriptions of European countries. In the 1630s India, Africa, Persia and China were added to the list. After 1636, when the privilege ran out, it was not renewed (why, nobody knows). So none of the following books were protected from piracy in this "series which is no series", or "a series without the title of a series".[10] Joannes de Laet's contributions to the development of the series in its first decade were very important; he edited 12 out of the 33 'republics' between 1628 and 1633 and to a great degree established a pattern. The Elzevier scheme of production turned out to be enormously successful. The 'republics', regardless of their respective forms of government, were described as regards geography, inhabitants, economy, form of government and history. The Latin text was legibly printed, though in the tiny format of 24th, and on high quality paper. Anyone interested could afford such a small book, and from the beginning they were presented as a group. This explains why not only the educated classes bought them, so did students, travellers, diplomats or merchants. All over Europe they became collectors' items of those who were looking for the complete list and did not much mind whether the volumes differed greatly from one another. Some volumes were specially written for the series, others were a compilation of existing texts by different authors with additions by the editor of what he had heard or read. After the end of the Thirty Years' War and of the Civil War in England the situation changed, particularly in the impoverished northern European countries. De Laet was now busy with work of his own and was no longer available; he died on 15 December 1649.

Varenius had to adapt his description of Japan, a country still missing in the series, to a more or less established scheme, which consisted

[9] J.A. Gruys, "De reeks 'Republieken' van de Elzeviers en Joannes de Laet," *Boekverkopers van Europa: Het 17de-eeuwse Nederlandse uitgevershuis Elzevier*, ed. B.P.M. Dongelmans et al., Zutphen 2000, 83.

[10] Gruys 84.

of a catalogue of questions, continuing along the lines to be found in the sixteenth century, and de Laet's pattern provided Varenius with examples of how to organise his material. Like him Varenius begins with the geographical description and then moves on to the inhabitants, to the monarchy past and present, and in between deals with themes of social history. At last he writes about trade and then about illnesses, which must have caught his particular interest as a student of medicine, but he was unable to find much source material and ended the book with a description of the funeral rites of the Japanese. Varenius had no time to lose in finishing the work from his notes, with all his parallel efforts to qualify as a doctor and establish himself in a recognised position. We need not be surprised that Varenius, an ambitious mathematician and medical man, a creative scientific thinker, was happy to be the author of a 'republic', but did not see his future in such writing. After all, as an expert remarked,[11] the compilation of such 'republics' required stamina and precision rather than originality or vision.

Vossius, who knew Louis Elzevier (as he usually called himself) very well, because he had been the publisher of all his works since 1645, very likely supported Varenius in his relations with him. Thus Varenius at last had his foot in the door at both publishing houses, in Leiden through his medical paper, and now in Amsterdam; although both Elzevier houses competed with each other, they never sacrificed their family relations to any hostile rivalry.[12] By this time it was Louis Elzevier, who had become the more important printer of scientific publications. He had been an apprentice in the Leiden firm of the Elzeviers, and on their behalf he had travelled to various countries in the north and south of Europe and had also been in England and France. In 1638 he left Leiden and settled in Amsterdam to establish his own printing and publishing business. Like Varenius he had been a student at Leiden, but of philosophy. His publishing programme was a great deal more progressive than that of his Leiden cousins, for among his authors were Descartes, Bacon, Hobbes, Clauberg, Comenius and Pascal. Not only was his business equal to that of Leiden, it was on the point of surpassing it, and so Varenius as a young author could not have found a better publisher.

[11] The Keeper of the special collections in The Royal Library of The Hague, Hans Gruys.

[12] For the Elzeviers see especially the contributions by O.S. Langhorst, J.A. Gruys, J. Kingma, B. Bennich-Björkman in Dongelmans et al.

VARENIUS AND THE WORLD OF LEARNING IN THE DUTCH REPUBLIC

Rienk Vermij

Abstract

Varenius und die gelehrte Welt in der Niederländischen Republik
Varenius lebte in den Vereinigten Provinzen, genauer: in Holland, als nach den politischen Auseinandersetzungen der vorhergehenden Jahre eine gewisse Stabilität eingekehrt war, die das goldene Zeitalter des Landes durch Wissenschaft, Kunst und Handel begründete. In diesem Beitrag wird das Umfeld untersucht, in dem Varenius arbeitete, und trotz schlechter Quellenlage können so seine Lebensumstände und Arbeitsbedingungen einsichtiger gemacht werden. Erörtert werden die Lehr- und Studienverhältnisse an den Universitäten (die Rolle Descartes') sowie die Möglichkeiten einer Karriere für einen jungen Gelehrten. Die Gründe, weshalb Varenius nicht am Athenaeum von Amsterdam eine feste Anstellung bekam, sind im Kontext der Stadtpolitik innerhalb der Konsuln zu suchen, nicht in der vorgeblichen religiösen Intoleranz, denn In- und Ausländer, die nicht der reformierten Kirche angehörten, konnten in den meisten Bereichen durchaus Karriere machen. Es gab Bedarf an Mathematikunterricht, besonders im Bereich der Seefahrt zur Ausbildung in Navigation einschließlich Astronomie, Arithmetik und Kartographie, noch dazu in einer Hafenstadt wie Amsterdam. Über die Vermittlung von praktischen Kenntnissen hinaus gab es auch Patrizier und Adlige, die an vertieften Einsichten in die Mathematik interessiert waren, aber diese konnte Varenius nicht erreichen, wenn er es denn gewollt hätte, denn er sah sich als Gelehrter, nicht Praktiker. In diesen Jahren des Umbruchs zu einer neuen Denkweise durch Descartes betrieb er als Humanist (wie Vossius) anscheinend eine ‚altmodische' Mathematik in der philologischen Tradition des Studiums antiker Autoren, wie es allerdings die meisten anderen Mathematiker der Zeit auch taten.

When Varenius arrived in the Dutch Republic in 1645, the country was still at war with its ancient overlord, the King of Spain. The republic had emerged as an unplanned by-product of the political and religious upheavals of the late sixteenth century, and a general revolt in the Low Countries against the pretensions of the Habsburg monarchy had led to the factual independence of the seven most northern provinces. Resulting from the emergencies of war, the new state had been rather haphazardly constituted. At a time when most European states witnessed a strengthening of monarchic rule, the Dutch arrived at a republican form of government. Sovereignty resided with the States, with the provincial assemblies, attended by representatives of the major towns, nobles and landed proprietors of each province. The Republic was *de jure*, though not always in fact, a coalition of seven independent republics. As a matter of fact, some of the provinces had more to say than others. The wealthiest and hence most powerful province was Holland, which contributed more than half of the Republic's funds. Holland owed its wealth largely to overseas trade. The provincial government was heavily dominated by the towns, which had eighteen votes in the provincial States against one for the nobility, and as a consequence, public life was dominated by urban oligarchies.

Although the struggle for independence continued, by 1645 it had almost been achieved. Negotiations were under way which shortly afterwards, in 1648, would lead to a peace treaty with Spain and to the formal recognition of Dutch independence by the European powers. During the course of the war, the Seven United Provinces had become a major player in European politics. Dutch shipping dominated the grain trade to the Baltic states. In the year 1645, a Dutch fleet sailed north to intervene in the war between Sweden and Denmark and defiantly entered the Sont in spite of a blockade by the Danish King—not so much on behalf of the Swedes, as on behalf of Dutch shipping and merchandising. Moreover, during the preceding half-century, the Dutch had become a first-rate colonial power, with colonies in both the East and the West Indies.

In the West, Dutch inroads in the Spanish and Portuguese colonies were largely checked after the reconquest of Dutch Brazil by the Portuguese in 1654. In the East, however, the Dutch became a dominant power. In 1602, the various trading companies had joined forces and established the VOC (Verenigde Oostindische Compagnie), the United East-India Company, which had been granted a charter and a formal monopoly by the States-General. From the VOC headquarters at Batavia on the isle of Java, the present Jakarta, Dutch power and influence were made felt over a wide area of Asia. As regards Varenius' later work in geography, it is certainly of interest that he came to live in a country and a town, Amsterdam, which were at the centre of a wide-ranging global network.

In this respect, special mention should be made of the Dutch trade with Japan, as Varenius was to write a book on this country. Trade with Japan was one of the VOC's main sources of bullion.[1] Its first Japanese trading post had been established in 1609 at Hirado. During this period, Japan initiated a gradually developed policy of isolation from the outside world. In 1635, Japanese merchants were forbidden to travel abroad. European merchants, on the other hand, were usually forbidden to enter the country. When in 1637 the Portuguese were formally excluded, the Dutch were the only Europeans left who were allowed to trade with Japan, although under very restricted conditions. In 1641, the VOC trading post was transferred to the tiny island of Deshima in the gulf of Nagasaki, where foreigners were kept under control and strictly isolated from the indigenous population. It appears that the Japanese allowed this "window to the west" to remain because they did not want to be completely dependent on the Chinese for their contacts with the outside world. The Dutch connection remained Japan's only western contact until 1853, when the U.S. Navy forced a new opening and demonstrated to the Japanese government the bankruptcy of its policy.

In retrospect, by 1645 the Dutch Republic had reached the acme of its power. After the peace treaty of Westphalia in 1648 had finally brought the war to an end, things could only decline. The Dutch 'Golden Age' had been brought about by industry, inventiveness, and daring. But to a large extent, the flowering of the Dutch Republic in the first half of the seventeenth century was possible only because of the temporary decline of its neighbours. With Germany, England and France devastated by war and civil war, the Dutch were left without serious competitors—and indeed made the most of it. But when peace was restored, the situation could not be maintained. Although Holland remained the dominant European trading power throughout the second half of the seventeenth century, it met with ever growing problems, a part of which arose from the renewed assertion of the other European powers, who were interested in developing their own economies. Increasingly, the Dutch were hampered by protectionist and mercantilist measures, the first being the English Act of Navigation of 1651. Whereas up to 1648 the Republic had followed an aggressive economic policy, after 1648 it was forced to go on the defensive.[2]

---

[1] See Reichert below.

[2] Jan de Vries and Ad M. van der Woude, *Nederland 1500–1815. De eerste ronde van moderne economische groei*, Amsterdam 1995. Engl. trsl.: *The First Modern Economy: Success, Failure, and Perseverance of the Dutch Economy, 1500–1815*, Cambridge 1997, 476f.

Politically too, the Dutch could not sustain the pretensions of a great European power. The first Anglo-Dutch naval war, which broke out in 1652, revealed the vulnerability of the Dutch even in their home waters. Moreover, the country which had victoriously beaten off the army of one of Europe's greatest powers two decades earlier needed French help in 1665 against the raids of a petty neighbour, the Prince-Bishop of Münster. The lowest point came in 1672, when the French King conquered about half of the Dutch territory during a single campaign. The Dutch Republic would remain a viable state and a wealthy economy for the next century and a half, but the outburst of creative energy was definitely over. This, of course, was not immediately clear. In 1645, Varenius arrived in a country that was powerful, wealthy, and self-confident, and where scholarship was flourishing. He looked around, he liked what he saw, and he decided to stay.

Learning and Scholarship

It is not amazing that Varenius felt attracted to this country. Many scholars did. No less than in the economy, the first half of the seventeenth century had been a 'Golden Age' for the Dutch in art and culture. The Dutch Republic was a highly urbanised society. The rate of literacy was unusually high and learning was generally held in high esteem. There were many book-sellers and printing houses that produced a wide range of books and pamphlets as well as maps and prints, with relatively little control by either state or university. There were also many teaching institutions. The University of Leiden, established in 1575 amidst the turmoil of the revolt, had grown into one of the foremost universities of Europe. It drew many foreign students, especially from Germany and the British Isles. Many Germans, Scotsmen, Frenchmen and Englishmen occupied chairs at Dutch universities. Besides Holland, other provinces, too, established their own universities, or, if that was not possible, at least an 'illustrious school'. Such schools offered a university curriculum, but lacked the right of granting doctoral degrees. Their stature depended on the number of inhabitants and on the wealth of the respective provinces. Even small towns supported their Latin schools.

Many foreign students and scholars liked this open and intellectually stimulating environment and took up residence in the Dutch Republic for a shorter or longer period. The Bohemian educational reformer

Jan Amos Comenius lived here for many years. Other scholars made short trips to the United Provinces in order to acquaint themselves with the intellectual life there, as did Pierre Gassendi in 1628–1629 and Marin Mesenne in 1630. By far the most important foreign scholar living in the Netherlands was the French philosopher René Descartes. Descartes lived in the Netherlands for some twenty years and used the opportunity to spread his ideas among Dutch scholars, both university professors and independent virtuosi. His *Discours de la méthode* and his *Essays* were published in 1637 at Leiden and his *Principia philosophiae* in 1644 at Amsterdam. Descartes' ideas were indeed well received in the Dutch Republic, even at Dutch universities.

Varenius' stay in Holland coincides with the rise of Cartesian philosophy in the Netherlands, and this development is clearly reflected in Varenius' correspondence with his teacher Jungius at Hamburg. After arriving in Leiden, Varenius wrote on 20 March 1646 that Mr Descartes had few followers, but still, he had some, although his opinions were not taught publicly. A year later, in May 1647, he reported on a disputation in which a corollary that attacked Descartes caused great turmoil. He predicted that Descartes would find many adherents and stated that Aristotle's philosophy was already in disrepute. In December of the same year, he told Jungius that the Cartesian philosophy was cultivated as much or more than the Aristotelian, and was taught openly at both Leiden and Utrecht, although much opposed by the theologians. Descartes' philosophy thus met not just with acclaim but also with fierce opposition.

The first university where his ideas were taught had been the University of Utrecht. Descartes' concepts were championed here by the professor of medicine Henricus Regius. He was attacked by the theologian Gisbertus Voetius. In the end, after much debate and manoeuvring in which Descartes himself also took part in 1645, the authorities imposed silence on all parties and forbade the further teaching of Cartesian theses. At Leiden, various professors were friends of Descartes: the mathematicians Golius and Van Schooten, and the professor extraordinary of medicine Walaeus (mentioned by Varenius). Descartes' main advocate, however, was the professor of philosophy Adriaan Heereboord, a rather quarrelsome character (who does not seem to have grasped the full extent of Descartes' ideas). Here, too, theologians were among the main opponents of Descartes, notably Jacob Trigland and Jacob Revius. The dispute was waged by means of public disputations. In 1647 the authorities at Leiden attempted to restore public peace by prohibiting further references to Descartes or Cartesian philosophy.[3]

Varenius was an interested observer of these disputes. He certainly was familiar with Descartes' work and he may have met him in person. But as a student of Jungius, his philosophical ideas had probably already been fixed by 1645—critical of Aristotle, but critical of Descartes as well. Varenius was certainly not impressed by Descartes' physics. In fact, in his *Geographia Generalis* (1650:180–190) he promised a discussion of Cartesian physics, but regrettably this has never been published. His attitude in the debates on Cartesianism remained that of an outsider and he does not appear to have taken sides.

Varenius did not live to see that the prohibitive measures of the Dutch authorities failed to be effective or to pacify the opponents. In the following years, especially around 1656, the Cartesian philosophy would become a main subject of public debate, discussed in disputations, pamphlets and at the synods of the Dutch Reformed Church.

Prospects of a Scholarly Career at Amsterdam

When Varenius matriculated at Leiden University in May 1645, he took rooms in the house of the cobbler Herman Bogemaecker at the Sint Pieterskerkgracht. The next year, however, he left Leiden for Amsterdam. For about a year, he tutored the children of an Amsterdam regent. In the seventeenth century, a tutorship was not a position deemed suitable for a scholar of note. It was rather a position for a promising student who in this way could obtain funding for his studies. Constantijn Huygens, secretary of the stadholder, appointed a tutor for his children in 1637. This was Abraham Mirkenius, a student of theology at Leiden. After a year, Mirkenius left his position and recommended a certain Henricus Bruno, another student. We may presume that for Varenius too, his tutorship was a way of earning a living while pursuing his studies.[4]

[3] Much has been written on the Cartesian debates at the Dutch universities. A good overview is offered by Theo Verbeek, *Descartes and the Dutch: Early Reactions to Cartesian Philosophy 1637–1650*, Carbondale 1992.

[4] From the eighteenth century, there are some instances of tutors either doing important research in their position (Abraham Trembley) or who became scholars of note afterwards (Adriaan Reland, J.H. Lambert). Even in the eighteenth century, this appears to have been exceptional. In the eighteenth century the position of tutors

Serving in the household of an influential person of course meant that one made useful acquaintances, which could further one's career. Surely Varenius hoped that a grateful patron would certainly help his former servant in his further career. However, a patron as a rule would only promote his protégés to a position considered befitting his stature. A position of note had to be deserved by other means as well—birth, family connections, qualifications, and so on. Varenius felt sufficiently qualified to set his hopes on the chair of mathematics of the Athenaeum, the illustrious school, at Amsterdam.

The Amsterdam Athenaeum had been established in 1632 by the town government. It had met with vehement opposition (there had even been a trial) from the University of Leiden, which claimed that the new institution constituted an infringement on Leiden's privilege as provincial university. Probably the fear of competition was not wholly unwarranted, but the almost complete absence of archival records of the Amsterdam Athenaeum for the early period makes it rather difficult to verify. It seems, however, that the Athenaeum was established not so much to provide for educational needs (although concern for the local youth certainly did play some role) as to create a sign of civic pride. Rather than to provide the land with ministers or lawyers, it was meant as a centre of higher learning which should enable the citizens of Amsterdam to be citizens of the world of learning as well, and thus embody the ancient ideal of the *mercator sapiens*. The Amsterdam Athenaeum was not the first illustrious school, but it was the first to represent purely local interests. As such, it was a new type of institution. Later in the century, all self-respecting Dutch towns would open some kind of institution of higher learning.[5]

What disciplines had to be taught was a matter of town policy, and thus what chairs had to be filled, or not filled, or created, was a matter of permanent debate. In general, propaedeutic disciplines, like history

in the household of their master often was somewhat ambiguous: not belonging to the servants, but not really being a member of the family either. This may well have held for seventeenth-century tutors as well. See Petrus J. Buijnsters, "De gouverneur/gouvernante tussen adel en burgerij: Enkele beschouwingen naar aanleiding van een vertoog van Justus van Effen uit 1734," *Nederlandse literatuur van de achttiende eeuw: Veertien verkenningen*, Utrecht 1984, 86–98; and his *Justus van Effen 1684–1735: Leven en werk*, Utrecht 1992, 48–56.

[5] Willem Frijhoff, "Het Amsterdamse Athenaeum in het academische landschap van de zeventiende eeuw," *Athenaeum Illustre: Elf studies over de Amsterdamse doorluchtige school 1632*–1877, ed. Eco O.G. Haitsma Mulier, Amsterdam 1997, 37–65.

or rhetoric, were deemed more essential than theology or law, which were taught only intermittently. Mathematics was such a propaedeutic discipline and a chair for mathematics was established very quickly. By Varenius' time, however, it had fallen out of favour, allegedly because it attracted only a few students. After Professor John Pell left for Breda in 1646, his chair was allowed to remain vacant for some time.

Not unreasonably, Varenius felt that the chair might fall to him. To qualify as a candidate, one needed strong arguments or powerful supporters, as Varenius himself realised. In this respect, he appears to have fared badly. As he explained to Jungius, his main protector, an Amsterdam burgomaster, died on an embassy to Russia. This clearly refers to the burgomaster Albert Coenraadsz. Burgh. Born in 1593, Burgh had held many offices in Amsterdam. Among other things, he was appointed curator of the Athenaeum on its establishment in 1632. Earlier he had been on an embassy to Russia in 1630–1631. He went a second time in the summer of 1647, but when still on his way to Moscow fell ill and died 24 December at Novgorod.[6] This left Varenius without a powerful patron. When he nevertheless sent his friends to the Amsterdam burgomasters to plead his cause, his request was turned down with the argument that he was a Lutheran.

The Religious Issue

This was a diplomatic way out for the regents, but it cannot have been the real cause of their refusal. Being a Lutheran was certainly an embarrassment for a would-be professor, but not an insurmountable problem. The Dutch Republic was rather atypical in its religious policy. Reaction against religious persecution had been among the main causes of the Dutch revolt. As a consequence, throughout the seventeenth century many people were critical of religious dogmatism and leading regents remained wary of ecclesiastical pretensions. Freedom of conscience was never in danger in the Dutch Republic. In the case of religious liberty, the issue was somewhat more complicated. The Catholic Church, compromised by its alliance with Spanish absolutism, had been formally

[6] Jacob Scheltema, *Rusland en de Nederlandenbeschouwd in derzelver wederkeerige betrekkingen*, vol. 1, Amsterdam 1817, 208–211; J.E. Elias, *De vroedschap van Amsterdam 1578–1795*, vol. 1, Haarlem 1903 (reprint Amsterdam 1963), 327f.

forbidden. Protestants had formed the core of resistance during the Revolt, and the Reformed Church had emerged as the official Church of the country. However, many inhabitants had remained Roman Catholic or indifferent to Reformed theology. In this situation, whereby the ruling Church represented only a small fraction of the population, toleration, at least *de facto*, of a wide variety of dissenting confessions (Catholics, Lutherans, Remonstrants and Mennonites) proved unavoidable.

Dutch reality was a curious mix between judicial rigor and practical flexibility. The Reformed Church was the 'public church' of the country. In principle, all members of government, anybody in a position paid by the government and all people with public offices had to be members. This would include professors at the universities or the illustrious schools. In practice, there were compromises. For professors of theology, the rule was strictly applied, of course. Professors of law also were regarded as important public functionaries, for whom open dissent could not be tolerated. However, professors of lower-ranking disciplines, like medicine, mathematics or philosophy, were dealt with more leniently. Catholics were not admitted and Remonstrants (or Arminians) also had a hard time for several years after the official ban on Arminianism was promulgated by the Synod of Dordrecht. In 1619, several Leiden professors were fired or suspended because they were Arminians.

However, at other times religious dissent proved no insurmountable obstacle. An interesting case is Burchard De Volder, originally a physician in Amsterdam, where he served the Remonstrant community. He himself was a Mennonite. This did not deter the curators of Leiden University from offering him in 1670, at the proposal of the Amsterdam burgomaster Johannes Hudde, the chair of mathematics and physics. Apparently, this decision raised some protests, for in their next meeting, the curators decided that De Volder should enter the Reformed Church before he could accept his chair. De Volder was amenable and joined the Walloon Church, the francophone sister-Church of the Dutch Reformed Church. Originally established by refugees from the southern Netherlands, the Walloon Churches had in the course of time developed into a rather fashionable, and decidedly less dogmatic, variant of the Dutch Reformed Church.[7]

---

[7] Gerhard Wiesenfeldt, *Leerer Raum in Minervas Haus: Experimentelle Naturlehre an der Universität Leiden, 1675–1715*, Amsterdam 2002, 57f.

Lutherans even seem to have been less of a problem than Mennonites. There are various instances of Lutherans who became professors at Leiden, such as the linguist Gronovius. Albertus Kyperus, also a Lutheran, converted to the Reformed Church when acting as a lector at Leiden. Later, he became professor of philosophy at the illustrious school at Breda and finally, in 1650, of medicine at the University of Leiden. Joining the reformed Church was not just advantageous; however, it made him vulnerable to the accusation of insincerity.

A change of confession was not even always necessary. In 1688, Govert Bidloo became professor of medicine at Leiden University, while being and remaining a Mennonite. There were some objections, but then, Bidloo was a protégé of the stadholder William III who took little heed of such protests. Another protégé of William III, the Lutheran Paul Hermann, was appointed to the chair of botany, also without being required to change his confession.[8]

As the last instances show, much depended upon the local backing of the candidate as well as upon specific circumstances: the alertness and influence of the local Church council and the leniency of the local governors. As regards the Amsterdam Athenaeum, Varenius could hardly have wished for better in this respect. Maintaining orthodoxy was not the main priority of the Amsterdam town government. The first professors at the Athenaeum, Vossius and Barlaeus, had been people very close to Arminianism and were suspected of unorthodoxy by the Amsterdam Church council. If the Amsterdam regents had decided that they wanted Varenius for the job, his Lutheranism would have been no problem. But as they did not want him, it offered a convenient argument.

Mathematics and Mathematicians in the Dutch Republic

After 1647, it becomes difficult to trace Varenius' activities. His request having been turned down, he had to find some means of supporting himself. It is not quite clear how he did this, but it is clear that he did

[8] L. van Poelgeest, "The stadholder-King William III and the University of Leiden," *Fabrics and Fabrication: The Myth and Making of William and* Mary, eds. Paul Hoftijzer and C.C. Barfoot, Amsterdam 1990, 115, 132f.

not give up hope of pursuing a scholarly career. His book on Japan appeared in 1649 and his *Geographia Generalis* in 1650. He obtained his doctorate in medicine at Leiden in 1649, but apparently he never practiced. His doctorate in medicine served Varenius mainly as proof of his scholarly qualifications.

As far as is known, he continued to live in Amsterdam. He did not obtain a post at another teaching institution, for instance a Latin school. In various other cases, Latin schools offered positions for scientifically-minded people. This was true of Nicolaus Mulerius, who in the end was called to a university chair, and for Isaac Beeckman, who remained a Latin school teacher until his death. As true scholars were rare, the position was open to foreigners. Andreas Cellarius, from the Palatinate, became rector of the Latin school at Hoorn in 1637, and Jacobus Cocceius, from Hanau, became, at about the same time, rector of the Latin school at Haarlem. It might be true, however, that for a schoolteacher to whom parents were to entrust their young children, the religious issue was of more consequence than for a professor at a university.

As we thus have no evidence on Varenius' sources of income, we may approach the question from the other side and ask what opportunities the Dutch Republic offered to a person with scholarly ambitions but little money. At first glance, the prospects would have appeared dim. Early modern scholars were either people of independent means or backed by wealthy patrons. Most important in this respect were the princely courts, where many poets and artisans found a living, but where also all kinds of scholars—mathematicians, astrologers, alchemists—gathered together. Local magnates or magistrates could fulfil similar roles on a smaller scale. The Dutch Republic, however, lacked a true princely court or powerful magnates. Dutch science was flowering by the middle of the seventeenth century, but, like Dutch trade, it was increasing primarily due to free competition and individual enterprise. And whereas at the beginning of the century, state-building and economic expansion still seemed to offer many possibilities for upward social mobility, with surveyors becoming burgomasters, by the middle of the century a social hierarchy had established itself.

However, in other respects some possibilities did exist. Although there was no princely court, the towns were rich and there was a burgeoning civic culture. Artists or scholars who knew how to answer the needs of the Dutch citizens certainly could make a living. In particular as a mathematician, Varenius had no need to despair of finding a suitable job. Mathematical knowledge was much in demand in the Dutch Republic and many mathematicians made a living by keeping private schools.[9] Broadly speaking, there were two kinds of mathematics teachers. On the one hand, there were those who mainly taught navigation, which included a lot of astronomical knowledge. The VOC and other large companies saw to it that their pilots were qualified for their jobs and understood the science of navigation. In 1619, the Amsterdam department of the VOC appointed its first examiner of the pilots, the mathematician Cornelis Jansz. Lastman. In 1633, it was ruled that anybody who wished to serve as a pilot on one of their ships, had to pass an exam. In the course of time similar measures were adopted by the other departments of the VOC, by other major companies and by the Dutch Admiralties. This led to a continual demand for instructions in navigation.

Teachers of navigation were mostly active in the big naval ports, in particular Amsterdam, and in the north of the province of Holland, where most sailors were recruited. A well-known teacher in this field was Klaas Hendriksz. Gietermaker, author of a famous handbook on navigation. Another was Dirk Rembrandtsz. van Nierop, who lived in the village of Nieuwe Niedorp, in Northern Holland. Originally a shoemaker, he became a prolific writer on all kinds of mathematical subjects. He calculated almanacs, published astronomical tables, and defended the theories of his friend Descartes. Many of these mathematicians belonged to religious minority groups, in particular Mennonites. As can be seen, some of these people were active in a wider field than just navigational instruction. Competition between teaching schools spurred individual mathematicians to distinguish themselves in fields like astronomy, cartography, instrument-building, or mathematical problem-solving. Important teachers could combine their schools with a position of pilot examiner.[10]

---

[9] J.A. van Maanen, *Facets of Seventeenth Century Mathematics in the Netherlands*, Diss. Utrecht 1987, 1–41; G. Alberts et al., "Mathematics in the Netherlands: A Brief Survey With an Emphasis on the Relation to Physics," *A History of Science in the Netherlands. Survey, Themes and* Reference, eds. Klaas van Berkel, A. van Helden and L.C. Palm, Leiden 1999, 370–377.

[10] Carolus A. Davids, *Zeewezen en wetenschap: De wetenschap en de ontwikkeling van de navigatietechniek in Nederland tussen 1585 en 1815*, Amsterdam 1986; and his "Ondernemers in kennis: Het zeevaartkundig onderwijs in de Republiek gedurende de zeventiende eeuw," *De zeventiende eeuw* 7 (1991): 37–48; Rienk H. Vermij, *The Calvinist Copernicans: The Reception of the New Astronomy in the Dutch Republic*, Amsterdam 2002, 192–202.—Davids (1986: 294–301, 312–333) discusses elaborately the development of the exam system and the navigational instructors.

Maybe more to the point, the science of navigation was not just practical knowledge, but also very much in fashion. The Dutch knew very well where their riches derived from. Rich merchants bought atlases and adorned their walls with world-maps. The Dutch Republic, in particular Amsterdam, was the world's centre for the production of maps, atlases and globes. Only parts of these were actually used at sea. According to van der Krogt, many of them were just show-pieces, bought by Amsterdam merchants to show off their riches and their interest in navigation and geography. Geography itself certainly had its share in this general interest. So Varenius' books, published in 1649–1650, may have been an attempt to draw attention to his scholarship, but they were certainly written with a keen eye on the market.

Navigation, however, was not the only application for mathematics. On the other hand, there were people who concentrated on teaching arithmetic, mainly with an eye to trade and commerce. A well-known mathematics teacher of this kind was Simon Hansz. Cardinael. Originally from Friesland, he established himself at Amsterdam in 1605. He managed a private school of mathematics, mainly for arithmetic as it seems from his books, besides being active as a surveyor and inspector of weights and measures. In the beginning of his career, he published a book on geometry which earned him some renown. At this time, the Dutch Republic was still in its initial stage and Cardinael eagerly seized the opportunities which the occasion presented to him. By 1617, he had joined the project of establishing a *Nederduytsche Academie* at Amsterdam, an institution for academic instruction and literature in the Dutch language, where he delivered lectures in arithmetic. However, as Sitters informs us, protests from the Dutch Reformed Church eventually brought the project to naught, and therewith Cardinael's hopes of attaining higher academic status.

As can be seen, it was practical mathematics that was most in demand. Apart from navigation and arithmetic, this also included surveying, cartography, engineering and the like. But in addition, many noblemen and patricians felt the urge to gain some theoretical knowledge. One of the most fashionable mathematics teachers was Jan Jansz. Stampioen, who in the 1640s taught at the homes of the notables of The Hague. Among his pupils, he counted the son of the Prince of Orange; Elizabeth of Bohemia (the daughter of the 'Winter King' who resided in exile in The Hague); and the sons of Constantijn Huygens. Teachers of the elite were not just teaching arithmetic or navigation. They were expected to know Latin and be well versed in the classical authors.

However, even those teachers who taught the aristocracy rather than common sailors were seldom persons of scholarly standing. Generally speaking, a career as a private teacher of mathematics would not be fitting for a university-educated person, other than as a temporary solution to bridge a certain period. Frans van Schooten Jr., the son of a mathematics professor at the Leiden engineering school, gave private lessons when he returned home from a stay in Germany. This was clearly a temporary move while waiting for an opportunity to obtain a more permanent position. Van Schooten was a great friend of Descartes, whose ideas he propagated. Descartes thereupon urged Constantijn Huygens to send his sons to take lessons with Van Schooten instead of Stampioen. Huygens also supported Van Schooten in taking over his father's chair at Leiden when the latter died in 1645.

Varenius' case appears in some ways comparable to that of the younger Van Schooten. He was no mean practitioner, but a well-read scholar, versed both in the classical sources and in the more recent philosophical and scientific authors. Like Van Schooten, he might have been giving private lessons as he could not do so officially at the Athenaeum. Amsterdam must have harboured a lot of people who were at least potentially interested. Moreover, as the position at the Amsterdam Athenaeum remained vacant for the following years, Varenius could have hoped that his situation would change for the better and that, when he acquired some renown as a teacher and author, the Amsterdam burgomasters might change their minds in the end. However, if he indeed entertained such hopes (which remains pure speculation), he must have been thwarted when professor Gerard Vossius, who had encouraged him and was a man of great influence, died in 1649.

The Turning of the Tide

Varenius attributed his lack of success to the utilitarian spirit of the Dutch ruling class, their lacking a love of true learning. At first sight, this reasoning seems somewhat obscure, as he decided to establish himself in the Dutch Republic exactly because of its attractive scholarly climate. It is true, however, that practical mathematics was most in demand and that institutional backing for higher scholarship was difficult to acquire. However, the Amsterdam Athenaeum was established as an institute of higher learning, not just a school for practical skills.

We might well ask, therefore, what was the kind of learning that Varenius would be expected to represent. Traditionally, mathematics at the universities was a scholarly undertaking, closely linked to humanist studies and philology, which were deemed the summit of learning at the time. The professor of mathematics at Leiden, Willebrord Snellius, had been a pupil of the famous humanist Joseph Justus Scaliger. Snellius reconstructed ancient mathematical texts. Although he did write on such practical topics as surveying and navigation, his research was based on a meticulous study of classical sources. A textbook on cosmography, formally a part of mathematics, was published not by Snellius but by his colleague Paul Merula, who occupied the chair of history. Snellius' successor, Jacob Golius, combined the chair of mathematics with that of eastern languages, in which he became particularly famous. (Frans van Schooten and his father, as stated, were not university professors, but taught at the Leiden engineering school, which was connected to the university.) At the University of Harderwijk, the chair of mathematics and philosophy was occupied by Johannes Isacius Pontanus (a pupil of Tycho Brahe), who was also historiographer of the States of Gelderland and wrote most of his works on history or classical philology. In Groningen likewise, curators wanted to put the professor of history, Ubbo Emmius, in charge of mathematics as well. In the end, Emmius got the chair of Greek instead, while mathematics went to his colleague for medicine, Nicolaas Mulerius. Mulerius again was a true humanist, with a special interest in mathematical astronomy and chronology.[11] It was only at Franeker, where Adriaan Metius was the first professor of mathematics, that mathematics had a more practical application from the start. In the year of the opening of the university, 1598, Metius asked and obtained permission to teach mathematics courses in Dutch, for the benefit of lay practitioners—surveyors and the like.

Therefore, it seems that in the first half of the seventeenth century, mathematics, as far as it was a scholarly undertaking, was regarded primarily as a kind of 'historical' or descriptive knowledge, as opposed to 'philosophical' knowledge (the knowledge of causes and essences). By the mid-seventeenth century, however, the landscape of learning was drastically changing. The 'new philosophy' was making rapid headway.

[11] Vermij 2002: 18–25; and Vermij, "Mathematics at Leiden: Stevin, Snellius, Scaliger," *Der "mathematicus": Zur Entwicklung und Bedeutung einer neuen Berufsgruppe in der Zeit Gerhard Mercators*, ed. Irmgard Hantsche, Bochum 1996, 75–92.

Up to then, humanist studies had been the core of learning. But under the influence of writers like Bacon and Descartes, the veneration of classical authorities as a source of all wisdom was rapidly declining. Freedom from prejudice and rational investigation were henceforth considered the keys of learning. Natural philosophy was coming to the fore as the new core of the arts curriculum. This was, however, no longer a philosophy based on the scrutiny of ancient texts, but a philosophy based on natural investigation and rational thinking.

It might seem that with this change, mathematics would profit. Formerly, mathematics had the support of a most respected discipline, classical philology, but it had little independence. For the new philosophers, mathematics became the paragon of rational thinking. Euclid's *Elements* became a model for structuring the whole of philosophy. It was therefore a key element in the world of learning. In practice, however, mathematics declined. The appeal to mathematics made by the new philosophers was mainly a rhetorical manoeuvre. The new physics was based on a rather elementary knowledge of mathematics; it did not require any advanced studies. Most philosophers who came to occupy chairs at Dutch universities in the second half of the seventeenth century had only a slight knowledge of mathematics. Thus, whereas mathematics was deemed all-important, the practice of mathematics itself seems to have lost much prestige. When its base in classical philology became irrelevant, it was difficult to separate it from more practical activities. Learned knowledge seemed no longer required. The calculation of astronomical tables, formerly the work of advanced scholarship, was by 1650 executed by a former shoemaker like Dirk Rembrandtsz. van Nierop.[12]

Varenius decidedly wanted to distance himself from practical work. He wanted to be a scholar. But what can we say of the kind of scholarship he practiced? We know little about his mathematical abilities. When applying for the post at the Athenaeum, he wrote a work on conics in order to demonstrate his competence. The work was never published and the manuscript has not survived. From what he tells his teacher Jungius, it appears to have been a rather unoriginal work, largely a compilation from classical authors like Archimedes, Apollonius and Pappus. Probably it contained nothing new, and most certainly Varenius

---

[12] G.H.W. Vanpaemel, "Gemengde wiskunde en experimentele fysica," *De zeventiende eeuw* 7 (1991): 49–57.

made no use of the new analytical techniques developed by Descartes. In all fairness, it should be said that very few mathematicians at this time made us of these techniques.

As regards learning in general, although Varenius was impressed by the new astronomical discoveries and highly critical of Aristotelian philosophy, it does not appear that he had been won over to the 'new philosophy', even on a rhetorical level. He simply ignored the fierce philosophical debates which were waged between Cartesians and Aristotelians and which were the talk of the day in the Dutch Republic at that time. He chose to go his own, distinct way. Actually, his work appears closer to the historical-descriptive method which was characteristic of humanist learning. That he was backed by the humanist Gerard Vossius is significant in this respect. With hindsight, it is easy to appreciate the empiricist and scientific approach Varenius applied in *Geographia Generalis*. But to his contemporaries, the work must have appeared rather old-fashioned.

We might well ask, therefore, whether Varenius actually had the right qualifications to be nominated at the Athenaeum. It might well be that the Amsterdam regents felt that learning in their town should be more in line with contemporary philosophical developments. As the kind of scholarship that Varenius practiced became overtaken by the fashion of the new mechanical philosophy, he must have felt that the love of learning had gone astray in Amsterdam. Not until after his death (1650) was mathematical instruction at the Amsterdam Athenaeum offered again, when in 1653 the Amsterdam regents gave permission to Alexander de Bie to teach mathematics privately at the Athenaeum. A few years later, de Bie, who, like Varenius, was not a follower of Descartes, was formally made professor. He was born in Amsterdam, which may have helped in gaining local backing. Moreover, he obtained the support of persons of note. Varenius, however, never succeeded in obtaining a position that would secure his livelihood.

II

VARENIUS AS A TRANSMITTING AGENT OF KNOWLEDGE—JAPAN

VARENIUS ALS WISSENSVERMITTLER EINER FREMDEN WELT—JAPAN

REISE- UND ENTDECKUNGSGESCHICHTLICHE GRUNDLAGEN DER *DESCRIPTIO REGNI IAPONIAE*

Folker Reichert

Abstract

The Sources of the Descriptio regni Japoniae

The *Descriptio regni Japoniae* is based on a number of sources in which Varenius found more or less suitable and up-to-date information on Japan. These works can be divided into three groups:

1. Surveys of geographical or Asian studies, which provided a geographical context for writing about Japan in the *Descriptio*,
2. Reports by Jesuit authors who were involved in the first Christian missionary work in Japan,
3. Reports by officials of the „Verenigde Oostindische Compagnie" (VOC) [United Netherlands Chartered East Indies Co.] in Japan.

In the first group Marco Polo's book (*Le divisament dou monde* 1298/99) is particularly noteworthy. Although the author had never set foot in Japan, the first news of those distant islands reached Europe through him. His *Zipangu* contains a number of fantastic aspects, but their meaning can be reconstructed. Here we find a blend of authentic information, folk traditions, European evaluations and literary attempts. From that time on, the island became an established part of the European world picture. Even drawings of regional maps were based on it.

After the discovery of Japan by Portuguese smugglers in 1543, merchants and missionaries regularly frequented the islands. Maps from the second half of the sixteenth century indicate that the knowledge of Japan had become more and more detailed and differentiated. It was particularly the missionaries of the Jesuit Order who wrote regular reports of their activities in an effort to obtain support. Owing to an elaborate missionary method, which was to a large extent adapted to local conditions, they acquired extensive knowledge of the mentality and the cultural traditions of the Japanese. The Portuguese Jesuit Luís Fróis even drew up a catalogue of differences (*Algumas contradições e diferencas de custumes antre a gente de Europa e esta provincia de Japão* 1585), which, while adopting the rhetorical commonplace of 'the world upside down' (Herodotus), was also intended to serve future missionaries as a cultural orientation guide. Varenius took over some observations in order to describe the extremely foreign character of the Japanese civilisation.

> In the early seventeenth century the Christians were persecuted and the Portuguese and Spaniards expelled, only the agents of the VOC still had regular, but strictly controlled access to the Japanese islands. Their contacts with the local people were organised in the form of "controlled cultural relations" (Urs Bitterli). Nevertheless, a man like François Caron, an experienced Asian traveller and competent linguist, could still collect rich and new information about Japan and the Japanese. Until 1641 he was in charge of the Dutch trading post as 'Opperhoofd' [head] near Nagasaki. His *Beschrijvinghe van het machtigh Coninckrijck Iapan* 1645 [Description of the Mighty Kingdom of Japan] became a European bestseller and remained unsurpassed until the eighteenth century and the work of Engelbert Kaempfer. It dealt with geographical questions and the topical situation in Japan at the time, but also described exotic rituals like cutting open the abdomen (*seppuku, harakiri*); thus Caron satisfied the interests of his readers.

Geographische Kompendien sind grundsätzlich Texte, die auf der Kompilation von Aussagen und Erkenntnissen anderer, also auf Fremdtexten, basieren.[1] Das gilt für Erdkunden so gut wie für Landeskunden, für die *Geographia Generalis* so gut wie für die *Descriptio regni Japoniae*, die beide von Bernhard Varenius verfasst wurden. Autoren von Landeskunden haben zwar eine bessere Chance, ihren Gegenstand persönlich und aus eigener Anschauung zu kennen. Aber auch sie sind in aller Regel auf die Informationen anderer, auf ältere Darstellungen, Kartenwerke und vor allem Reiseberichte als authentische Informationen, angewiesen. Das gilt auch und gerade für Varenius, der sein Leben zwischen Hitzacker, Uelzen, Helmstedt, Hamburg, Königsberg, Leiden und Amsterdam verbrachte und den Gegenstand seines Japanbuches nur aus Büchern und Bibliotheken oder—wenn man so will—nur vom Hörensagen kannte.[2]

Daran lässt er auch nie einen Zweifel: Schon das Titelblatt kündigt an, dass die *Descriptio regni Japoniae* 1649 aus den Aussagen verschiedener

[1] Markus Friedrich, „Chorographica als Wissenskompilationen: Probleme und Charakteristika," *Sammeln, Ordnen, Veranschaulichen. Zur Wissenskompilatorik in der Frühen Neuzeit*, Hg. Frank Büttner, M. Friedrich u. H. Zedelmaier (Pluralisierung und Autorität 2), Münster 2003, 83ff.

[2] Zur Biographie vgl. S. Günther, *Varenius* (Klassiker der Naturwissenschaften. 4), Leipzig 1905. Nachdruck Amsterdam 1970, 12ff, 23ff; Klaus Lehmann, „Das kurze Leben des Bernhard Varenius (1622–1650)," *Bernhard Varenius: Der Beginn der modernen Geographie*, Hg. Wolfgang Griep u. Susanne Luber, Begleitband zur Ausstellung 2001 (Veröffentlichungen der Eutiner Landesbibliothek 5), Eutin 2001, 9ff; sowie die Beiträge (siehe oben) von Schuchard und Lehmann.

Autoren zusammengestellt wurde (ex variis auctoribus collecta), und der Beschreibung selbst ist ein Katalog der Quellenschriften vorangestellt, auf die der Autor sich vornehmlich stützte: „Autores & Scriptores, ex quibus haec desumta sunt" (3*7^{r-v}; 1974 dt 33f). Darüber hinaus wies er auch im Einzelnen nach, woher er seine Informationen bezog. Nur wenige seiner Aussagen blieben ohne Beleg. Obwohl sich die *Descriptio* an eine breitere Öffentlichkeit wandte und wie die vorherigen 34 Bände der „Elzevirschen Republiken" in gedrängter Form über politische, historische und geographische Fragen informieren wollte,[3] erhob sie doch den Anspruch gelehrter Gründlichkeit und wissenschaftlicher Solidität.

Man kann die von Varenius herangezogenen oder wenigstens erwähnten Autoren in drei Gruppen einteilen:

1. eine kleine Zahl von erd- oder asienkundlichen Gesamtdarstellungen, die dem Gegenstand des Buches den Rahmen gaben und eher Informationen von allgemeiner Bedeutung enthielten, im übrigen aber einen höheren Bekanntheitsgrad beanspruchen konnten und deshalb auch nur gelegentlich expressis verbis angesprochen sind;
2. eine Anzahl von (mehrheitlich portugiesischen) Autoren aus dem Jesuitenorden, die in der Mehrzahl an dem ersten christlichen Missionierungsversuch in Japan teil nahmen und das Land aus eigener Anschauung kannten;
3. eine ähnliche Zahl von Berichten, die von Funktionsträgern der niederländischen Vereinigten Ostindischen Compagnie (Verenigde Oostindische Compagnie, kurz: VOC) stammten und Japan ebenfalls aus eigenem Erleben, aber aus der Sicht der Kaufleute beschrieben.

Ich werde im Folgenden an einigen Beispielen zu zeigen versuchen, worin der Wert und die Bedeutung dieser Texte denn lagen, wie sie das geographische Weltbild ihrer Epoche beeinflussten und was sie einem gelehrten Autor wie Bernhard Varenius zu geben vermochten.

---

[3] Georg Frick, *Die Elzevir'schen Republiken* (Hallesche Abhandlungen zur Neueren Geschichte 30), Halle 1892, 6, 20f.

1. *Marco Polos* Divisament dou monde *und andere klassische Texte zur Geographie*

Die geographischen Gesamtdarstellungen, die Varenius zitierte, geben das geographische Weltbild wieder, das ihm vor Augen stand. Japan kam darin vor, allerdings nicht bei jedem der genannten Autoren. Wenn Philipp Clüver (1580–1622) im frühen 17. Jahrhundert davon spricht, Peter Apian (1495–1552) ein Jahrhundert zuvor aber nicht,[4] dann werden damit geographie- und entdeckungsgeschichtliche Entwicklungen deutlich, von denen noch zu sprechen sein wird. Erst recht von Ptolemaios von Alexandria, dessen Kosmographie im 2. Jahrhundert nach Christus entstand, konnte niemand erwarten, dass er Japan erwähnte. Denn die Weltsicht der Antike reichte im Osten bestenfalls bis Südchina und nicht weiter.[5] Trotzdem galt der Text seit seiner Übersetzung ins Lateinische im frühen 15. Jahrhundert als Grund- und Basistext der Geographen. Zwar wurde das Kartenbild, das sich aus ihm ergab, nach den Entdeckungsfahrten der Portugiesen und Spanier seit dem späten 15. Jahrhundert von Auflage zu Auflage verändert und durch so genannte Tabulae novae erweitert.[6] Doch der Autor und sein Werk blieben kanonisch, so dass auch Varenius dort nachschlagen musste—ohne Erfolg, wie er sagt.[7]

Ergänzungen ergaben sich im Westen, wo eine Neue Welt vor den Augen der Kartographen sich abzeichnete, schon vorher aber im Osten Asiens, den Marco Polo bereist und durch sein Werk, den *Livre dou divisament dou monde*, bekannt gemacht hatte. Bis weit ins 16. Jahrhundert hinein galt es als eine taugliche Unterlage für die Geographen, danach als Klassiker und schließlich—für Goethe zum Beispiel—immer noch als literarisch bedeutend, weil es in der Lage sei, „das Gefühl des Unendlichen, Ungeheuren in uns aufzuregen".[8] Auch Varenius

---

[4] Varenius 1649: 8f, dt 40f.

[5] Folker Reichert, *Begegnungen mit China: Die Entdeckung Ostasiens im Mittelalter* (Beiträge zur Geschichte und Quellenkunde des Mittelalters 15), Sigmaringen 1992, 15–64.

[6] Zur Ptolemäischen Kosmographie vgl. *Lexikon zur Geschichte der Kartographie*, bearb. v. Ingrid Kretschmer, J. Dörflinger u. F. Wawrik, 2 Bde., Wien 1986, 644ff.

[7] Varenius 1649: 9, dt 41.

[8] Johann Wolfgang von Goethe, „Marco Polo" in *Noten und Abhandlungen zum Westöstlichen Diwan*.—Zu Marco Polos Rezeption und Bedeutung vgl. Reichert, „Marco Polos Buch: Lesarten des Fremden," *Fiktion des Fremden: Erkundung kultureller Grenzen in Literatur und Publizistik*, Hg. Dietrich Harth, Frankfurt a. M. 1994, 180–202; John Critchley, *Marco Polo's Book*, Aldershot 1992; John Larner, *Marco Polo and the Discovery of the World*, New Haven 1999.

schätzte den Autor und nannte ihn an fünfter Stelle seines Kataloges der benützten Literatur. „M. Paulus Venetus" heißt es dort ganz lapidar unter „Autores" (dt 33). Was sich dahinter verbirgt, zeigt ein Blick in das Buch Marco Polos.

Marco Polo war im Jahre 1271 im Alter von siebzehn Jahren mit Vater und Onkel von Venedig aufgebrochen, traf vier Jahre später in Khanbaliq bei Peking ein und verbrachte dann siebzehn Jahre seines Lebens am Hofe des Großkhans der Mongolen in China. Er gehörte zu den zahlreichen Ausländern, die am Hofe Khubilais lebten und—in zum Teil sehr hohen Ämtern—die mongolische Reichsverwaltung mittrugen. Wie hoch Marco Polo dabei rangierte, ist schwer zu bestimmen. Seine eigenen Angaben sind sichtlich übertrieben. Auch die Behauptung, als Sondergesandter Khubilais das chinesische Reich bereist zu haben, wirkt nicht sonderlich glaubwürdig. Marco Polo wollte damit seine Leser beeindrucken und sein breites landeskundliches Wissen unterstreichen. Ihm aber umgekehrt jede Augenzeugenschaft abzusprechen, hat sich als abwegig erwiesen. Vor wenigen Jahren tobte eine öffentliche Debatte um die uralte Frage, ob Marco Polo jemals in China gewesen sei. Angeblich habe er in Persien auf Horchposten gesessen und lediglich die Angaben anderer verwertet. Die Debatte trug absurde Züge und wurde am Ende eindeutig (wenn auch offenbar nicht endgültig) zugunsten Marco Polos entschieden. Meines Erachtens kam er von Khanbaliq/Peking bis ins südliche China und hatte Gelegenheit, die dazwischen liegenden Provinzen und Städte kennenzulernen und wenigstens oberflächliche Eindrücke von der chinesischen Zivilisation zu gewinnen.[9]

Darüber hinaus behandelte er die umliegenden Länder, soweit sie mit den Herrschaftsansprüchen der mongolischen Großkhane konfrontiert wurden. Dazu gehörten Burma, Java und auch Ceylon, vor allem aber Japan. Drei umfangreiche Kapitel beschäftigen sich mit jenem fernen Reich im äußersten Osten der Erde. Es heiße Zipangu, liege 1.500 Meilen östlich von China und werde von schönen weißen Menschen mit guten Manieren bewohnt. Allerdings beteten sie vielarmige Götzen an und seien idres/idolatri, also Bilderverehrer, Götzendiener. Sogar

[9] Ausgelöst wurde die Debatte durch Frances Wood (*Did Marco Polo Go to China*, London 1995), entschieden durch Igor de Rachewiltz („Marco Polo Went to China," *Zentralasiatische Studien* 27, 1995: 34–92). In der Presse wird das Ergebnis der wissenschaftlichen Diskussion freilich nicht zur Kenntnis genommen und Marco Polos Glaubwürdigkeit nach wie vor in Zweifel gezogen (vgl. etwa Ulli Kulke, „In vierzig Jahren durch die Welt", *Die Welt*, 18. Dezember 2004; Ralf-Peter Märtin, „Angeber auf Geschäftsreise", *Die Zeit*, 22. Dezember 2004, 30).

Anthropophagie, Kannibalismus also, komme dort vor. Die Insel sei reich an roten Perlen, die sogar größeren Wert als die weißen besäßen. Toten stecke man eine davon in den Mund. Auch Gold finde man auf Zipangu in Mengen. Der Palast des Herrschers sei mit goldenen Fußböden ausgestattet und mit einem Dach ebenfalls aus purem Golde bedeckt. Dieses Reich übe keine Herrschaft über andere Länder aus, es dulde aber auch keine Herrschaft über sich. Ein Versuch des Großkhans Khubilai, Zipangu zu unterwerfen, sei in Meeresstürmen gescheitert. Die Insel liege im Chinesischen Meer, dem Meer von Cin, und sei eine von insgesamt 7.448 indischen Inseln, auf denen man duftende Bäume, Aloëholz und alle Arten von Gewürzen, namentlich schwarzen und weißen Pfeffer, finden könne. Die Seeleute kämen von den ost- bzw. südostchinesischen Hafenstädten Quinsai (Hangzhou) oder Zaitun (Quanzhou) mit dem Sommerwind herüber und segelten mit dem Winterwind wieder zurück. Die Reise nach Zipangu dauere also ein ganzes Jahr.[10]

An Marco Polos Bericht war manches phantastisch. Denn der Verfasser war nie auf Zipangu, nur vom Hörensagen kannte er das Land und seine Bewohner.[11] Von dem, was man ihm mitteilte, hat er manches nur halb verstanden, anderes wurde bei der Entstehung und in der Überlieferung seines Buches entstellt oder verzerrt. Aber einiges erwies sich als zutreffend:[12] Schon der Name kam dem Original näher als der gegenwärtig in den westlichen Sprachen gebräuchliche Begriff und gibt die seit dem 7. Jahrhundert in Japan vorherrschende

[10] Marco Polo, *Il Milione, a cura di Luigi Foscolo Benedetto*, Firenze 1928, 163ff. c. 160–162; und *Il Milione: Die Wunder der Welt*, Übers. u. Nachw. Elise Guignard, Zürich 1983, 277ff; Peter Kapitza, *Japan in Europa: Texte und Bilddokumente zur europäische Japankenntnis von Marco Polo bis Wilhelm von Humboldt*, Bd. 1 München 1990, 45ff (allerdings nach der mitteldeutschen Fassung). Zur Biographie und zum Buch Marco Polos vgl. Critchley; Larner; Reichert 1992: 114ff; Reichert 1994; sowie Leonardo Olschki, *Marco Polo's Asia: An Introduction to His* Description of the World *Called* Il Milione, Berkeley 1960; Marina Münkler, *Marco Polo: Leben und Legende*, München 1998; und Münkler, *Erfahrung des Fremden: Die Beschreibung Ostasiens in den Augenzeugenberichten des 13. und 14. Jahrhunderts*, Berlin 2000, 102ff.

[11] Vgl. Polo 1928: 167 Z.40: „Or desormés ne vos conterai plus de ceste contree, ne de cestes ysles, por ce que trop [sunt] des[v]oiables et encore que nos ne i somes estés."

[12] Zum Folgenden vgl. Kazuo Enoki, "Marco Polo and Japan," *Oriente Poliano: Studi e conferenze tenute all' Is. M.E.O. all' occasione del 7 centenario della nascita di Marco Polo (1254–1324)*, Hg. Etienne Balasz, Rom 1957, 23–44; Reichert, „Zipangu: Marco Polos Japan und das europäische Weltbild zwischen Mittelalter und Neuzeit," *Engelbert Kaempfer (1651–1716) und die kulturelle Begegnung zwischen Europa und Asien*, Hg. Sabine Klocke-Daffa, J. Scheffler u. G. Wilbertz (Lippische Studien 18), Lemgo 2003, 147–168.

Selbstbezeichnung (Nihon/Nippon[-koku], nämlich „das Land der aufgehenden Sonne") durch chinesische Vermittlung wieder. Das portugiesische Japão, dem die meisten europäischen Sprachen folgten, stellt davon eine Verballhornung dar und erscheint 1513 erstmals in einer europäischen Quelle.[13]

Marco Polo irrte, wenn er von einer einzigen Insel nur sprach. Er meinte Kyushu und vergaß Honshu, Shikoku und die kleineren Inseln ringsum. (Hokkaido wurde erst seit dem 17. Jahrhundert von Süden her besiedelt und gehörte lange Zeit noch nicht wirklich zu Japan.) Doch die Entfernung vom Festland gab er mit 1.500 Meilen ganz zutreffend an. Denn er meinte chinesische Meilen (li), ein Längenmaß, das nach der Beschaffenheit des Geländes variierte (also bergauf kürzer, bergab länger wurde), im Durchschnitt aber 500 Metern entsprach. Erst Marco Polos Leser, Kartographen und Seefahrer, rechneten seine Angaben in italienische Seemeilen um und kamen auf diesem Wege zu ganz anderen Distanzen. Dass sie sich mit Hilfe zweier Winde, des Südwestmonsuns im Sommer, des Nordwestmonsuns im Winter, leicht überwinden ließen, beruhte ebenfalls auf der Erfahrung der Seeleute und beschrieb die Grundlagen der Schifffahrt im Chinesischen Meer. Eher europäischen Vorstellungen folgte Marco Polo, wenn er Zipangu zu den indischen Inseln rechnete und 7.448 weitere in seiner Nähe erwähnte. Denn für die mittelalterlichen Kosmographen reichte Indien vom mittleren bis in den äußersten Osten und meinte dort alles Land beim Äquator.[14] Dort durfte man den Ursprung der Gewürze und Dufthölzer vermuten, nach denen Europa begehrte. Auf diese Weise wurde Zipangu in europäischen Augen zur hinterindischen Gewürzinsel, obwohl es nichts von alledem besaß.

Gold gab es auf Zipangu, und während des 13. Jahrhunderts wurden regelmäßig größere Mengen nach China exportiert. Die Relation zum chinesischen Silber war für die Käufer so günstig, dass die Vorstellung entstand, auf der Insel stünden riesige Mengen zur Verfügung. Der Volksmund machte daraus eine Geschichte von goldenen Dächern und Fußböden. Marco Polo sprach es ihm nach. Seine Informanten

---

[13] Vgl. Hirosato Iwai, "A Study of the Varied Names for 'Japan'," *Cultural Nippon* 8 (1940): 21–48, 113–144; Armando Cortesão, "The First Account of the Far East in the Sixteenth Century: The Name 'Japan' in 1513," *Comptes rendus du Congrès International de Géographie Amsterdam 1938*, vol. 2, 7, Leiden 1938, 146–152.

[14] Zum Indienbegriff der mittelalterlichen Geographie vgl. Helmut Gregor, *Das Indienbild des Abendlandes (bis zum Ende des 13. Jahrhunderts)*, Wien 1964, 11ff.

kamen offenbar aus dem chinesischen Süden und erzählten ihm Dinge, die sie selbst nur vom Hörensagen kannten. Auch der Hinweis auf die guten Sitten der Insulaner entsprach einem chinesischen Urteil, und die Existenz rot schimmernder Perlen ist immerhin für das 17. Jahrhundert bezeugt. Marco Polo seinerseits hat wohl auch selbst einige Missverständnisse erzeugt und die Erzählungen seiner Gewährsleute um eigene Urteile ergänzt. Zum Beispiel der Brauch, Toten etwas besonders Wertvolles (allerdings nicht Perlen, sondern Jade) in den Mund zu legen, war nur in China zu finden und aus der Beschreibung vielarmiger (buddhistischer) „Götzenbilder" auf der Insel spricht der Abscheu des Christen. Kein Chinese oder Mongole hätte sich ähnlich geäußert.

Am befremdlichsten wirkt jedoch die Erzählung von der gescheiterten Eroberung Zipangus durch die Flotte des Großkhans. Zweimal, 1274 und 1281, hat Khubilai den Versuch unternommen. Zweimal kamen den Verteidigern starke Stürme, Taifune, zur Hilfe, die die mongolische Flotte vor der Westküste Kyushus aufrieben. Mongolisch-chinesische Waffen und Rüstungen, die vom Meeresgrund gehoben wurden, legen bis heute Zeugnis davon ab.[15] Die Verteidigungsanlagen, die vor dem zweiten Unternehmen in der Bucht von Hakata errichtet und teilweise in und bei der heutigen Großstadt Fukuoka ergraben wurden, hielten zwar stand, erwiesen sich aber nicht als entscheidend. Die hilfreichen Winde gingen deshalb als „Götterwinde" (kamikaze) in das mythische Repertoire des Landes ein und wurden in der japanischen Geschichte noch öfters memoriert (zuletzt während des Zweiten Weltkrieges), waren aber nicht immer so wirkungsvoll wie damals.

Marco Polo beschreibt die Dramatik des Geschehens, verschweigt nicht das Desaster und nennt auch die Verantwortlichen bei Namen.[16] Er spricht aber nicht von zweien, sondern von einem Unternehmen und beschreibt außerdem die kühne Tat eines Trupps von 30.000 Mann, die den Insulanern in den Rücken gefallen seien, sich in den Besitz der Hauptstadt (mitsamt einigen schönen Frauen) gesetzt und schließlich ein Bleiberecht für sich selbst ausgehandelt hätten. Dies alles ist ebenso frei erfunden wie die grausame Bestrafung der beiden Feldherren: Der

---

[15] Vgl. dazu Otto Franke, *Geschichte des Chinesischen Reiches: Eine Darstellung seiner Entstehung, seines Wesens und seiner Entwicklung bis zur neuesten Zeit*, 2. Aufl. Berlin 2001, Bd. 4: 432ff u. Bd. 5: 210ff; Morris Rossabi, *Khubilai Khan: His Life and Times*, Berkeley 1988, 99ff, 207ff. Zu den im „Invasion Memorial Museum" in Fukuoka gezeigten Fundstücken vgl. Torao Mozai u. Nakamura Koji, "The Lost Fleet of Kublai Khan," *National Geographic Magazine* (1982: 11): 635–649.

[16] Paul Pelliot, *Notes on Marco Polo*, 3 Bde., Paris 1959–1973, 1ff, 871f.

eine sei enthauptet, der andere auf einer einsamen Insel ausgesetzt worden, wo er mit in Büffelhaut eingenähten Händen Hungers habe sterben müssen. Marco Polo, der den zweiten Invasionsversuch nicht als Augen-, wohl aber als Zeitzeuge in China erlebte, hatte offenbar Gründe dafür, das Geschehen so und nicht anders zu schildern. Den Großkhan Khubilai, einen der beiden Helden seines Buches, wollte er nicht als einen gütigen, sondern als strengen und gerechten Herrn zeigen, und die mongolischen Invasoren als intelligentes, in Kriegen listiges und nur vor den Naturgewalten kapitulierendes Volk zu charakterisieren, lag ihm am Herzen. Siebzehn Jahre hatte er am Hofe Khubilais verbracht und sich dabei die Weltsicht, Maßstäbe und Wertvorstellungen der mongolischen Oberschicht zu eigen gemacht. Sein Bericht von Zipangu sollte eine schmähliche Niederlage beschönigen und das Prestige des Großkhans Khubilai vor den Wechselfällen der Geschichte bewahren. Aufs Ganze gesehen, erweist sich der Text als ein eigentümliches Pasticcio aus authentischer Information, volkstümlichen Traditionen, europäischen Urteilen und literarischen Zwecken.

Dennoch legte er den Grund für das europäische Japanbild in den nächsten 200 bis 250 Jahren. Christoph Columbus wollte über Zipangu zum ostasiatischen Festland segeln und hoffte, auf der Insel einen Zwischenstop einlegen zu können und dabei auch etwas von deren sagenhaften Goldschätzen mitnehmen zu können.[17] Kartographische Zeugnisse wie vor allem die Werke des deutsch-italienischen Kartenmachers Heinrich Hammer (Henricus Martellus) zeigen, was das späte Mittelalter von Japan/Zipangu dachte. Zumal die erste topographische Karte des Landes von der Hand eines europäischen Kartographen vermittelt davon einen anschaulichen Eindruck (Fig./Abb. 7). Es galt als Solitär am östlichsten Rande der Welt und war deshalb nicht leicht zu erreichen, aber der Zugang schien auch nicht unmöglich. Nach wie vor hielt man Japan für eine einzige Insel, die regelmäßig in Form eines „Ziegels" dargestellt wurde.[18] Im Landesinneren gab es Ebenholz, Muskatnuss und andere Reichtümer, die dazu berechtigten, Zipangu zu jenen mythischen Inseln zu rechnen, die man schon immer im östlichen Ozean nahe beim irdischen Paradies vermutet hatte.[19]

[17] Reichert 2001: 207ff; 2003: 151ff.

[18] Zur frühesten europäischen Japan-Kartographie Reichert 2003: 153ff. Zu Henricus Martellus vgl. *Lexikon zur Geschichte der Kartographie* 467f.

[19] Reichert, „Mythische Inseln," *Mittelalter-Mythen*, Hg. Ulrich Müller u. W. Wunderlich, Bd. 5, St. Gallen 2007.

Fig. 7. Henricus Martellus Germanus: Zipangu, c. 1490

Varenius kannte seinen Marco Polo genau und berief sich mehrfach auf ihn, allerdings in bezeichnender Weise: Früher habe Japan Zipangu geheißen, und früher habe es dort Gold in rauen Mengen gegeben, doch heute nicht mehr. Wo er von der Region Anian spricht, deren Kenntnis ebenfalls auf Marco Polo zurückgeht, bezog sich dies nicht auf Japan, sondern auf ein mythisches Land im hohen Norden, das kein Europäer jemals besucht hatte.[20] Als Japan-Autor dagegen war Marco Polo durch die Ereignisse überholt worden und konnte nur mehr als ein geographie- und entdeckungsgeschichtlicher Klassiker dauernde Geltung beanspruchen.

2. *Portugal und die Mission der Jesuiten*

1543 landeten portugiesische Schmuggler auf der Insel Tanegashima südlich von Kyushu, seitdem entwickelte sich ein ebenso reger wie regelmäßiger Schiffs- und Handelsverkehr mit Macao in China und Malakka auf der Malaiischen Halbinsel. Er war in Form eines hoch profitablen Monopolhandels organisiert, bei dem japanisches Barrensilber gegen Gold, Porzellan und Seide aus China, europäische Feuerwaffen und indische Luxuswaren getauscht wurden. In einer Art Kreishandel oder Ringtausch wurden Waren zwischen Indien (Goa), der Malaiischen Halbinsel, China und Japan gehandelt.[21] Portugiesische Schiffe fuhren nach Hirado, Funai (Oita) und zu anderen Häfen, bis sich—seit 1571—der Handel auf die Ortschaft Nagasaki im südlichen Kyushu konzentrierte.

Gleichzeitig, Hand in Hand mit den Kaufleuten und später sogar durch den portugiesisch-japanischen Handel finanziert, kamen christliche Missionare aus dem Jesuitenorden ins Land; allen voran Francisco de Xavier (Franz Xaver, der heilige Apostel der Japaner, 1506–1552), der zweieinhalb Jahre an verschiedenen Orten in Japan verbrachte und bis nach Kyoto gelangte, ohne allerdings zum Tenno vorgelassen zu werden.[22] Da das Land im Inneren zerstritten war und zu diesem

[20] Varenius 1649: 1, 2, 16; dt 35, 36, 46. Polo 1928: 126; 1983: 210f. Vgl. Reichert 1996: 133ff.

[21] Hierzu und zum Folgenden Wolfgang Reinhard, *Geschichte der europäischen Expansion*, Bd. 1: *Die Alte Welt bis 1818*, Stuttgart 1983, 76ff; Peter Kirsch, *Die Barbaren aus dem Süden: Europäer im alten Japan 1543–1854* (Expansion. Interaktion. Akkulturation 6), Wien 2004.

[22] Georg Schurhammer, *Franz Xaver: Sein Leben und seine Zeit*, Bd. 2, Freiburg 1971, 3.

Zeitpunkt keine zentrale Vormacht besaß, schienen die portugiesischen Kaufleute und Missionare durchaus willkommen zu sein. Denn die Aussicht auf Bündnispartner und Waffenhilfe (zumal mit Feuerwaffen) trieb einige Daimyo („Regionalfürsten") dazu an, zum Christentum überzutreten und westliche Taufnamen anzunehmen. Japan schien für mehrere Jahrzehnte ein geradezu offenes Land zu sein, zumal im Vergleich mit den nachfolgenden Jahrhunderten seiner fast vollständigen Abschließung nach außen.

Francisco de Xavier konnte nur die Grundlagen für die Evangelisierung der japanischen Bevölkerung legen, aber seine zahlreichen Nachfolger aus der Societas Jesu trieben die Mission auf den japanischen Inseln mit beispiellosen Erfolgen voran. So wie gleichzeitig in China befolgten sie eine ausgearbeitete Missionsmethode, die darauf hinauslief, durch die bewusste Anpassung (Akkommodation) an die kulturellen Verhältnisse, an Religiosität und Mentalität der Bevölkerung, die Adressaten der Mission unmittelbar ansprechen zu können.[23] Nur dadurch gelang ihnen der Aufbau einer christlichen Kirche in Japan. Um 1580 bestanden circa 200 Kirchen mit 150.000 Gemeindemitgliedern im ganzen Land, 65 jesuitische Missionare waren auf den japanischen Inseln tätig, zuerst vor allem im Süden, dann auch auf den nördlichen und östlichen Landesteilen. Zunächst waren ihre Bemühungen auf die Fürsten und die Oberschicht gerichtet, daran anschließend wurde aber auch die einfache Bevölkerung erreicht. Sogar Toyotomi Hideyoshi (einer der drei großen Reichseiniger) soll gesagt haben: wenn nicht das Polygamieverbot wäre, könnte er sich zur Konversion zum Christentum entschließen.[24] Aber auch so stellte die Mission in Japan den größten Erfolg der Jesuiten in Asien dar. Die westliche Forschung spricht sogar von einem „christlichen Jahrhundert" in der japanischen Geschichte.[25]

Mit den sogenannten „Litterae annuae", jährlichen Briefen, hatten die Missionare regelmäßig über ihr Vorgehen an die Ordenszentrale in Rom zu berichten. Diese Briefberichte wurden teilweise veröffentlicht und erregten kein geringes Aufsehen in Europa. Übertroffen wurden sie noch durch den spektakulären Besuch einer Gesandtschaft von vier

[23] Kirsch; Reinhard, „Gelenkter Kulturwandel im 17. Jahrhundert: Akkulturation in den Jesuitenmissionen als universalhistorisches Problem," In: Wolfgang Reinhard, *Ausgewählte Abhandlungen* (Historische Forschungen 60), Berlin 1997, 358ff.

[24] Horst Gründer, *Welteroberung und Christentum: Ein Handbuch zur Geschichte der Neuzeit*, Gütersloh 1992, 237.

[25] Gründer 210ff; Charles R. Boxer, *The Christian Century in Japan 1549–1650*, Berkeley 1951, VII.

jungen, vornehmen christlichen Japanern, die im Jahre 1585 nach Rom, Venedig, Mailand, Madrid und Lissabon kamen und in ganz Europa eine Welle von Publikationen auslösten. Insgesamt 55 Druckschriften wurden über dieses Ereignis publiziert und auch in Japan profitierten Kunst und Kartographie von den Erfahrungen der Gesandtschaft.[26] Ihr Unternehmen gehörte zu den sogenannten umgekehrten Entdeckungsreisen, die erst in neuerer Zeit vermehrte Aufmerksamkeit von der reise- und entdeckungsgeschichtlichen Forschung erfahren.[27]

Auf all diesen Wegen gelangte eine Fülle neuer und genauerer Informationen über Japan nach Europa. Nachrichten über Geographie und Geschichte, Städte und Dörfer, Lebensweise und Brauchtum, den Kaiserhof in Kyoto, das Shogunat in Edo (Tokyo), den Adel in den Provinzen und die einfachen Leute allenthalben fanden Eingang in Medien, Bibliotheken und die gelehrten Diskurse in Europa. Auch die Kartographen nahmen daran Anteil und schufen Kartenwerke, die den gewachsenen Kenntnissen von Japan Rechnung trugen.[28]

Fernão Vaz Dourado steht mit seiner Karte von 1568 (Fig./Abb. 8) sichtbar noch im Anfang der sich weitenden Kenntnisse von Japan. Die „Barbaren des Südens" (nanban/namban), wie die (Süd-)Europäer von den Japanern genannt wurden, wussten zunächst nur auf der südwestlichen Insel Kyushu Bescheid und erwarben erst allmählich ein Wissen von den anderen Inseln. Hirado, Kagoshima und Nagasaki spielten im Kartenbild noch gar keine Rolle. Auch die eigenartige, nach Süden statt nach Norden abknickende Form von Nord-Honshu lässt vorerst nur schemenhafte geographische Kenntnisse erkennen. Später wurde Vaz Dourado durch den portugiesischen Kartographen Luis Teixeira (SJ) abgelöst, dessen Darstellung der japanischen Inseln in Abraham Ortelius' *Theatrum orbis terrarum* (1595) Eingang fand und dadurch kanonisch werden sollte (Fig./Abb. 9). Zwar war der Verfasser niemals in Japan gewesen, aber er konnte auf japanische Informationen

---

[26] Kapitza 1: 153ff, 199ff; Donald F. Lach, *Asia in the Making of Europe*, part 1, Chicago 1965, 2, 688ff.

[27] Michael Harbsmeier, Hg., *Stimmen aus dem äußersten Norden: Wie die Grönländer Europa für sich entdeckten*. Stuttgart 2001.

[28] Zur Entwicklung der Japan-Kartographie vgl. Hugh Cortazzi, *Isles of Gold: Antique Maps of Japan*, New York u. Tokyo 1983; Walter Lutz, Hg., *Japan mit den Augen des Westens gesehen: Gedruckte europäische Landkarten vom frühen 16. bis zum 19. Jahrhundert*, München u. New York 1993; Kazutaka Unno, *Chizu ni miru Nippon—Japan in Maps: Wakoku, Zipangu, Dainippon*, Tokyo 1999 [japan.].—Vgl. Boxer 133f über die Abhängigkeit der portugiesischen Karten von japanischen Vorbildern.

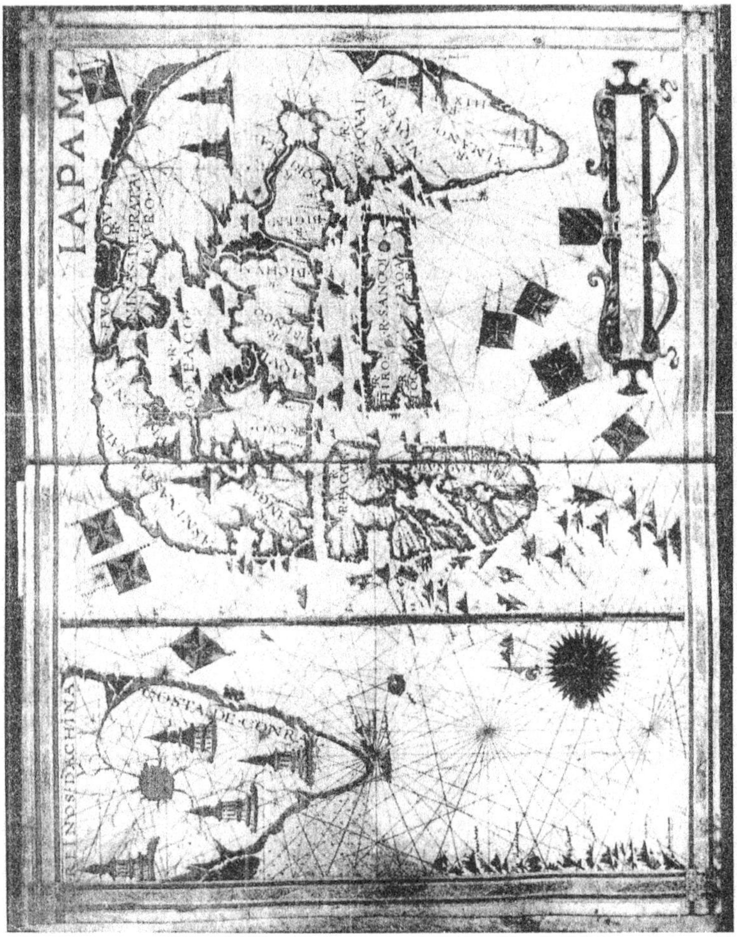

Fig. 8. Fernão Vaz Dourado: Iapam

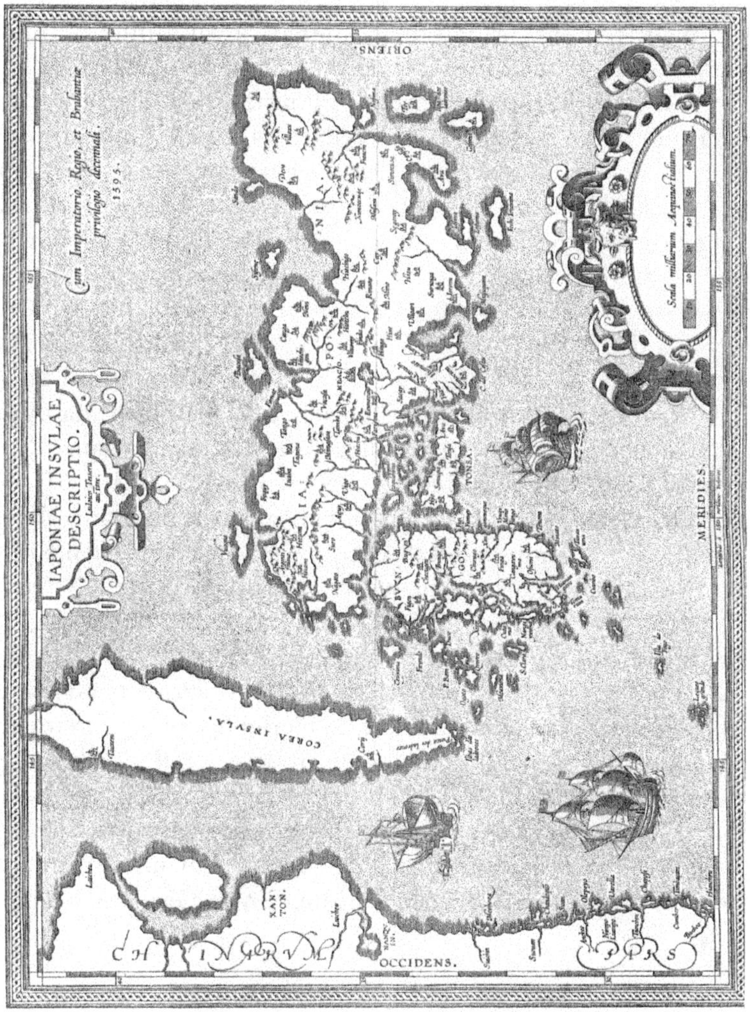

Fig. 9. Abraham Ortelius/Luis Teixeira: Iaponiae insulae descriptio, 1595

zurückgreifen und dadurch ein genaueres und dichteres Kartenbild erstellen. Die Umrisse der Inseln kommen der Wirklichkeit näher, und zahlreiche Ortsnamen können identifiziert werden. In Kyushu und West-Honshu sind sie nach wie vor dichter, nach Nordosten hin dünner gestreut, aber Tohoku (Nord-Honshu) ist zutreffend orientiert. Korea ragt keilförmig in die Karte hinein, von Hokkaido ist so wenig zu sehen wie auf den japanischen Karten der gleichen Epoche.[29]

Aber nicht nur geographisches, sondern auch ethnographisches Wissen wurde durch die Missionare erhoben und durch ihre regelmäßigen Berichte nach Europa kolportiert. Informationen über die örtliche Bevölkerung, deren Lebens- und Verhaltensweisen, Brauchtum und religiöse Vorstellungen waren für den Orden sogar besonders wichtig, da die jesuitische Missionsmethode darauf basierte. In allen gedruckten und ungedruckten Berichten werden solche Fragen daher mehr oder weniger ausführlich behandelt.

Ein besonders anschauliches und in vielerlei Hinsicht bezeichnendes Beispiel gibt der Portugiese Luís Froís (1532–1597).[30] Als junger Mann war er der Societas Jesu beigetreten, um in der Asienmission tätig zu werden. Sein Weg führte ihn zuerst nach Goa in Indien, dann nach Malakka, schließlich (1562) nach Japan, wo er sich über 34 Jahre hinweg an verschiedenen Orten aufhielt, zunächst in Miyako (Kyoto), dann in Sakai bei Osaka, schließlich in Bungo und Nagasaki auf Kyushu. Der jesuitischen Missionsmethode entsprechend suchte er den Kontakt zu den Mächtigen und unterhielt ein Verhältnis zu Oda Nobunaga (dem starken Mann hinter dem Shogun und ersten der drei großen Reichseiniger), das nach außen sehr eng wirkte.

Froís wird noch heute in Japan geschätzt, da er die erste Grammatik der japanischen Sprache konzipierte. Außerhalb Japans waren jedoch seine landeskundlichen Einsichten von größerem Interesse. 1585 verfasste er einen Traktat über die Kulturgegensätze zwischen Japan und Europa: *Algumas contradições e diferencas de custumes antre a gente de Europa e esta provincia de Japão.* Es handelt sich dabei um einen ausführlichen und in jeder Hinsicht charakteristischen Katalog von Gegensatzpaaren in

---

[29] Zu Vaz Dourado und Teixeira vgl. *Lexikon zur Geschichte der Kartographie* 627, 806f.

[30] Zu Biographie und Werk: Boxer 56ff; Kirsch 60ff; Engelbert Jorißen, *Das Japanbild im „Traktat" (1585) des Luis Froís* (Portugiesische Forschungen der Görresgesellschaft: 2. Reihe, 7), Münster, Westf. 1988, 22ff, 134ff.

Form kurzer Sentenzen[31] nach dem Muster: bei uns ist es so, in Japan aber ganz anders, und dies kann auf allen Ebenen gezeigt werden: beim Hausbau, im Theater, bei Lebensmitteln, Kindererziehung, Stellung der Frau usw. Insgesamt stellte Fróis 609 solche kontrastierenden Vergleiche zusammen und kann schon durch die schiere Menge beeindrucken. Aber auch die einzelnen Aussagen sind ebenso anschaulich wie unterhaltsam zu lesen. Die folgende Auswahl[32] soll davon einen Eindruck vermitteln:

> [107] Wir grüßen, indem wir die Kopfbedeckung abnehmen; die Japaner erweisen Höflichkeit, indem sie sich die Schuhe ausziehen.
> Wir gebrauchen bei Trauer die schwarze Farbe, die Japaner die weiße.
> Wir betreten das Haus mit Schuhen an den Füßen; in Japan ist das unhöflich, und man muß die Schuhe an der Pforte lassen.
> [125] In Europa gehen die Männer voran, die Frauen hinterher, in Japan gehen die Männer hinterher und die Frauen voran.
> [173] Die Leute in Europa lieben gebackenen und gekochten Fisch, die Japaner essen ihn viel lieber roh.
> [175] Bei uns kühlt man den Wein, in Japan wärmt man ihn an.
> Wir sitzen, wenn wir trinken, auf Stühlen; sie hocken auf den Knien.
> [179] Bei uns gilt es als dreckiges Benehmen, das Essen sehr laut zu kauen und den Wein bis zum letzten Tropfen auszuschlürfen; die Japaner unter sich halten das eine wie das andere für Feinheit.
> [231] Unsere Kirschbäume geben sehr schmackhafte und schöne Kirschen, die japanischen geben sehr kleine und bittere Kirschen und schöne Blüten, die die Japaner hochschätzen.
> [259] In Europa sucht man Klarheit in den Worten und meidet die Zweideutigkeit, in Japan sind zweideutige Ausdrücke die beste Sprache und am meisten geschätzt.
> [263] Wir reinigen die Nasenlöcher mit dem Daumen oder Zeigefinger; die Japaner tun es, weil sie kleinere Nasenlöcher haben, mit dem kleinen Finger.

Luís Fróis hat es verstanden, mit seinem Traktat über die Kulturgegensätze zwischen Japan und Europa Alterität zum Ausdruck zu bringen, und in der Tat scheint vieles zutreffend, was er berichtet: Sake (Reiswein), Sashimi (roher Fisch) und Sakura (Kirschblüte), japanische Tischsitten und Sprechweisen werden charakterisiert, um kulturelle Differenz im Spiegel von Alltag, Verhaltensstilen und Lebensformen sichtbar werden zu lassen. Sogar die Angaben zum Nasenbohren bei Europäern und

[31] Die Forschung spricht von Distichen ohne Versform, Aphorismen o.ä. und zieht auch japanische Vorbilder in Erwägung (Jorißen 217ff).

[32] Alle zitiert nach Kapitza 1: 133ff.

Japanern treffen nach wie vor zu. Die Vorstellung schließlich, anders als alle anderen zu sein, entspricht dem japanischen Selbstverständnis bis auf den heutigen Tag. Das Verzeichnis basierte somit nicht nur auf Beobachtungen im Alltag, sondern auch auf der langjährigen Auseinandersetzung des Verfassers mit der japanischen Kultur.

Die sprachliche Form, die er dabei wählte, wirkt jedoch keineswegs neu. Schon Herodot gab das gültige Muster mit der Beschreibung Ägyptens: „Wie der Himmel bei den Ägyptern anders ist als in anderen Ländern, der Strom sich anders verhält als die anderen Flüsse, so stehen auch die Sitten und Bräuche der Ägypter größtenteils in allen Stücken im Gegensatz zu denen der übrigen Völker." Es folgt dann eine Serie von Beispielen, die um den Topos der verkehrten Welt kreisen.[33] Spätere Autoren hielten sich an Herodots Beispiel und betonten auf ähnliche Weise die kulturellen Kontraste, wenn sie glaubten, eine verkehrte Welt vor sich zu haben. Denn immer wieder fanden sich Kulturen, die dem europäischen Betrachter ganz anders als die eigene gestaltet und nach völlig anderen Grundsätzen geordnet zu sein schienen.

Dies galt für die Muslime, deren Kultur man als eine Gegenwelt begriff, für China, das den Europäern in fast jeder Hinsicht konträr zu sein schien, und erst recht für die japanischen Inseln, deren Lage am gegenüberliegenden Rand der eurasiatischen Landmasse die Verwendung des Topos geradezu provozierte. Auch andere Missionare vor und nach Luís Fróis wiesen auf die zahlreichen kulturellen Unterschiede hin, und Alessandro Valignano (1539–1606), der drei Jahre lang als Visitator auf den Inseln wirkte und sich besonders intensiv mit den Perspektiven der Mission auseinander setzte,[34] sprach ausdrücklich und wörtlich von einer verkehrten Welt, die er in Japan erblickte, einer Welt, in der alles anders läuft als in Europa (... Japón es un mundo al revés de como corre en Europa). Wer dort etwas ausrichten wolle, müsse sich an die japanische Lebensweise anpassen.[35] Die von ihm vorgeschlagene Missionsmethode auf dem Wege der Akkommodation konnte er damit begründen, und wahrscheinlich ist auch Fróis' Liste vor diesem Hintergrund zu verstehen. Vielleicht war sie als Handreichung oder Anleitung für angehende Missionare gedacht.

---

[33] Herodot, *Historien* II 35–36.

[34] Vgl. Joseph Schütte, *Valignanos Missionsgrundsätze für Japan* (Storia e letteratura 36), 2 Bde., Rom 1951–1958; Joseph Moran, *The Japanese and the Jesuits: Alessandro Valignano in Sixteenth-Century Japan*, London 1993.

[35] Jorißen 111.

Varenius dagegen verfolgte keine praktischen Zwecke, als er die Berichte der Jesuiten von Japan exzerpierte und wie Froís wenigstens auf zwei Seiten die für ihn auffälligsten kulturellen Unterschiede zusammentrug.[36] Sein Interesse an Japan war geographisch-landeskundlich orientiert, und eine solche Liste konnte dazu beitragen, einen besonders sinnfälligen Eindruck von der gänzlichen Andersartigkeit und Exotik des Landes zu vermitteln.

3. *Berichte aus dem Umkreis der Vereinigten Ostindischen Kompagnie*

Die Berichterstattung der Vereinigten Ostindischen Kompagnie (VOC) resultierte aus einer völlig veränderten politischen und ökonomischen Konstellation.[37] Denn seit dem ausgehenden 16. Jahrhundert hatte ein allgemeiner Stimmungsumschwung zu Ungunsten Portugals und der Jesuiten stattgefunden. Deren Mission hatte von der inneren Zersplitterung Japans profitiert und Handlungsspielräume zwischen den konkurrierenden Parteien zu nutzen verstanden. Regionale Machthaber hatten am portugiesischen Handel partizipiert, portugiesische Waffenlieferungen zu schätzen gewusst und deshalb das Christentum toleriert.

Mit der fortschreitenden Einigung und Zentralisierung Japans durch Oda Nobunaga, Toyotomi Hideyoshi und Tokugawa Ieyasu entfiel sozusagen die Geschäftsgrundlage für das alles. Nach der Schlacht von Sekigahara, in der Ieyasu seine Gegner überwand (1600), bestand kein Anlass mehr, mit den Jesuiten zu kooperieren. Vielmehr wurden Verdächtigungen laut, dass die Missionare nichts weiter als eine Fünfte Kolonne der portugiesischen oder spanischen Kolonialmacht darstellten und das Bekenntnis zum Christentum fremde Loyalitäten begünstige. Auch Anschuldigungen durch die holländische Konkurrenz, die religiös indifferent auftrat und sich als alternativen Handelspartner anbot, spielten dabei eine Rolle. Ein Engländer in Diensten der VOC, William Adams (1564–1620), erwarb sich auf diesem Wege als Anjin-san bzw. Miura Anjin einen prominenten Platz in der japanischen Geschichte.[38] Seit 1587 wurden zunächst regionale, dann landesweite

[36] Varenius 1649: 9, dt 41.

[37] Zum Folgenden vgl. Boxer 308ff; Reinhard 1983: 84f; Kirsch 157ff.

[38] Vgl. Giles Milton, *Samurai William: The Englishman Who Opened Japan*, New York 2004 (zuerst 2002).

Edikte gegen die Tätigkeit der jesuitischen Missionare erlassen, die sich schließlich zu regelrechten Christenverfolgungen auswuchsen. 1597 wurden in Nagasaki 26 Christen, davon 6 Missionare, gekreuzigt (die heute als heilige Märtyrer verehrt werden),[39] 1622 wurden mehr als 120 Missionare und Konvertiten exekutiert. Insgesamt dürften in den Verfolgungen des 16. und 17. Jahrhunderts an die 300.000 japanische Christen ums Leben gekommen sein, von denen 3.000 als katholische Märtyrer anerkannt wurden.

Als schließlich 1637 auf der Halbinsel Shimabara (östlich von Nagasaki) ein Aufstand ausbrach, der als christlich inspiriert interpretiert werden konnte, war der Anlass (oder auch: Vorwand) zu durchgreifenden Maßnahmen gegeben, die das vorläufige Ende des Christentums in Japan und die mehr als 200 Jahre dauernde Abschließung des Landes (sakokurei) zur Folge hatten: Sämtliche Grenzen wurden hermetisch geschlossen, Fremden war die Einreise, Japanern die Ausreise untersagt, jede Zuwiderhandlung wurde mit dem Tode bestraft. Die Verwendung christlicher Symbole war strengstens verboten, zur Überwachung wurde eine besondere religiöse Behörde geschaffen. Sämtliche Einwohner hatten sich in den buddhistischen Tempeln registrieren zu lassen. Portugiesen und Spanier wurden des Landes verwiesen, allein auf der winzigen, in Form eines geöffneten Fächers künstlich aufgeschütteten Insel Deshima im Hafen von Nagasaki wurde eine von holländischen Kaufleuten bewohnte Handelsstation noch geduldet. Deren minimale Ausmaße (600 × 200 Schritt) konnten sie allerdings eher als ein Gefängnis denn als Handelsstützpunkt erscheinen lassen. Abgesondert vom Festland, mit einem Zaun als Barriere umgeben und von japanischem Wachpersonal beaufsichtigt, lebten die holländischen Kaufleute ohne ihre Familien, unterhielten nur eng begrenzte Kontakte zur japanischen Bevölkerung, und auch ihr Handel wurde permanent argwöhnisch kontrolliert. Mit anderen Worten: in der von Urs Bitterli entworfenen Typologie der europäisch-außereuropäischen Kulturbeziehungen stellte Deshima den Musterfall einer „kontrollierten Kulturbeziehung" dar. Die Portugiesen hatten mit ihren Handels- und Missionsstationen immerhin noch ein „System der begrenzten Kontakte" mit Japan unterhalten.[40] Betrieben wurde die Faktorei durch die VOC, eine private Kaufmannsgesellschaft

[39] Twenty Six Martyrs Museum, *Catalogue*, Nagasaki 1987 (japan.).

[40] Urs Bitterli, *Alte Welt—neue Welt: Formen des europäisch-überseeischen Kulturkontakts vom 15. bis zum 18. Jahrhundert*, München 1986, 55ff, 152ff.

zur Gewinnmaximierung im Asienhandel, die Stützpunkte entlang der Handelsrouten zwischen Europa und Süd-, Südost- und Ostasien unterhielt.[41] An der Spitze der Faktorei stand ein Opperhoofd (Oberhaupt), der die Interessen der VOC gegenüber den japanischen Behörden vertrat und dazu jährlich einmal mit (zunächst kleinem, dann erweitertem) Gefolge zum Shogun nach Edo (Tokyo) reisen musste. Sein Weg führte quer durch das Land und gab durchaus auch Gelegenheit zu landeskundlichen Beobachtungen und Erkundigungen. Was ein aufmerksamer Beobachter daraus machen konnte, zeigt vor allem das große Japan-Buch Engelbert Kaempfers aus dem westfälischen Lemgo, der zwischen 1690 und 1692 als „Oberchirurg" auf Deshima lebte und zweimal an der Hofreise teilnahm. Der Opperhoofd dagegen durfte nur ein Jahr in Deshima bleiben und wurde dann durch einen anderen Vertreter der VOC abgelöst.

Der erste, den diese Regelung betraf, war François Caron (1600–1673). Kind hugenottischer Flüchtlinge und schon als Küchenjunge im Dienst der VOC, gelangte er 1619 nach Hirado vor Kyushu. Dort lernte er Japanisch und war als Dolmetscher auch bei Reisen nach Edo dabei. Zeitweilig lebte er als Geisel in Japan (1629), schließlich aber als Kaufmann in Hirado. 1639–1641 war er Opperhoofd der dortigen Faktorei, dann auf Deshima. Er wohnte in einem japanischen Haus, war mit einer Japanerin verheiratet und schätzte sogar das japanische (extrem) heiße Bad (ofuro).[42] Mit einem Wort: kein anderer Europäer wusste mehr über Japan als Caron.[43] Später bekleidete er noch höhere Ämter in der VOC, aber auf Dauer bedeutend war er nur durch seine *Beschrijvinghe van het machtigh Coninckrijck Iapan*, die zuerst in Amsterdam 1645 erschien und danach zahlreiche Ausgaben und Übersetzungen in Deutsch, Englisch, Französisch, Latein, Schwedisch und Italienisch erlebte. Sie war ein internationaler Bestseller, der die Zeitgenossen

---

[41] Zur VOC und ihrer Rolle in Japan vgl. Eberhard Schmitt, Thomas Schleich u. Th. Beck, Hg., *Kaufleute als Kolonialherren: Die Handelswelt der Niederländer vom Kap der Guten Hoffnung bis Nagasaki 1600–1800*, Bamberg 1988 (dieselben zu Deshima: 171ff); Eberhard Schmitt, Hg., *Wirtschaft und Handel der Kolonialreiche* (Dokumente zur Geschichte der europäischen Expansion 4), München 1988; Ken Vos, „Dejima und die Handelsbeziehungen zwischen den Niederlanden und dem vormodernen Japan," *Japan und Europa 1543–1929*, Hg. Doris Croissant u. Lothar Ledderose, Berlin 1993, 72–82; Grant Koon Goodman, *Japan and the Dutch 1600–1853*, Richmond 2000; Leonard Blussé, W. Remmelink and I. Smits, Hg., *Bridging the Divide: 400 Years The Netherlands—Japan*, Leiden 2000.

[42] Lach part 3:1, 459.

[43] Caron 2000: 22ff; Kirsch 150ff.

faszinierte. Auch Varenius zitierte François Caron gerne, namentlich insgesamt 21 Mal.

Dreierlei war Carons Buch zu entnehmen: Erstens hielt es den Stand des geographischen Wissens über die japanischen Inseln fest. Dazu gehörten recht genaue Kenntnisse der Hauptinseln Kyushu, Honshu und Shikoku, dazu gehörte aber auch die Feststellung, dass man nicht genau wisse, wie die Küsten im Norden verliefen. Zwar hatte sich der Jesuit Girolamo degli Angeli (1568–1623) einige Zeit auf Yezo/Hokkaido aufgehalten und schon Froís hatte manches von Land und Leuten berichtet,[44] aber unklar blieb 1. ob Hokkaido eine Insel oder eine Halbinsel, also Teil des Festlandes sei; 2. ob sich zwischen Hokkaido und Honshu eine schiffbare Durchfahrt oder nur eine Meeresbucht befinde. Auch Caron konnte die Frage nicht lösen, und spätere Auflagen enthalten eine Karte, die die Unsicherheit aller Ortskundigen, auch der japanischen Gewährsleute, spiegelt (Abb./fig. 10). Varenius ließ die Frage ebenfalls offen und folgte Luís Froís, um Carons Angaben zur Urbevölkerung Hokkaidos, den Ainu, zu ergänzen.[45] Erst durch die Erkundungsfahrten des Holländers Maarten Gerritszoon Vries (1643) und des Franzosen Jean-François Galaup de La Pérouse (1787) wurde Klarheit über den Küstenverlauf im nordwestlichen Pazifik gewonnen.[46]

Zweitens war Caron bemüht, seinen Lesern ein aktuelles Bild von Japan um die Mitte des 17. Jahrhunderts zu vermitteln. Zu diesem Zweck gab er seiner Darstellung eine 25-seitige Aufstellung über die Erträge der Provinzen und die Einkommen der regionalen Fürsten bei.[47] Dem heutigen Leser scheint sie kaum noch erträglich, aber dem Bestseller war sie damals nicht abträglich. Varenius, der Auszüge aus ihr abdruckte,[48] gab sie reichhaltiges und vor allem authentisches Zahlenmaterial an die Hand, um seiner Leserschaft einen Begriff von den ökonomischen Verhältnissen der Provinzen und den Machtmitteln des Shogunats in Edo zu vermitteln.

Drittens schließlich enthielt Carons Beschreibung des Königreichs Japan auch reiche ethnographische Information. Zahlreiche Details, die Varenius über Lebensweisen, Mentalitäten, Rechtsformen oder

---

[44] Dietmar Henze, *Enzyklopädie der Entdecker und Erforscher der Erde*, 5 Bde., Graz 1978–2004, hier: 1: 78f, 2: 302.
[45] Caron 2000: 85ff.
[46] Henze 5: 430ff, 3: 128ff.
[47] Caron 2000: 89.
[48] Varenius 1649: 68ff, 1974: 87ff.

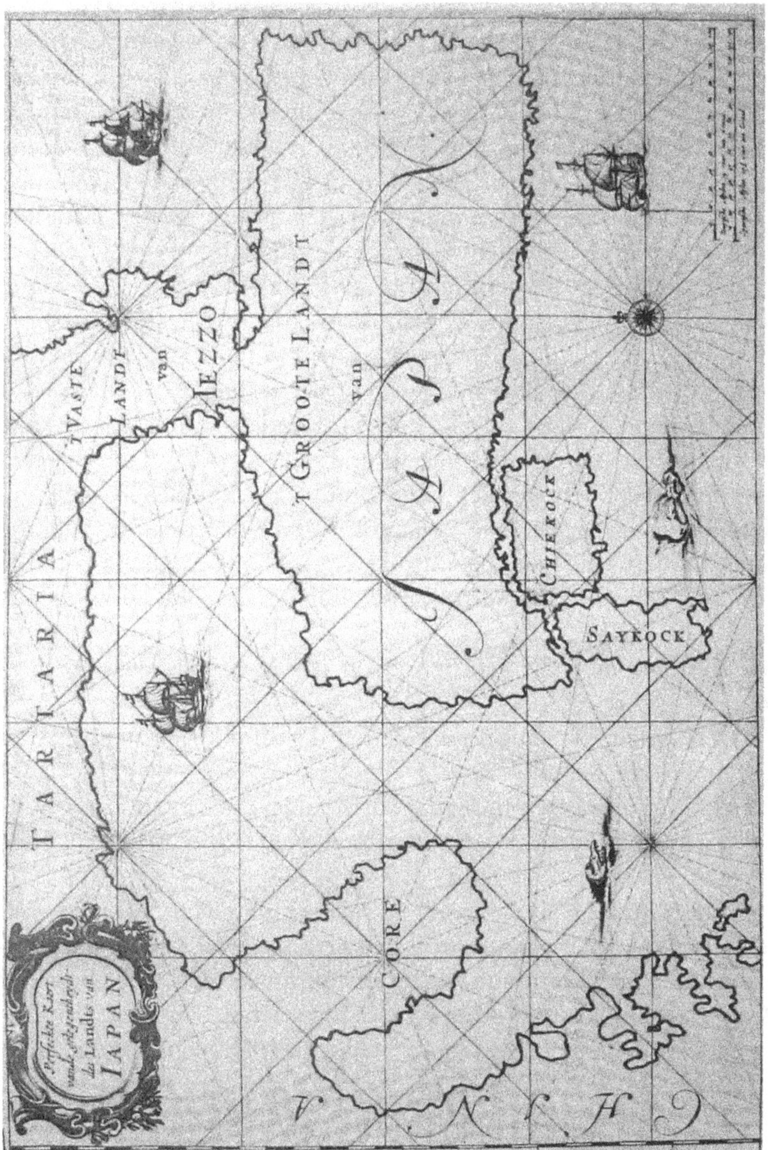

Fig. 10. Perfeckte Kaert van de gelegentheydt des Landts van Iapan, 1661

die Stellung der Frau in Japan mitteilt, sind Carons Darstellung entnommen, darunter auch etwas so Spektakuläres wie das japanische Seppuku (vulgo Harakiri), das Ritual der Selbsttötung bei drohendem Ehrverlust, zur Wiederherstellung von Ehre oder zur ehrenvollen Milderung der Höchststrafe.[49] Luís Fróis hatte in seinem Verzeichnis kultureller Gegensätze nur beiläufig davon gesprochen,[50] wahrscheinlich weil es ja schon aus theologischen Gründen nicht zur Nachahmung empfohlen werden durfte und bei der Vorbereitung der Missionare keine Rolle spielen konnte. Aber Caron hat mehrfach davon gehandelt und aufmerksam beschrieben, wie, auf welche Art und Weise und bei welcher Gelegenheit man sich den Bauch aufschlitzt und eine „Selbstentdärmung"[51] vornimmt. Auch eine Abbildung durfte nicht fehlen und trug sicher zum Verkaufserfolg des Buches etwas bei.[52] In der Ausgabe von 1662 drang sie sogar bis ins Titelblatt vor (Fig./Abb. 11). Denn was in der Berichterstattung der Jesuitenmissionare immer nur nebensächlich blieb und allenfalls am Rande erwähnt wurde, konnte in einer am Buchmarkt orientierten und die Interessen eines breiteren Publikums berücksichtigenden Darstellung eine Hauptrolle spielen und in deren Wirkungsgeschichte vollends in den Vordergrund treten.

Auch Varenius ließ sich beeindrucken und hat Carons Beschreibung des Seppuku übernommen, um anschaulich darstellen zu können,

- wie es ist, wenn man sich den eigenen Bauch in Kreuzesform aufschneidet,
- dass es fünfzig verschiedene Formen davon gibt (bei Caron waren es nur vierzig Arten),
- dass beim Heraustreten der Gedärme der Tod eintritt,
- dass manche Leute sich zuvor noch die Kehle durchstechen können.

So gut hat ihm eine Stelle gefallen, dass er sie gleich zweimal in fast gleichlautenden Worten zitierte.[53] Die in Europa bis auf den heutigen Tag anhaltende Faszination eines aristokratischen, sonst aber wenig praktizierten Rituals und seine regelmäßige Überschätzung durch den europäischen Beobachter ist hier mit Händen zu greifen. Varenius nutzte

[49] Caron 2000: 89.
[50] Jorißen 364.
[51] Ulrich Pauly, *Seppuku: Ritueller Selbstmord in Japan*, Tokyo 1995.
[52] Caron 2000: 118, 120, 122, 125, 126f.
[53] Varenius 1649: 119, 203; dt 119, 187.

Rechte Beschryvinge
Van het Machtigh Koninghrijck van
IAPPAN,
Bestaende in verscheyde Vragen, betreffende des selfs Regiering, Coophandel, maniere van Leven, strenge Justitie &c. voorgestelt door den Heer *Philips Lucas*, Directeur Generael wegens den Nederlandsen Staet in India, ende door de Heer
FRANCOYS CARON,
President over Comp. ommeslach in Iappan, beantwoort inden Iare 1636.
Welcke nu door den selven Autheur oversien / vermeerdert en uytgelaten is de Fabuleuse aenteekeningen van Hendrick Hagenaer, soo dat nu alles met zijn voorige Origineel komt te accorderen / en met Kopere Figueren verrijckt.

In 's Gravenhage, by *Iohannes Tongerloo*, Boeckverkooper,

Fig. 11. François Caron, *Rechte Beschryvinge van het Machtigh Koninghrijck van Iappan*, 1662: Seppuku

somit seine Möglichkeiten, aktuelle und authentische Informationen zu verwerten. Er steht aber auch in den Anfängen eines orientalisierenden Bildes von Japan, das sich später (zumal nach der Abschließung des Landes) noch weit phantasievoller ausdrücken konnte.

Resümee

Zusammenfassend ist festzuhalten, dass Bernhard Varenius aus einer überschaubaren, aber gut sortierten Anzahl von Quellen seine Informationen zusammengetragen hat. Ein Klassiker wie Marco Polos Buch erinnerte an das früheste Wissen, das Europa von Japan besaß. Durch die Briefe und Berichte der Jesuiten erfuhr man von den Bedingungen und Chancen der Missionsarbeit, aber auch von den Hemmnissen, die sie erschwerten. François Caron schließlich bot aktuelle Informationen und brachte das geographische und ethnographische Wissen der Europäer auf einen Stand, den die Berichte der Jesuiten nicht erreicht hatten oder nicht erreichen wollten. All diese Texte besaßen Vorzüge wie Nachteile und konnten unter verschiedenen Gesichtspunkten gelesen, exzerpiert und ausgewertet werden. Bernhard Varenius erstellte aus ihnen ein landeskundliches Kompendium, dessen Qualität in erheblichem Maße auf der Qualität seiner Vorlagen beruhte. Ob und inwieweit die Leistung des Verfassers über die eines bloßen Kompilators hinausging, wie eigenständig oder abhängig er mit den Grundlagen seines Wissens umging, bleibt zu diskutieren.

DIE *DESCRIPTIO REGNI JAPONIAE* IN DER LITERARISCHEN TRADITION DER EUROPÄISCHEN ‚DESCRIPTIONES'

Reinhard Düchting

Abstract

The Descriptio Regni Japoniae *in the Literary Tradition of European 'descriptiones'*

Varenius' book on Japan is a literary example of a geographia specialis, even though the maxims developed by its author in the *Geographia Generalis* (1650, 3f) as regards the so-called 'affectiones' (8+1 affectiones caelestes, 10 terrestres, 10 humanae) were not strictly, not even approximately observed. Instead, for the description of Japan, the traditionally open literary form is used, in which the initial required dates and information are followed by various memorabilia about the properties of a country (more or less as listed in the 10 points under the affectiones humanae).

My remarks in the following 11 sections are those of a philologist and might be entitled "On rereading Varenius" or "*Descriptio Regni Japoniae* reconsidered":

1. descriptio specialis (and affectiones III: 10 affectiones humanae)
2. *Descriptio Regni Japoniae* (1649)
3. de-scriptio (Greek ek-phrasis) alicuis rei
4. describere, descriptio in the "Dedicatio" and "Praefatio"
5. de quo plura videantur in libro De religione (= *Tractatus* 1649)
6. Japonia *illustrata*?
7. Varenius in German (1974); citations, corrigenda
8. *Descriptio Regni Japoniae* as the last of the Elzevier 'republics'
9. usus and recreatio animi
10. Remarks on the literary quality
11. David Chytraeus (Varenius's great-grandfather), *De Creichgoia* [1558], Wittenberg 1561 and M.F. Wendelinus, *Admiranda Nili* (1623)

Das Ergebnis dieser Betrachtungen nehme ich vorweg, indem ich sage, dass das Japan-Buch (1649) des Varenius der literarische Fall einer Geographia specialis ist, ohne indes die vom Verfasser selbst in seiner *Geographia Generalis* (1650, 3f) entwickelten Maximen der sogen. affectiones (8+1 affectiones caelestes, 10 affectiones terrestres; 10 affectiones humanae) strikt oder auch nur annähernd zu beachten; vielmehr nutzt die Japan-Beschreibung die traditionell offene literarische Form, welche einigen notwendigen ersten Daten und Nachrichten allerlei Memorabilia nach dem Proprium eines Landes und am ehesten noch entlang den zehn Punkten der affectiones humanae folgen lässt. Die Betrachtungen stimmen in diesen Teilen weitgehend mit der Studie von Blanke[1] überein.

Die Bemerkungen verstehen sich als Beobachtungen eines Philologen: Beim Wiederlesen des Varenius oder: Descriptio regni Japoniae—reconsidered; die Vortragsform eines Kommentars zu den folgenden 11 Punkten ist beibehalten.

1 descriptio specialis (und affectiones III: A 10 humanae). 2 *Descriptio Regni Japoniae* (1649). 3 descriptio alicuius rei. 4 describere, descriptio in der Dedicatio und Praefatio. 5 de quo plura videantur in libro De religione (= *Tractatus* 1649). 6 Japonia *illustrata*?. 7 Varenius deutsch (1974); Zitate und Corrigenda. 8 *Descriptio regni Japoniae* als letzte der Elzevierschen Reichs-Beschreibungen. 9 usus und recreatio animi. 10 Bemerkungen zur Literarität. 11 David Chytraeus: *De Creichgoia* [1558] (1561) und M.F. Wendelinus: *Admiranda Nili* (1623)

1. *descriptio specialis (und affectiones III: C 10 humanae)*

Die drei Arten „Beziehungen" (A terrestrische, B caelestische und C Human-Bezüge) sind auch von Schwind in der Einleitung der dt. Ausgabe XXIIIf diskutiert.[2] Die sogen. (B) 8 caelestes (unten unter Testimonia 1 zitiert) sind: Entfernung des Ortes vom Äquator und Pol; Neigung der Gestirn-Tagesbahnen gegen den Horizont; Tagesdauer; Klima und Zone; Hitze, Kälte und alljährliche Unwetter (tempestates; nicht, wie übersetzt, Jahreszeiten), meteorologische Erscheinungen; Aufgangszeit und Dauer der Gestirne; Transit der Gestirne durch den Meridian; ihre Bewegungsgeschwindigkeit iuxta Copernicam hypothesim. Eine 9. affectio, die von manchen Astrologen einzelnen Regionen zugeordneten Zeichen des Zodiacus, schien Varenius schon immer

[1] Horst Walter Blanke, „Marco Polo, Bernhard Varenius und Engelbert Kaempfer. Oder: Vom Hörensagen über die gelehrte Recherche zum Autopsiebericht: Drei Stationen der europäischen Japan-Kunde," *Bernhard Varenius (1622–1650): Der Beginn der modernen Geographie*, Hg. Wolfgang Griep u. Susanne Luber, Begleitband zur Ausstellung 2001 (Veröffentlichungen der Eutiner Landesbibliothek 5), Eutin 2001, 36–49. Siehe auch unten Blankes Beitrag in diesem Band.

[2] Zum Begriff der affectiones vgl. Staffhorsts Beitrag und Volkmanns Übersetzung in der Appendix; ihr Vorschlag zur Übersetzung lautet ‚Eigenschaften' oder ‚Merkmale'.

ein Hirngespinst (vana doctrina). Varenius geht nur ganz knapp auf einige dieser affectiones caelestes im 1. Kapitel seiner *Descriptio* ein (dt. 36). Die sogen. (A) 10 terrestres, d. h. Land- und Bodenbezüge sind (s.u. Testimonia 2): Grenzen und Umfang, Gestalt, Größe; Gebirge, Wasserverhältnisse, Wälder und Wüsten, Fruchtbarkeit und Dürre, Vorkommen von Mineralien, Tierwelt; auf einige Punkte kann Varenius im 1. und 2. Kapitel seiner Descriptio eingehen. Diese affectiones einer Geographie specialis (aus der *Geographia Generalis*) bilden ein generelles Raster für eine spezielle Landbeschreibung wie die gewählte von Japan. Wie also wenn von Wüsten (deserta) oder von Bergbau in irgendeinem anderen Land nichts zu berichten ist? Varenius spricht in der *Geographia Generalis* nicht von seiner *Descriptio regni Iaponiae* als einer geographia specialis; eine solche hat die „Konstitution einzelner Länder der Erde" zum Gegenstand (singularum regionum telluris constitutio), und zwar eine zweifache, nämlich die Deskription einer chora (Land, Region), also: choro-graphia und die Deskription, kleinteiliger, einer Gegend (tractus) oder eines Ortes (topos, locus) also: topo-graphia. Das Japan-Buch ist also eine Chorographie, eine Landbeschreibung, sie ist ausschließlich aus der Lektüre (und für manche Punkte nach möglichen eigenen Recherchen) geschrieben; die gesammelte Lektüre redigiert und ordnet Varenius am ehesten nach Gesichtspunkten, die sich für eine solche Beschreibung immer anbieten und die sich unter den sogen. 10 affectiones humanae finden (s.u. Testimonia 3) Die Überlegungen, ob Varenius die Raster seiner Geographie generalis für eine geographia specialis womöglich noch nicht entwickelt hatte, als er das Japan-Buch zusammenschrieb, sind müßig, Vorhaltungen, dass er seinen Maximen selbst nicht gefolgt sei, gegenstandslos—beides ist zu trennen: der große Entwurf einer Geographia generalis (inklusive geographia specialis) und das wenig frühere, doch gleichzeitige Editionsprojekt einer Descriptio Japoniae. Drei „Prinzipien" hatte er für eine der Wahrheit nahekommende Geographie genannt: erstens und zweitens die Kenntnis der zu berechnenden und zu vermessenden Verhältnisse auf der Erde und am Himmel (und durch den Himmel) und drittens: experientia und solche besonders für eine geographia specialis: „sie kann und muss sich auf die Erfahrung allein und Beobachtungen von den Menschen stützen, welche die je einzelnen Länder beschrieben haben" (geographia inprimis specialis sola experientia et observatione hominum nititur, qui regiones singulas descripserunt). Varenius kann fast nur literarischen Nachrichten folgen trotz gelegentlicher Bedenken bei divergierender Überlieferung: fides sit penes auctorem, d.h. die Güte des hier Kolportierten garantiert allein der es erzählt. Die Nachrichten erstmals gesammelt, geordnet und zu großem Teil detailreich, interessant und lebendig erzählt zu haben, ist wohl die immer deutlicher werdende Leistung dieser ersten Japan-Monographie überhaupt.

2. *Descriptio Regni Japoniae (1649). Eine bibliographische Beschreibung*

[Titelkupfer] DESCRIPTIO / REGNI IAPONIAE / Cum quibusdam / affinis materiae, / Ex Variis auctoribus collecta / et in ordinem redacta per / BERNARDVM VARENIVM / Med· D· / Amstelodami, / Apud Ludovicum Elzevirium. / Anno M· DC· XLIX. (ungez. S. 48+) 210 (mit Appendix 287) 24°. [s. Willems; Gruys][3]

Dedicatio consulibus et senatoribus reipublicae Hamburgensis ac fautoribus [Dabam Amstelodami Calendis Iulii, Anno MDCXLIX] (*2r–*4r [3–7])

Praefatio ad lectorem (*4v–*7r [8–13])

Dissertatio de Rebuspublicis in genere [Quoniam Descriptio haec mea Iaponiae numerum rerumpublicarum claudit, non ingratum iuvenibus fore putavi, si quaedam cognitu necessaria de Rebuspublicis in genere et earum numero praemisero]: definitio, divisio, optima forma reipublicae, incommoda regiae potestatis absolutae, incommoda aristocratiae, incommoda democratiae, mutatio rerumpublicarum in aliam formam, de numero rerumpublicarum, Asiae regna [caesar Iaponiae multis vasallis regio titulo gaudens], Africae regna., Americae regna. (*7v–3*6v [14–44])

Autores et scriptores, ex quibus haec desumta sunt (3*7r–v [45–46])

Tituli capitum libri I [I–XXVIII]... Finis (3*8r–v [47–48])

Descriptio regni Iaponiae (1–210)

1. De [nomine] situ Iaponiae, magnitudine, divisione et primo Europaeorum de ea notitia 1
2. De terrae Iaponiae proprietatibus, aquis, montibus, herbis, animalibus 10
3. De origine incolarum Iaponiae 19
4. De supremo Iaponiae imperatore et de dairis 23
5. De caede Cubi, quae anno MDLXV facta est, et eius successore usque ad LXXIII annum 32

---

[3] Alphonse Willems, *Les Elzevier: histoire et annals typographiques*, Bruxelles 1880, 1095; Jan Albert Gruys, „De reeks Republieken van de Elzeviers en Joannes de Laet," *Boekverkopers van Europa: Het 17de–eeuwse Nederlandse uitgevershuis* Elzevier, Hg. Berry P.M. Dongelmans et al., Zuphen 2000, 96.

Appendix (211–278) ad Descriptionem regni Iaponiae continet haec: 1 De dairi Iaponici ad caesarem adventu 213, 2 De regno Siamensi [= Jodocus Schoutenius, Descriptio regni Siam, 1636 *belgico sermone*] 229. Collecta ex aliis scriptoribus 273, 3 (Appendix secunda) Historiam quandam Persicam (279–[287])

3. *descriptio alicuis rei*

de-scriptio (ganz äquivalent zur griechischen ek-phrasis) meint die Beschreibung, Ausmalung „irgendeines Gegenstands, irgendeines Themas" (descriptio alicuis rei), also von Dingen, auch etwa auf ihnen dargestellten Szenen: berühmt und unerreichtes Vorbild die Descriptio des Schilds des Achill (clipeus Achillis) aus der *Ilias*; die Beschreibung des Sattels von Enites Pferd (aus dem *Erec* des Hartmann von Aue), einer Haube im *Meier Helmbrecht* sind weitere beliebige Beispiele. Neben imaginär-literarischen Kunstobjekten tritt die Beschreibung von bezeugten oder noch erhaltenen Kunstwerken wie Wandmalereien,

Tapisserien oder Kirchenbauten. In den letzten Jahren hat man diesem reizvollen Sujet erhöhte Aufmerksamkeit gewidmet; ich nenne nur eine Würzburger Ringvorlesung 1991/1992 über „Ekphrasis. Kunstbeschreibung und Beschreibungskunst von der Antike bis zur Gegenwart", die Monographie von Christine Ratkowitsch, *Descriptio picturae* (1991), nämlich anhand von Texten des Baudri von Bourgueil, Walter von Chatillon, Alanus ab Insulis, Johannes von Hauvilla und Joseph von Exeter; die Arbeit von Haiko Wandhoff, *Ekphrasis* (2003), sowie, ebenfalls 2003, die monumentale Arbeit von Ulrich Schlegelmilch mit dem Titel *Descriptio templi*, also von programmatischen Kirchenbauten und mit weit ausgreifenden Darlegungen und Texten aus unserer Zeit. Der Terminus descriptio gehört der Rhetorik und zwar dem genus demonstrativum der Lob- und Prunkrede an. Ursprünglich Bezeichnung für eine anschauliche Redeweise, wird descriptio nahezu eine Gattungsbezeichnung; sie kann auch den Bericht und die literarische Summe von Nachrichten über ein kennengelerntes Land meinen. Das Geschichtswerk des Herodot ist reich an oft genug sich verselbständigenden Einlagen, Exkursen von Deskriptionen erlebter Länder von Nordafrika bis Persien, die Griechenland und Athen um 450 mit einer Fülle von ethnographischen Nachrichten in Kenntnis setzten. Die Frage, ob der Begriff descriptio autorisiert ist oder nicht, ist nicht entscheidend; gerade die Reise- und Länderbeschreibungen der Spätantike und des Mittelalters können unter anderen Bezeichnungen gehen und allemal Be-schreibungen bieten: das Iter (des Horaz nach Brindisi), die Itinera (Itinerarien), die Navigationes; oft ist das Interesse freilich nicht auf eine Descriptio des Ziellandes gerichtet, sondern auf die einzelnen Reisestationen oder auch auf das Zielland als religiöse Sehnsucht. Die literarische Beschreibung ist dann nicht selten eine bloße Reihung von Orten mit kultisch-liturgischen Memorabilien und natürlich von archäologisch und historisch interessanten Nachrichten; Varenius hätte für diese Fälle wohl von einer Topographie, mit Bemerkungen von topos zu topos, gesprochen. Auch die literarische Isolierung eines Topos, einer Urbs, wie sie Hochmittelalter und Humanismus und Renaissance in den zahlreichen Laudes urbis (urbium) geübt haben, liefert, in Prosa und oft Versen, natürlich Beschreibungen geschichtlicher und literarischer Nachrichten, die von dem jeweiligen Fall der gewählten Stadt bestimmt und sonst literarisch offen sind; nicht anders die sogen. Hodoeporica,[4]

[4] Hermann Wiegand, *Hodoeporica: Studien zur neulateinischen Reisedichtung des deutschen Kulturraums im 16. Jahrhundert*, Baden-Baden 1984.

oft viele hundert Verse zählende Reisedichtungen deutscher und nördlicher junger Akademiker und Diplomaten über romanische Länder; die Neuigkeiten, die sie beschrieben, waren interessant genug, dass sie noch zeitgenössisch als literarisches Genus gesammelt gelesen wurden. Die hochmittelalterliche *Topographia Hibernica* ist eine veritable Descriptio Irlands, der topos hier also eine chora, das irische Inselland, und der Waliser Giraldus Cambrensis beschreibt darin im Jahr 1187 in den drei Büchern (die am englischen Hof Heinrichs II. begeistert aufgenommen wurde und etwa vor dem Klerus in Oxford vorgelesen werden musste) Irlands Beschaffenheit, Geschichte und Kulte mit allerlei Kuriositäten und Fabulösem. Unter den zwölf von Varenius exzerpierten lateinischen, portugiesischen und niederländischen Autores et scriptores des 16. und seines, des 17. Jahrhunderts ist einzig M. Paulus Venetus (Marco Polo) mittelalterlicher Autor (vor 1300). Nichts kann die in den letzten Jahren untersuchte *Deutsche Landesgeschichtsschreibung im Zeichen des Humanismus* (hier in Wolfenbüttel im September 1999 vorgetragen und diskutiert),[5] beitragen für den Fall Varenius, sie behandelt die humanistische Historiographie und Landeschronistik der politisch von den jeweiligen Gegenseiten beanspruchten Territorien wie Elsass, Schweiz und Teilen Bayerns, behandelt auch die Bistumsgeschichtsschreibung und das ehrgeizige, literarische Sorgenkind einer Descriptio Germaniae durch Päpste und Humanisten nach 1450; auch die *Erkundung und Beschreibung der Welt*, die Vorträge eines Symposiums (Gießen 2000) sammelt, bietet interessante Beiträge zur Poetik der Reise- und Länderberichte,[6] ist aber für Varenius allenfalls für die Tradition und Beobachtung von Phänomenen und Strukturen hilfreich.

Die *Descriptio Japoniae* des Varenius hat als Buchwerk keinen Vorgänger und kann sich auch zur literarischen Gestaltung auf kein Modell stützen: es sammelt geschickt verschiedenste Nachrichten, gruppiert einige Kapitel und bricht abrupt ab; allein darin, dass Varenius mit der Namensbezeichnung und geographischen Beschreibung des Landes beginnt, beachtet er eine Tradition der Landesbeschreibungen.

---

[5] Franz Brendle, Dieter Mertens, Anton Schindling u. Walter Ziegler, Hg., *Deutsche Landesgeschichtsschreibung im Zeichen des Humanismus* (Contubernium 56), Stuttgart 2001.

[6] Xenja von Ertzdorff und Gerhard Giesemann, Hg., *Erkundung und Beschreibung der Welt: Zur Poetik der Reise- und Länderberichte*, Amsterdam 2003.

4. *describere, descriptio in der Dedicatio und Praefatio*

Varenius folgt keineswegs den Titeln der Elzevierschen Respublicae; die wenigsten (etwa ein Sechstel) der kleinen Monographien haben descriptio oder umschreibend eine Form von describere im Titel: Afrika, Griechenland, Rhätien-Graubünden, Wallis, Böhmen und China. Varenius sichert das Wort durch häufige Verwendung in der Dedicatio und Praefatio (s.u. Testimonia 4); diese Eindeutigkeit der Begriffsverwendung wird durch die Rückverweisungen in den beiden späteren Bänden noch gestützt.

5. *de quo plura videantur in libro De religione* (Tractatus *1649*)

Die Verweisungen verbinden einzelne Kapitel und erinnern den interessierten Leser an bereits woanders Ausgeführtes oder kommende Partien (s.u. Testimonia 5 und dt. 218 Anm. 109). Verwiesen wird nicht auf die *Geographia Generalis* (1650), wohl aber, Japan-bezogen, auf den: liber alter, l. secundus de Religione, also auf die elf Kapitel des *Tractatus* (1649).

6. *Japonia* illustrata?

Allein zu fragen Japonia *illustrata*?, evoziert ein ganzes, der descriptio-Literatur und -tradition sehr nahekommendes Genus, nämlich den von Flavius Blondus (Biondo) nach 1450 begründeten Typ einer il/lustrierten, also er/leuchteten, aus dem Dunkel der Geschichte durch Zeugnisse er/hellten Land- oder Stadtbeschreibung. Biondo hatte dies durch seine achtteilige *Italia illustrata* (1453 beendet und Rom 1474 gedruckt) demonstriert: eine historische Topographie des antiken und mittelalterlichen, nach Regionen geordneten Italiens (noch ohne Süditalien und Sizilien). Der hochgelehrte päpstliche Notar und Diplomat hat mit seinem Titel das Stichwort gegeben für andere humanistische Antiquare, welche Regionen und Städte mit dem Abdruck literarischer Quellen illustrieren: Inschriften und Urkunden, Briefen und Gedichten. Das charakteristische illustrata (fast immer so weiblich zu den weiblichen Regionen und Städten) hatte sich längst von Biondo gelöst und Karriere gemacht. Zu erinnern ist nur an die berühmte (Urbs) *Norinberga illustrata* des deutschen Humanisten und

Dichterkönigs Helius Eobanus Hessus. Varenius muss Titel und diese Beschreibungsart gekannt haben; in der Praefatio schreibt er: „ich habe gedacht, dass meine schriftstellerische Mühe sich dann lohne, wenn ich das, was ich über das Regnum Japonicum gelesen habe, sammelte, in eine Ordnung bringe und bestimmten Kapitel zuteile und [alles in allem] bis dahin Dunkles, Unbekanntes erhelle und schließlich publiziere" (operae itaque pretium me facturum putavi, si illa, quae de eo legi, colligerem, in ordinem et certa capitula distribuerem, obscura illustrarem tandemque publici iuris facerem).

Descriptio regni Iaponiae—gegen eine Iaponia illustrata mit freilich fast nur zitierten und übersetzten Autoritäten spräche sonst wenig. 1987 zeigte die Herzog-August-Bibliothek eine Ausstellung und präsentierte den Katalog unter dem Titel (des gleichnamigen Werks von Athanasius Kircher, 1667): *China illustrata: Das europäische China-Verständnis im Spiegel des 16. bis 18. Jahrhunderts.*

7. *Varenius deutsch (1974): Zitate und Corrigenda*

Die deutsche Ausgabe von 1974[7] mit Einleitung von Schwind und Kommentar und Anmerkungen von ihm und H. Hammitzsch ist in jeder Hinsicht höchst verdienstvoll; auch die Übersetzung von Volkmann ist vorzüglich. Nur weniges bleibt nachzutragen und zu verbessern.

Zitate:

Das Zitat am Anfang der Dedicatio aus Terenz (dt 3) ist nicht nachgewiesen: „Der Versuch, von anderen zu erfahren, was dir nützlich werden kann, ist schlau" stammt aus der *Andria* IV 5 Vers 811f.; wie jedes Zitat ist es von Varenius argumentativ (wegen: usus) ausgehoben worden. Terenz war Schulautor, vielleicht ist der Vers auch sekundär gefunden worden. Ebenfalls nicht nachgewiesen ist das Zitat (dt 67): „Wenn man das Recht verletzen muss, dann [nur] um der Herrschaft willen", das Varenius Cicero zuschreibt; dort steht es *De officiis* III 21/82. Das Dictum hat eine kleine Geschichte: Sueton zitiert es in seinem Leben des Julius Caesar 30.5 unter Berufung auf Cicero und in dem Wissen,

[7] Bernhardus Varenius (Bernhard Varen), *Descriptio regni Japoniae: Beschreibung des japanischen Reiches*, Hg. Martin Schwind u. Horst Hammitzsch, Übers. Ernst-Christian Volkmann, Darmstadt 1974. Nachdruck München 2000.

dass man ihm, Cicero, diese lateinische Fassung nach den *Phoenissae* (Vers 524) des Euripides verdankt (si violandum est ius, regnandi gratia / regni causa violandum est). Auch die Herkunft von: summum ius summa iniuria (höchstes Recht höchstes Unrecht, dt 67) bleibt unerklärt. Merkwürdigerweise ist es auch im *Heauton Timorumenos* (Vers 796) des Terenz belegt und wird von Cicero wiederum in *De officiis* I 10/33 zitiert, es hat dann natürlich als staatspolitisches Rechtfertigungsmuster eine lange Geschichte.

Im 6. Kapitel (dt 67) heißt es, „aber man hatte [damit] den Bock zum Gärtner gemacht" (hircus olitor erat constitutus). Mir ist nicht ganz klar, ob die Partie (mit dem Proverbium) schon in der Vorlage steht oder Text des Varenius ist; jedenfalls scheint das Proverbium unrömisch zu sein (die Römer sagen in gleichem Sinn etwa, dass man Tauben einem Habicht zugesellt habe o. ä.). Den Bock zum Gärtner machen ist nicht lateinisch-romanisch, sondern seit dem 16. Jh. (Hans Sachs) und zunehmend gegen 1650 belegt in niederdeutschen Quellen.[8] Auch noch zu überprüfen bleibt die sprichwörtliche Formulierung wenig später (dt 68): „In der Erwägung aber, dass ein toter Löwe nicht beiße, [...]".

Zu korrigieren und supplieren ist dies: zweimal ist von inanis maiestas (lat 28, dt 55 und Anfang Kap. 17) die Rede; statt von: inhaltloser und machtloser Würde der Dairi zu sprechen, würde ich nach den weiteren Ausführungen inanis als „eitel" übersetzen. Die (etwa dt 41) aufgezählten navigationes der Chinesen Richtung Japan sind eher mit den eben unternommenen See-Reisen oder—fahrten (statt See-Wegen) zu übersetzen. Unter den bibliographischen „Autores et scriptores" ist ein: Lex(icon) Iaponicum nicht übersetzt (dt 34). Am Ende von Kap. 1 (dt 41) ist ein Satz ohne Anschluss und unvorbereitet: „Diese nämlich pflegten [...] auf Schiffen des Handels wegen dorthin zu kommen.", „diese" bezieht sich auf die Chinesen im vorangehenden Satz des Originals; doch dass Japan „Chinensibus semper notissima" gewesen ist, fehlt in der Übersetzung.

Ein Stück, gerade aus der für die Studien so wichtigen programmatischen Praefatio, ist in der Übersetzung (dt 6) ausgefallen: [viri quidam docti, qui...] et me de latina descriptione cogitare cognoscerent (...und also wussten, dass ich über eine lateinische Beschreibung nachdachte); und die enge Verbindung von Medizin und Physik „hinkt" (dt 7) ohne

[8] Lutz Röhrich, *Lexikon der sprichwörtlichen Redensarten*, 5 Bände, 5. Aufl. Freiburg 2001, hier: 1: 226f.

gegenseitige Hilfe (claudicare) und schließt sich keineswegs aus (als stünde dort eine Form von: claudere).

Ob die Entscheidung, Belga, belgicus des lateinischen Texts als Belgier, belgisch stehenzulassen und nur zu konstatieren (dt 204 Anm. 109), sie meinten wie immer: Holländer, holländisch, sinnvoll und nötig war, ist zumindest fraglich; in der Dissertatio (dt 20) selbst ist schon alles richtig gesehen: in Indien nennt man die „Belgier" (Belgae) Holländer (Hollandi), und nicht nur dort sind die lateinischen Belgae die Niederländer.

8. Descriptio regni Japoniae *als letzte der Elzevierschen Reichs-Beschreibungen*

Die *Descriptio regni* [rei publicae] *Iaponiae* des Varenius schließt die Zahl der Elzevierschen sogen. Republiken ab (s. Dissertatio: Quoniam...). 33 europäische und dann außereuropäische waren von 1625 bis 1641 erschienen bei den Elzeviers in Leiden; wie gesagt, haben die wenigsten Büchlein Descriptio (oder eine Form davon) im Titel sondern Libri oder Commentarius u.ä. Nach mehrjähriger Unterbrechung erscheint 1649 als letzte die Japan-Monographie des Varenius, nunmehr in Amsterdam bei Ludwig [Lodewijk] Elzevier; Freunde hatten wohl vermittelt. Varenius aber hatte gewusst, dass eine Beschreibung Japans in der Serie der Republiken noch aussteht (venit mihi in mentem Iaponici Regni, quod solum ex omnibus Regnis latino sermone nondum explicatum esse deprehenderam)—ein Indienbüchlein war 1631, eines über China als vorletztes 1639 erschienen. Varenius kann seine angelesenen, übersetzten, geordneten Materialien anbieten, ermuntert von gelehrten Freunden. Über der Drucklegung ist er mit dem Drucker ständig in Verbindung. Eine Notiz vor dem zweiten oder dritten Stück der Appendix (vor der Historia Persica) gibt einen schönen Einblick in diesen begleitenden Prozess der Herstellung (s.u. Testimonia 6) zwischen Autor und Drucker-Verleger, Käufer und Leser.

Ein paar Bemerkungen zur Einrichtung der Ausgabe sind vielleicht willkommen. Der Satz der lateinischen Ausgabe ist sehr sorgfältig, wie aus einem Guss, ohne Anmerkungen; umso mehr fallen die wenigen Angaben unter dem Text auf (die zahlreichen vergleichbaren Bemerkungen stehen sonst alle im Text): lat 1 Maffeus zu: ab urbe enim Liampo, 2 Macceu, Macao zu: ab Amaccano, 161 Parvum fasciculum, 178 Epist. Die Formel, wonach für die Richtigkeit des Gesagten die

Quelle bürgt (fides sit penes auctorem, penes quos fides sit), steht einmal im Text, einmal als Anmerkung darunter.[9] Diese Handvoll Notate unterm Text ist offensichtlich nach dem abgelieferten Manuskript des Varenius so zu stehen gekommen; die Ausgabe kann insgesamt als eine Leseausgabe ohne jegliche Anmerkungen bezeichnet werden.

9. *usus und recreatio animi*

Käufer- und leserorientiert musste das Japanbüchlein sein; das war das Interesse des Verlegers wie des Autors. Die Serie der „Republiken" konnte nun auch eine Beschreibung Japans vorlegen, die als erste lateinische Monographie höchst nützlich war (da bisher „von niemandem vollständig beschrieben": animadverti a nemine hactenus illud regnum integre descriptum esse, Dedicatio; ähnlich in der Praefatio), weil Beschreibungen verschiedener Reiche und Republiken einen großen Nutzen für ein zu bildendes und herrschaftsfähiges Gemeinwesen haben (magnus usus tam in formanda quam regenda societate). Der augenscheinlich ausgezeichnete Nutzen empfiehlt jegliche solche historiographische Bemühung (insignis utilitas praecipue studium hoc commendat); dem dient auch das hier in der Dedicatio angebrachte Zitat aus Terenz.

Das Büchlein soll rekreativ wirken, aber auch delektieren. Die Nützlichkeit (prodesse, usus) scheint Varenius höher zu gelten, allein für eine „erfreuliche Lesbarkeit und Erquickung des Geistes" (iucunditas et animi recreatio) ist zu sorgen. Man darf die Hörer oder Leser nicht einschläfern, somnolent machen; darum muss man den dürren, knochigen Zahlenkolonnen und Daten der affectiones caelestes und terrestres Humana von menschlichen Riten und Gewohnheiten, Sitten und Unsitten beimischen. Formal heißt das, dass Varenius die ethnographischen Nachrichten mit kleinen Narrationes und Exempla anreichert. Gelegentliche Anführungen einheimischer Begriffe (gebroken land dt 37, schoendragers dt 85) verlebendigen die Darstellung ebenso wie sparsame Vergleiche. Zu nennen sind etwa japanische Einwohner in einer Tunica, wie sie wohl die Polen tragen (fere ut Poloni lat 164 dt 156), oder japanische Klosterkollegien (coenobia) verglichen mit unseren Akademien (veluti Academiae nostrae) oder zeitgenössische Erinnerungen an den

[9] Vgl. lat 138, dt 135; lat 194, dt 180f.

Papst (pontifex Romanus lat 28, dt 55; lat 63, dt 83), Königin Christine von Schweden (dt 26), der dann der Religions-Tractat vom selben Jahr 1649 gewidmet ist, oder die kürzliche (nuper) Hinrichtung König Karls I. von England am 30. Januar 1649 (dt 22, 25f).

Dass Varenius sich mit dem Japanbuch aus einem Erschöpfungszustand nach „diffizilsten" mathematischen Studien freischreibt, rekreiert und mit der schnellen Publikationsfolge der literarischen Arbeiten gleichsam zu „zivileren" und gefälligeren Arbeiten wechselt, um nach der Veröffentlichung ganz frei zu sein für den Arztberuf und ein auskömmliches Leben, beschreibt die Praefatio in sehr persönlichen, besorgten und hoffnungsvollen Worten.

10. *Bemerkungen zur Literarität*

Die Descriptio ist zum allergrößten Teil ein Übersetzungswerk. Die Nahtstellen zwischen den ausgehobenen Textpartien und verbindendem Text des Varenius sind noch nicht untersucht, auch nicht fallweise; zur Beurteilung aber der Leistung und des Gestaltungswillens des Autors wäre dies eine Voraussetzung.

Varenius weiß, wenn zwei Überlieferungen divergieren, und sucht ein Urteil; gelegentlich bleibt ihm ein Zweifel. Man müsste eben in all diesen Fällen Gewissheit suchen, „Erfahrung einholen und allein den Erzählungen derer, die dort gelebt haben, folgen" (Kap. 2: lat 10, dt 42 tutius itaque est de hoc experientiam consulere et eorum, qui in illis [regionibus] vixerunt, narrationes sequi). Das schriftstellerische Übel der Übertreibungen (Hyperbeln) kennt er. Über jesuitische Nachrichten (aber wohl nicht nur über sie) schreibt er einmal (lat 166, dt 158): aber man muss wissen, dass bald in dieser, bald in jener Schrift über entfernte Länder recht häufig übertrieben wird (sed hyperbolas non raro tum in hisce tum in aliis scriptis de longinquis regionibus adhiberi scito). Wenn der angelesene und exzerpierte Stoff zu einem Thema wenig hergibt, belässt er es dabei (und muss es wohl dabei belassen); aber er thematisiert es auch (Kap. 24 De vestitu, am Schluss), Testimonia 7—er bedenkt also die notwendige Kürze (brevitas) eines Kapitels und die literarisch mögliche Ausführlichkeit (prolixitas) eines anderen.

Ganz als eigene Leistung gehört Varenius die Untersuchung und Darstellung (dissertatio) über Staatswesen im Allgemeinen und die bekannte Staatenwelt, die er der ethnographischen Beschreibung (descriptio) des einzelnen Reichs Japan vorausschickte; die Originalität dieser Dissertatio und ihre Nähe zur zeitgenössischen Diskussion staatstheoretischer und politischer Entwürfe zu untersuchen, gerade in den Niederlanden um 1650, ist noch eine Aufgabe. Für die prinzipiellen Teile leiten Varenius die Axiome aus der *Politik* des Aristoteles und *De officiis* des Cicero.

Zweimal beteiligt sich Varenius an einem philosophisch-ethischen Diskurs und nimmt Position über Tugenden und Laster (einleitend Kap. 20) und, damit zusammenhängend, über das sittliche Problem des Schicklichen, dessen was sich gehört und nicht gehört (Kap. 21), Testimonia 8. Der zweite Fall liegt jenseits rechtlicher und zu entscheidender Prinzipien, ist vielmehr bestimmt durch Einschätzung, Vorlieben und Gewohnheit eines Volks; zahlreiche Beispiele demonstrieren Eigentümlichkeiten eines japanischen Geschmacks. Die einleitenden Bemerkungen dagegen zum Kap. 20 stellen der Eindeutigkeit von Lastern (vitia) tiefsinnige Gedanken zu Virtutes gegenüber, nicht ohne Kenntnis und Aufnahme von Gedanken Platos und Plotins, des Paulus und Augustinus: nur Schein und Schatten der wahren und vollen Idee von Virtus können die Menschen nachjagen und täuschen sich dabei noch meistens und gern. Die lateinische Erzählkunst oder Übersetzungs- und Nacherzählkunst des Varenius ist beträchtlich, sie ist mit den jeweiligen Vorlagen zu überprüfen und zu konturieren. Kapitel 5 und 16 sind nichts als Erzählung.[10] Glänzend, als Ereignisse ungeheuerlich und lapidar wie in Marmor erzählt (nach Caron, 1645) sind die je zwei Exempel für Treue (fidelitas) und Schamgefühl (pudor, verecundia) von Frauen in Kapitel 14 (lat 95, dt 101 duo fidelitatis exempla totidemque verecundiae seu pudoris annotat Caronius... lat 98, dt 103 pudoris exemplum Caronius refert hoc).

11. *Chytraeus,* De Creichgoia *[1558] 1561 und Wendelinus,* Admiranda Nili *(1623)*[11]

Von den Schul- und Lehrjahren des Varenius wüssten wir gern noch mehr als wir wissen, von der Schulzeit in Uelzen und auf den Hohen

[10] Siehe aber auch Kap. 12 dt 96, 13 dt 100f, 18 dt 115f. 117f.

[11] David Chytraeus, *De Creichgoia—Kraichgau*, Facs. von 1561, übers. u. kommentiert v. Reinhard Düchting u. Boris Körkel, Ubstadt-Weiher 1999; Markus Friedrich Wendelinus, *Wunder des Nil: Admiranda Nili 1623*, Faks.-Ausgabe, Hg. Reinhard Düchting u. Boris Körkel, Heidelberg 2000.

Schulen in Helmstedt und Hamburg, an dessen Gymnasium (wie er in der Widmung des Japanbuchs an die Hamburger Ratsherren erinnert) er „die ersten Fundamente in der Philosophie und besonders Mathematik und Physik gelegt hatte", dann für nur noch fünf Jahre in Leiden und Amsterdam. Zum Bildungshorizont gehört auch die weitverzweigte Gelehrtenfamilie Varenius und David Chytraeus, der Urgroßvaters des Japanologen. Der Melanchthonschüler, der ein halbes Jahrhundert in Rostock ausharrte, war auch ein bedeutender Historiker. Unter anderem hat er eine kleine Landesgeschichte, die erste seiner Heimat: des Kraichgaus 1558 in Rostock vorgetragen. Doch ist sie in nichts mit dem Japanbuch zu vergleichen. Thema und erforderliche Darstellung aller Proprietäten der Regionen sind notwendig verschieden; einzig im Anfang wohl der meisten Descriptiones: mit der Größen- und Formatbeschreibung des Landes und dem Etymon des Ländernamens liegen literar-strukturelle Gemeinsamkeiten. Das Kraichgaubüchlein des David Chytraeus löst eine doppelte Dankesschuld ein: an Peter von Menzingen, seinen Patron, der ihn bis nach Rostock stipendiert hat, und an Philipp Melanchthon, der ihn wie einen Sohn (filii loco) aufgenommen, gefordert und gefördert hatte—insgesamt ein schönes und bedeutendes Zeugnis auch für die Kraichgauer Ritterschaft, die Leistung der Schulen und besonders Lateinschulen der Region, für die ersten Jahre der reformatorischen Bewegung, die für sie gewonnenen bedeutenden Prediger und Lehrer, die Heidelberger Universität und vor allem seinen Landsmann Melanchthon. Das übrige Geschichtswerk des Chytraeus sollte gleichwohl auf seinen Urenkel Varenius hin in den Blick kommen.

Der reformierte Schulmann und Theologe Wendelinus, ein Erz-Kurpfälzer (archipalatinus) aus der Nähe Heidelbergs ist von 1612 bis zu seinem Tod 1652 Direktor des berühmt gewordenen Gymnasium illustre in Zerbst (Sachsen-Anhalt), das zeitweise mit der lutherischen Universität in Wittenberg konkurrieren wollte. Als Anfang des 30jährigen Kriegs auch das Zerbster Gymnasium eine schwierige Phase durchmacht, erlaubt ebendiese dem Wendelinus, in den Hundstagen des Sommers 1622, sich literarischen Arbeiten zu widmen. Wie Varenius mit Japan, hatte er sich aus eigentlich nicht näher genannten Gründen (oder doch: ihm ist das Zerbster Wasser der Nuthe und das berühmte Bier der Stadt der „heimische Nil") mit Ägypten und besonders dem Nil beschäftigt. Wie Varenius nicht in Japan, so ist Wendelinus nicht in Ägypten gewesen; beide unternehmen literarische Kopfreisen und legen eine Gesamtdarstellung aus literarischen Quellen und ihrer Lektüre vor.

Wendelinus kann sich dabei auf noch umfangreichere und schon antike Literatur stützen bis zu zeitgenössischen Titeln und Kartenmaterial. Auf dem Titelblatt ist bereits vermerkt, dass der „philologische, geographische, historische und hieroglyphische Kommentar" aus 318 griechischen und lateinischen, alten und jüngeren Autoren gezogen ist und die *Admiranda Nili* dadurch „illustriert" (Admiranda Nili illustrata). Auch seine Nil-Monographie beginnt mit den Namenserklärungen des Menschheitsstroms, auch ihr ist eine gelehrte Dissertatio (auch wenn sie sich als Praefatio dedicatoria gibt) vorausgeschickt. Wie Varenius in der *Descriptio regni Iaponiae* in der Dissertatio auf die Herrschaftsformen von Reichen (regna) zu sprechen kommt, so Wendelinus von all den Wundern des Nil prinzipiell auf das Wasser als den Anfang allen Lebens. Die Lebensspuren des Varenius verlaufen sich anders als die des Wendelinus; doch auch sein Werk, konzipiert vor den ägyptologischen, beim Erscheinen so sehr gefeierten Werken des Athanasius Kircher, bleibt trotz einer zweiten Auflage unbekannt bis auf unsere Tage—und das war wenigstens im Fall Varenius anders.

* * *

Testimonia

1 affectiones *caelestes*: elevatio poli, distantia loci ab aequatore et a polo; obliquitas motus diurni stellarum supra horizontem illius loci; quantitas diei longissimi ac brevissimi; clima et zona; calor, frigus atque anni tempestates, item pluvia, nix, venti et alia meteora; stellarum ortus, apparentia et mora supra horizontem; stellae per verticem loci transeuntes; quantitas vel celeritas motus

2 affectiones *terrestres*: limites et circumscriptio; figura; magnitudo; montes; aquae, nempe fluvii, fontes, sinus maris; silvae et deserta; fertilitas et sterilitas, item fructuum genera; mineralia sive fossilia; animalia; longitudo loci

3 affectiones *humanae*: incolarum statura, conformatio, color, vitae diuturnitas, origo, cibus, potus; quaestus et artes, quibus occupantur incolae; mercatura et merces, quas mittit illa regio in alias; virtutes, vitia, eruditio, ingenium, scholae etc.; consuetudines circa puerperia, nuptias, funera; sermo seu lingua, qua utuntur incolae; regimen politicum; religio et status ecclesiasticae rei; urbes et loca celebriora; historiae memorabiles; viri illustres, artifices et inventa a singularum regionum incolis

4 Descriptio regni Iaponiae (Titelblatt). Aus der Dedicatio: quod diversorum regnorum atque rerumpublicarum descriptiones magnum praebeant tam in formanda quam regenda societate usum / etenim qui terrarum topographias, situm atque connexum ignorant, illi regnorum descriptiones neque apte et iucunde describere neque descriptas feliciter apprehendere possunt / quae cum

colligerem, animadverti a nemine hactenus illud regnum integre descriptum esse / Aus der Praefatio: cum omnes fere Respublicas... excepta sola Iaponia descriptas latino sermone scirem / viri quidam docti, qui... me de latina descriptione cogitare cognoscerent / publici iuris facere decrevi nimirum descriptionem hanc Iaponiae / accipe igitur benevolo animo, amice lector, hanc descriptionem Aus der Dissertatio: Quoniam Descriptio haec mea Iaponiae numerum rerumpublicarum claudit...

5 cap 2: lat 16, dt 46 sed de his plura capite de Cibo et Potu (=23) dicemus / cap 6: lat 40, dt 65 [Rückverweisung auf Kap. 4]; cap 10: lat 65, dt 84 de quibus capite de Religione (des Tractatus) dicetur / cap 11: lat 82, dt 91 de quo plura videantur in libro de Religione / cap 19: lat 123, dt 123 ut capit VIII diximus; lat 124, dt 124 capite XI dictum / cap 20: lat 131, dt 129 ut libro de Religione (= Tractatus), capite de Relig. Christ. exstirp. (= 11) dicemus / cap 25: lat 166, dt 158 dicemus autem de illorum Theologia altero libro; lat 170, dt 161 ut libr. II (= Tractatus) dicetur; lat 171, dt 162 ut capite XV in fine dictum; lat 173, dt 163 quod plenius cap I libr. seq. explicabimus, ubi... / cap 26: lat. 188, dt 176 dicemus libro secundo (= Tractatus); lat 189, dt 176 ut cap. I dictum est

6 (Descriptio 1649:279) Cum Typographus finem libri pervenisset, vidit aliquot paginas vacuas fore ideoque rogavit, ut aliquid lectu dignum in gratiam emtorum adiicerem. Legeram tunc forte in Belgicis navigationibus de caede multorum Magnatum in Persia historiam. Eam itaque lectoribus clausulae loco exhibere volui. Sic autem se habet:...

7 de reliquis ad vestitum pertinentibus nihil reperio annotatum apud scriptores; itaque huius capitis brevitatem compensabit sequentis (= De educatione et artium conditione) prolixitas.

8 cap 20 (lat 129, dt 127) Vitiorum moralium in omnibus terrae regnis largam licet metere segetem: sed Virtutum veram et genuinam Ideam, qualem vel Ethici mentis rationisque argumentationibus utentes vel Theologi revelatae innitentes doctrinae requirunt, frustra quaesieris: speciem tantum et umbram quasdam Virtutum sectantur mortales et admirantur, atque hisce augustum splendidumque Virtutis nomen attribuunt, sibi ipsis vel aliis adulantes. Unde fit, ut in explicandis diversarum gentium moribus scriptores Virtutis nomen non in angusta illa Philosophorum et Theologorum significatione sed de qualicunque laudabili actione Virtutum seminibus aut pedissequis accipiant, quod et nos hoc capite imitamur. cap 21 (lat 140f, dt 137) Multa in vita communi pro indecoris, multa pro decoris habentur, quorum pleraque Ethici docent esse indifferentia, magis opinione gentium quam iuris et aequi principiis inniti ostendunt. Unde non immerito honesta et turpia distinguuntur in duo genera, quorum alterum contineat ea, quae revera talia sunt et virtutum vitiorumque nominibus appellantur, alterum ea, quae ex consuetudine talia censentur et apud diversos populos diversa ratione observantur. Über den ersten Bereich war im vorigen Kapitel (= 20) die Rede...

DIE ANEIGNUNG UND STRUKTURIERUNG VON WISSEN IN DER POLYHISTORIE. EIN FALLBEISPIEL: BERNHARD VARENIUS

Horst Walter Blanke

Abstract

Knowledge and the Transfer of Knowledge in Polyhistory. A Case Study: Bernhard Varenius' Description of the Kingdom of Japan (1649)

Varenius' description of Japan (*Descriptio regni Japoniae*) provides a case study for the transfer of knowledge from overseas to Europe about the specific situation abroad, and it also provides an insight into what kinds of filters were in use in the process of this transmission. The information was mainly derived from texts, from scattered books or essays, and every now and then the state of knowledge was periodically updated and summarised. Varenius' *Descriptio Regni Japoniae* is such a compilation. (For the sources see Reichert's contribution.)

Knowledge can, as it were, be acquired en passant, rather by accident, but it can also be the result of a carefully prepared expedition, as was Cook's, which followed detailed and exact instructions. Varenius' book is based on different materials: on travel reports, which follow the natural chronology, and in particular on more or less systematic descriptions of countries. As far as we can see, Varenius made no use of the rich apodemic literature.

In his *Geographia Generalis*, published in 1650, Varenius developed a systematic model of analysis, which was, however, not yet the standard for his book on Japan, published the year before. Varenius' description of Japan was evidently written with the aim of appealing to a wider reading audience, and his book was meant to be entertaining reading for the educated classes.

As regards the knowledge of Japan in Europe, it is however a pioneer achievement, because until then there had been no description of Japan. Varenius' work is in fact a forerunner; without attempting a systematic order, his book presents a mosaic of Japan at the end of the sixteenth and the beginning of the seventeenth centuries as seen through European eyes. We find this illustrated, for instance, in his description of the Japanese system of government and the various social groups.

The *Descriptio Regni Japoniae* consists almost entirely of a written text in which, with one exception, no maps or pictures have been integrated.

The exception is the engraved title page in two parts (see fig. 1). The upper part (covering about one fourth of the page) shows nine persons kowtowing to the Japanese Shogun; thus the symbolic sign of unconditional submission, which, as described at length in the text, is typical of absolutist government, and is presented here iconographically. The larger part below shows three Japanese women holding up a map of Japan to a young European man with a broad-rimmed hat (next to the engraved title). He is writing in a book; this is probably a portrait of Varenius.

Varenius had never even come near Japan; he had to rely entirely on available literary reports and successfully made the most of them. In fact, he is regarded as a scholar who rejected any unconditional belief in authorities and mythological speculation, directing his interest instead towards empiricism with its preference for unbiased inductive thinking. Although Varenius' report is not even in parts based on autopsy, he relies on reported autopsies and even mentions his sources. We must grant him that for his survey he consulted the most important texts, available in print at the time. In respect to his influence, Varenius' *Geographia Generalis* justifies his fame, but for a long time his *Descriptio Regni Japoniae* was hardly known. Here it is a question of why Varenius was forgotten. Was it because of the Latin language? In my view that is not a convincing argument. Perhaps it was due to his essayistic style, which met the expectations of his contemporary readership, but was ultimately inferior to the work of a man like Kaempfer.

Einleitung: Wissen und seine Herkunft

Im Anschluss an einen früheren Beitrag, in dem der Forschungsstand über Varenius' Japan-Buch zusammengefasst sowie ein Beitrag zu Varenius als Reiseschriftsteller gegeben wurde,[1] geht es im Folgenden um die Fragen von Wissen und Wissenstransfer im Zeitalter der Polyhistorie.[2]

[1] Horst Walter Blanke, „Marco Polo, Bernhard Varenius und Engelbert Kaempfer. Oder: Vom Hörensagen über die gelehrte Recherche zum Autopsiebericht: Drei Stationen der europäischen Japan-Kunde," *Bernhard Varenius (1622–1650): Der Beginn der modernen Geographie*, Hg. Wolfgang Griep u. Susanne Luber, Begleitband zur Ausstellung 2001 (Veröffentlichungen der Eutiner Landesbibliothek 5), Eutin 2001, 29–67.

[2] Zum Begriff Polyhistoria/Polymathie vgl. Wilhelm Schmidt-Biggemann, *Topica universalis: Eine Modellgeschichte humanistischer und barocker Wissenschaft* (Paradeigmata.1),

Varenius' Japan-Bericht und sein Hauptwerk, die *Geographia Generalis*, sollen als Fallbeispiel dienen: Es geht mir vor allem um die Fragen, inwieweit Kenntnisse über die spezifischen Verhältnisse in Übersee nach Europa gelangten, welche ‚Filter' bei diesem Transfer-Prozess eine Rolle spielten.

‚Wissen', Wissensbestände wurden—zumindest im 17. Jahrhundert—vor allem in Form von Büchern und Aufsätzen einer Öffentlichkeit zugänglich gemacht. Daneben gibt es noch die Kommunikationsformen des gelehrten Briefwechsels und die zirkulierender Manuskripte, die aber, soweit überschaubar, für meine heutige Fragestellung keine bedeutende Rolle spielen:[3] Varenius hat, ganz anders als etwa Kaempfer (1651–1716), offenbar keinen ausgedehnten Briefwechsel mit anderen Gelehrten geführt. (Das mag an seinem jugendlichen Alter gelegen haben.) Das Hauptinformationsmittel bildeten Bücher.

Informationen über fremde Welten waren zwar prinzipiell zugänglich, lagen aber, aufs Ganze gesehen, in recht verstreuter Form vor. Infolgedessen wurde in zeitlichen Abständen immer wieder versucht, das zwischenzeitlich verstreut vorliegende Wissen zu sammeln und zu bündeln—eine Aufgabe, die besonders Handbüchern und Enzyklopädien zukam.[4] Aber diese Aufgabe wurde auch, und dies nicht zuletzt, von Reiseberichtkompilationen wie die von Pieter van der Aa (1659–1733), später von der *Allgemeinen Historie der Reisen* vorgenommen.[5]

Hamburg 1983; und seine „Polyhistoria/Polymathie", *Historisches Wörterbuch der Philosophie*, Hg. Joachim Ritter u. Karlfried Gründer, Bd.7, Basel 1989; Arpad Steiner, "A Mirror for Scholars of the Baroque," *Journal of the History of Ideas* 1 (1940): 321–34; Conrad Wiedemann, „Polyhistors Glück und Ende: Von Daniel Georg Morhof zum jungen Lessing", *Festschrift Gottfried Weber*, Hg. Heinz Otto Burger u. Klaus von See (Frankfurter Beiträge zur Germanistik 1), Bad Homburg v.d.H. 1967), 215–35. Vgl. das Hauptwerk von Daniel Georg Morhof, *Polyhistor literarius, philosophicus et practicus*, Hg. Johann Joachim Schwabe, 3 Bde, 4. Aufl. Lübeck 1747 (Reprint: Aalen 1970).

[3] Hans Erich Bödeker, „Lessings Briefwechsel," *Über den Prozeß der Aufklärung in Deutschland im 18. Jahrhundert*, Hg. H.E. Bödeker u. Ulrich Herrmann (Veröffentlichungen des Max-Planck-Instituts für Geschichte 85), Göttingen 1987, 113–38.

[4] Zur Frühgeschichte der Enzyklopädie siehe He(i)nrich Martin Gottfried Köster mit den reichhaltigen kommentierenden Randbemerkungen: *Encyclopädie* (1783), Nachdruck: *Historische Encyclopädie: Gesammelte Artikel über Historik und Didaktik aus der* Deutschen Encyclopädie, Hg. Horst Walter Blanke u. Dirk Fleischer, Bd. 2 (Wissen und Kritik 9.2), Waltrop 2003, 35–103.

[5] Siehe u.a. Pieter van der Aa, Hg., *Naaukeurige versameling der gedenk-waardigste zee en land-reysen na Oost- an West-Indiën*... 28 Bde. [mit 127 jeweils separat paginierten Teilen],

Bevor ich auf Varenius zu sprechen komme, noch eine weitere Vorbemerkung: woher stammt das ‚Wissen', von dem im Folgenden die Rede sein wird? Nun, im wesentlichen von Autoren, die entweder ihre Reiseerfahrungen in Form von meist der naturalen Chronologie folgenden Tagebüchern festgehalten haben oder die systematische Landesbeschreibungen der von ihnen bereisten Länder verfasst haben. Als Beispiele seien hier nur genannt: Georg Forsters (1754–1794) *A Voyage round the World* einerseits und Johann Reinhold Forsters (1729–1798) Buch über eben diese zweite Cooksche Reise andererseits. In den meisten Fällen gibt es Mischformen aus beiden, z.B. Kaempfers Japan-Buch. Nicht selten werden die beiden Varianten der Wissensvermittlung im literarischen Produkt hintereinander gestellt, z.B. bei Hans Staden (1525–1576?) in seiner *Warhaftigen Historia* von 1557.[6] Die unmittelbaren Erfahrungen sind dabei mehr oder weniger bearbeitet, unter Umständen werden sie mit anderen bereits vorliegenden Texten abgeglichen.

‚Wissen' kann quasi nebenbei, d.h. eher zufällig gewonnen werden; es kann aber auch das Ergebnis einer detailliert vorbereiteten und mit genauen Instruktionen versehenen Forschungsreise wie die Cookschen Expeditionen sein.[7]

Das Japan-Buch und seine Gliederung: Zur Herkunft des immanenten Kategorien-Rasters aus der apodemischen Literatur

Im Jahre 1649 erschien im Amsterdamer Verlag von Ludwig Elzevi(e)r folgende Schrift: *Descriptio Regni Iaponiæ. Cum quibusdam affinis materiæ, ex variis auctoribus collecta et in ordinem redacta.* Bei dieser Publikation handelte es sich um ein Sedez-Bändchen von insgesamt 311 Seiten Umfang [24

Leiden 1706–07; *Allgemeine Historie der Reisen [...]*, Übers., Hg. u. Red. Johann Joachim Schwabe, Abraham Gotthelf Kästner et al., 21 Bde., Leipzig 1747–74. Über die *AHR* bereite ich z. Zt. eine Monographie vor. Dort weitere bibliographische Nachweise.

[6] Vgl. Michael Harbsmeier, *Wilde Völkerkunde: Andere Welten in deutschen Reiseberichten der Frühen Neuzeit*, Frankfurt (M.) 1994, 97–121.

[7] Siehe auch Düchting in diesem Band. Vgl Stefan Fisch, „Forschungsreisen im 19. Jahrhundert," *Der Reisebericht: Die Entwicklung einer Gattung in der deutschen Literatur*, Hg. Peter J. Brenner, Frankfurt (M.) 1989, 383–405; ferner auch grundsätzlich die Typologien in Blanke, *Politische Herrschaft und soziale Ungleichheit im Spiegel des Anderen: Untersuchungen zu den deutschsprachigen Reisebeschreibungen vornehmlich im Zeitalter der Aufklärung* (Wissen und Kritik 6), Waltrop 1997, Bd.1: 77ff, 161ff, *passim.*

unpag., 287 gezählte Seiten]. Verfasser war Bernhard Varenius, der bereits kurze Zeit später jung starb.[8]

Diese Schrift steht offensichtlich im Zusammenhang mit einer anderen Publikation, nämlich seiner berühmten *Geographia Generalis*, die kurz darauf, nämlich im Jahre 1650, veröffentlicht wurde.[9] Die Frage ist freilich, wie sich dieser Zusammenhang konkret gestaltet. Denn eine Geographia Generalis, der Entwurf einer allgemeinen Erdkunde, macht nur Sinn bei einer gleichzeitigen Entgegensetzung einer Geographia Specialis, die darin in der Tat auch von Varenius programmatisch entwickelt und in einem Faltblatt verdeutlicht wurde (siehe Appendix, Fig./Abb. 14 u. Übersetzung).[10] Infolgedessen ist Varenius' *Descriptio Regni Japoniae* immer wieder als Musterbeispiel einer solchen Geographia Specialis angesehen worden.

Eine Inhaltsanalyse der *Descriptio* ergibt allerdings, dass der Gliederungsentwurf einer solchen Geographia Specialis bei der Niederschrift der Japan-Beschreibung noch gar nicht konzipiert war.[11]

Sachlogisch ist die Japan-Beschreibung also nicht als das wissenschaftliche Werk konzipiert, wie sie die in der *Geographia Generalis* entwickelte Gliederung für eine Geographia Specialis verlangt, sondern verfolgt ganz andere Gesichtspunkte. Es gilt auch nicht so sehr als wissenschaftliches Werk; es richtete sich nicht an die späthumanistische Gelehrtenwelt, sondern an einen viel größeren Personenkreis: eine interessierte Öffentlichkeit, ein gebildetes Publikum, das als eine

---

[8] Auszüge auch in Peter Kapitza, *Japan in Europa: Texte und Bilddokumente zur europäischen Japankenntnis von Marco Polo bis Wilhelm von Humboldt*, München 1990, Bd. 1: 573a–83a, Kap. 12–14 u. 25. Die deutsche Ausgabe beinhaltet auch einen japanologischen Kommentar: von Schwind u. Hammitzsch: „Anmerkungen zum Text der *Descriptio*", 203–35.—Zur *Descriptio* vgl. auch S. Günther, *Varenius*, Leipzig 1905. Nachdruck Amsterdam 1970: 30–36; Martin Schwind, „Die älteste Japanbeschreibung in europäischer Sprache: *Descriptio Regni Japoniae* von Bernhard Varenius, 1649," *Chirigaku Hyoron* [Geographical Review of Japan] 46 (1973): 81–91.—Allgemein über Varenius vgl. Günther sowie Friedrich Ratzel, „Varenius", *Allgemeine Deutsche Biographie* 39 (1895): 487–90.

[9] Dazu Günther 44ff; Kastrop 1982; Gottfried Lange, „Das Werk des Varenius: Eine kritische Gesamtbibliographie," *Erdkunde* 15 (1961): 1–18; Manfred Büttner, *Die Geographia Generalis vor Varenius: Geographisches Weltbild und Providentialehre* (Erdwissenschaftliche Forschung 7), Wiesbaden 1973 sowie seine „Begründung der wissenschaftlichen Geographie durch Bernhard Varenius (1622–1650)?: Gedanken zu den Defiziten der Varenius-Forschung aus Anlaß des Varenius-Jahres 2000," *Berichte zur Wissenschaftsgeschichte* 24 (2001): 237–254.

[10] Varenius 1650: 4, Faltblatt gegenüber S. 9. Vgl. Günther 59f; Schwind 1973: 84 u. in Varenius 1974: xxiii.

[11] Blanke 2001: 37f.

eigene soziale Gruppe erst noch im Entstehen war. Varenius verfolgte mit seiner *Descriptio Regni Japoniae* eindeutig publizistische Absichten; ihn leiteten finanzielle Erwägungen. Deshalb ist auch wohl kein Zufall, dass die Schilderungen von politischen Vorgängen und von den Sitten, die Europäern merkwürdig genug erscheinen mussten, im Vordergrund stehen.[12]

Varenius' Japan-Beschreibung ist nicht ohne Wert—im Gegenteil. Zwar kommt ihr nicht eine derart überragende wissenschaftshistorische Rolle wie der *Geographia Generalis*, die einen Paradigmenwechsel markiert, zu. Im Hinblick auf die Japan-Kunde ist sie aber eine Pionierleistung, gab es doch damals—von Caron (1600–1673) einmal abgesehen—noch keine Landesbeschreibung Japans:[13] „Unter Verzicht auf Systematik erstellt sie ein mit europäischen Augen gesehenes Mosaikbild für das Japan im ausgehenden 16. und beginnenden 17. Jahrhundert."[14] In seiner Widmungsrede an den Bürgermeister und Senat der Stadt Hamburg weist Varenius selbstbewusst darauf hin, „daß bisher noch niemand dieses Reich im Zusammenhang beschrieben hat."[15]

Varenius schloss hier eine Forschungslücke. In diesem Sinne ist Varenius' Werk als Ergänzung zu den Elzevirischen *Republiken*, einer Sammlung von insgesamt 34 wissenschaftlichen Staatsbeschreibungen, die eine gleiche äußere Ausstattung und einen ähnlichen Inhalt aufwiesen und die in lateinischer Sprache geschrieben waren, anzusehen. Zwar gehört Varenius' Band eigentlich nicht dazu, weil es nicht in Leiden (wie alle anderen Bände), sondern in Amsterdam erschienen ist, sachlich schließt es allerdings die Reihe ab. Seine Japan-Beschreibung ist so ein Band der Reihe neben anderen (nämlich der 35. bzw., in anderer Zählung, der 49. Band), aber er kann sich durchaus sehen lassen: Er ist zwar nicht überragend, aber er erfüllte alle Standards seiner Zeit.[16]

---

[12] Schwind 1973: 86 u. in Var. 1974: xxvi, xxxi.

[13] Zum Quellenmaterial Kapitza 1990 Bd.1: 47a–573a (von Marco Polo bis Varenius); Michael Cooper, *They Came to Japan: An Anthology of European Reports on Japan, 1543–1640*, Berkeley 1965. Vgl. Engelbert Jorißen, *Das Japanbild im* Traktat *(1585) des Luis Frois* (Portugiesische Forschungen der Görresgesellschaft: 2. Reihe, 7), Münster (Westf.) 1988; Oskar Nachod, *Die Beziehungen der Niederländischen Ostindischen Kompagnie zu Japan im siebzehnten Jahrhundert*, Leipzig 1897; Michael J. Huissen, "England Encounters Japan: English Knowledge of Japan in the Seventeenth Century," *Terrae incognitae* 5 (1973): 43–59.—Über Caron s.u.

[14] Schwind in Varenius 1974: xxxix.

[15] Varenius 1649: *3^v (1974: 4). Ähnlich in seiner Vorrede an den Leser: 1649: *4^v (1974: 5).

[16] Dazu Frick, dort 8–12 Verzeichnis der 35 Bde. (vgl. auch 24f), Zusammenfassung

Die Elzevirischen *Republiken* sind bei gleicher Ausstattung durchaus heterogen, auch von sehr unterschiedlicher Länge (von 283 bis zu 583 Seiten). In den wenigsten Fällen sind sie von einem Autor geschrieben (wie von Laet 1630 und Emmius 1632), in der Mehrzahl sind sie Sammelbände mit zum Teil sehr unterschiedlichen Beiträgen.[17] Der Band über Moskau besteht aus zwölf Beiträgen; er handelt über die Geographie, Genealogie und Religion sowie das Staats- und Militärwesen Russlands; er gibt in der Tat einen guten Überblick.[18] Mitunter werden verschiedene Staaten in einem Band beschrieben.[19] In den wenigsten Fällen handelt es sich um Werke aus einem Guss. Die *Respublica Hollandiæ* versammelt beispielsweise Beiträge unterschiedlicher Autoren, u.a. von Hugo Grotius[20]—sie gehört aber, streng genommen, nicht zu den Elzevirischen *Republiken*; sie erfährt gewissermaßen eine Fortsetzung durch die *Belgii confœderati Respublica*, in der die einzelnen holländischen Provinzen nacheinander beschrieben werden.[21]

Im Mittelpunkt seiner Darstellung—darauf wurde bereits hingewiesen—stehen die Lebensumstände der einzelnen sozialen Gruppen des japanischen Volkes. Diese Schwerpunktsetzung ist Programm (*2ᵛ, dt 3), das auch in der Gliederung des Stoffes deutlich wird. An die bereits erwähnte allgemeine Abhandlung über die verschiedenen Staatsformen schließt sich eine Auflistung seiner Quellen an (3*7ʳf, dt 33f). Darauf beginnt die eigentliche Darstellung der Beschreibung Japans in insgesamt 28 Kapiteln (vgl. dazu auch Düchting in diesem Band):

des Inhalts 21f, 28.—Ich habe folgende *Republiken* eingesehen: *Respublica regni Galliæ* 1626, *Poloniæ* 1627, *Scotiæ et Hiberniæ* 1627; [Johan de Laet:] *Belgii confœderati Respublica* 1630, *Respublica Moscoviæ* 1630; Emmius, *Graecorum Respublicae Descripta* 1632; *Respublica Imperii Romano-Germanici* 1634; *Respublica regni Hungariæ* 1634.

[17] Die *Respublica Galliæ* von 1626 versammelt folgende Autoren: Claude Seyssel (5–134), Jean de Tillet (135–381), Vincent de la Loupe (382–504), Julius Bellus (505–565), Giovanni Botero (685–606) und enthält darüber hinaus eine Regentenliste (606–13).

[18] Dort 3–24 Genealogie der russischen Herrscher. Dazu vgl. Mechthild Keller, Hg., *Russen und Russland aus deutscher Sicht 8.–17. Jahrhundert* (Westöstliche Spiegelungen: Reihe A, 1), München 1985.

[19] *Respublica Poloniæ etc.* 1627; *Respublica Scotiæ et Hiberniæ* 1627. Die Ausführungen über Schottland sind William Camdens *Britannia* (Frankfurt 1590 u.ö.) entnommen.

[20] *Respublica Hollandiæ, et Urbes*, Leiden 1630. Der Band erschien zuerst 1610 bei Johannes Maire, nicht bei Elvezir. Dort als Wiederabdruck 3–120: Hugo Grotius, *De antiquitate Reipublicæ Batavicæ liber singularis*.

[21] Laet 1630. In dem mir zugänglichen Exemplar sind die beiden Bände zusammengebunden: LB Eutin 138 IV f 10. Vgl. Wolfgang Griep u. Susanne Luber, Hg., *Reiseliteratur und Geographica in der Eutiner Landesbibliothek* (Kataloge der Eutiner Landesbibliothek), 2 Bde., Heide (Holstein) 1990. Bd. 1: 382f, 2: 580f.

- Kap.1. Die Lage, Größe, Gliederung Japans und die ersten Nachrichten der Europäer darüber;
- Kap.2. Eigenart des japanischen Landes, seine Gewässer, Berge, Pflanzen und Lebewesen;
- Kap.3. Über den Ursprung der Einwohner Japans;
- Kap.4. Der oberste Herrscher Japans und die Dairi;
- Kap.5. Die Ermordung des Kubo im Jahr 1565 und seine Nachfolger bis zum Jahr 1573;
- Kap.6. Herkunft des gegenwärtig regierenden Kaisers;
- Kap.7. Über die Regierungsform;
- Kap.8. Die kaiserliche Burg Edo und andere Burgen;
- Kap.9. Die hoheitliche Würde des Kaisers und die Ehrerbietung des Adels dem Kaiser gegenüber;
- Kap.10. Rangabstufungen;
- Kap.11. Jahreseinkünfte des Adels, Ausgaben des Kaisers und der Erwerb der übrigen Bevölkerung;
- Kap.12. Ehe und Lebensbedingungen des weiblichen Geschlechts;
- Kap.13. Dirnentum und Ehebruch;
- Kap.14. Treue und Schamgefühl der Frauen;
- Kap.15. Kindererziehung und Erbschaft;
- Kap.16. Bericht über das, was sich in der Ehe des Kaisers Chunagon ereignete;
- Kap.17. Die Ehefrauen des Dairi und das bei der Wahl einer Amme gebräuchliche Verfahren;
- Kap.18. Über die Rechtsstrenge und die Strafen für Missetäter;
- Kap.19. Über das Wehrwesen;
- Kap.20. Tugenden und Laster der Japaner;
- Kap.21. Über das, was man nach Volksmeinung für ehrenhaft oder für schändlich hält;
- Kap.22. Über Städte, Gebäude und Hausrat;
- Kap.23. Über das Essen und Trinken bei den Japanern;
- Kap.24. Über die Kleidung;
- Kap.25. Über den Entwicklungsstand der Wissenschaften, der freien und handwerklichen Künste bei den Japanern;
- Kap.26. Über den Handel;
- Kap.27. Über die Münzen;
- Kap.28. Krankheiten und Beerdigungen.

Woher stammt diese Gliederung? Womöglich aus der apodemischen Literatur?

Ein Vergleich mit der apodemischen Literatur, etwa mit dem *Methodus describendi regiones, urbes et arces* (1587) Heinrich Rantzaus (1526–1598) oder dem *Examen Rerumpublicarum potiorum totius orbis* (1660ff) Hermann Conrings (1606–1681), Varenius' Lehrer, zeigt, dass Varenius auf eine strenge Systematik verzichtet hat[22]—wie übrigens die anderen Autoren der Elzevirischen *Republiken* auch.

Die Apodemik—ars apodemica oder auch prudentia peregrinandi, deutsch Reisekunst oder Reiseklugheit genannt—zielte darauf ab, das Reisen einer reflektierten Methodik zu unterwerfen. Sie ist eine eigenständige Textgattung, die, indem sie verschiedene Vorläufer—hier seien nur die ärztlichen ‚Reiseregimina' genannt—verarbeitete,[23] mit Beginn des Späthumanismus (um 1570) aufkam, sich bis zum Ende des 18. Jahrhunderts hielt, offensichtlich zunächst viel gelesen wurde und die so auch für die Praxis der Bildungs- und der Forschungsreisen in Europa bedeutsam wurde. Sie ist neben den mündlichen Berichten von Reisenden und den Reisebeschreibungen, die zum Teil eine vorbildhaft prägende Wirkung besaßen, die dritte Quelle für die Reisevorbereitung. Dabei behielt sie ihre im letzten Viertel des 16. Jahrhunderts gewonnene Gestalt bis Ende des 18. Jahrhunderts bei, verflachte aber im ausgehenden 17. und frühen 18. Jahrhundert: In dem selben Maße, in dem beim Reisen—insbesondere bei der Kavaliersreise—die ursprüngliche Bildungsabsicht durch vordergründige Repräsentationsaspekte überdeckt wurde, verkam die apodemische Literatur, obwohl sie sich scheinbar weiterhin an die Reisenden richtete, zu einer Art Fachliteratur von Hofmeistern für Hofmeister. Seit der Mitte des 18. Jahrhunderts, mit dem Einsetzen bürgerlicher Bildungsreisen im großen Stil, kam es dann nochmals zu einer Blütezeit der Apodemik, was aber hier nicht weiter interessieren soll.

Die Fragestellungen, die in der apodemischen Literatur behandelt wurden, umfassen neun Themenbereiche, nämlich: (a) Definitionen des

---

[22] Dazu Justin Stagl, „Die Apodemik oder ‚Reisekunst' als Methodik der Sozialforschung vom Humanismus bis zur Aufklärung", *Statistik und Staatsbeschreibung in der Neuzeit vornehmlich im 16.–18. Jahrhundert*, Hg. Mohammed Rassem u. Justin Stagl, Paderborn 1980, 131–204; und Blanke 1997: 169 Anm. 233.—Eine Anthologie einschlägiger Texte bieten Mohammed Rassem u. Justin Stagl, Hg., *Geschichte der Staatsbeschreibung: Ausgewählte Quellentexte 1456–1813*, Berlin 1994, die beiden erwähnten Texte 160–81, 227–69.

[23] Vgl. bes. Stagls Abgrenzung der Apodemik von anderen Textgattungen, an die sie anknüpft: Itinerarien, Reisebeschreibungen, mittelalterliches "Städte- und Fürstenlob", Volks- und Landeskunde, Direktorien für Pilger, Reiseinstruktionen, Reiseratschläge, ärztliche Reiseregimina (1980: 137–41).

Reisens, (b) Reflexionen auf die verschiedenen Reisetypen, (c) Argumente und Exempel für und das gegen das Reisen, (d) religiöse und moralische Verhaltensregeln, (e) hygienische und diätische Ratschläge für den Reisenden, (f) lebenspraktische und technische Ratschläge unterschiedlicher Art (die so vielfältig und heterogen sind, dass sie sich nur schwer systematisieren lassen), (g) Anweisungen zur Benutzung der diversen Hilfsmittel des Reisens (z.B. der Routenbücher), (h) Anweisungen zum Anstellen eigener Beobachtungen auf Reisen und zur Niederschrift dieser Beobachtungen, schließlich auch (i) konkrete Informationen über einzelne Reiseländer (etwa zur politischen Verfassung oder zu einzelnen Sehenswürdigkeiten).

Was den ersten Punkt angeht, wird die der Erlangung neuer Erkenntnisse sowie der Vervollkommnung der Persönlichkeit dienende Reise (peregrinari) scharf und entschieden vom sinn- und zwecklosen Herumvagabundieren (vagari) abgegrenzt: Der Reisetyp, für den die apodemische Literatur mit Verve plädierte und den sie methodisch entwickelte, ist die Bildungsreise. Die (im achten Punkt aufgeworfene) Fragestellung, wie man die gewonnenen Informationen festhält, systematisiert und auswertet, resultiert in dem Vorschlag, der Reisende möge zum einen ein Reisetagebuch führen, in das alle Beobachtungen detailliert in chronologischer Form eingetragen würden, zum anderen aber auch ein systematisch angelegtes ‚Hauptbuch', in dem die einzelnen Informationen unter einzelne Rubriken zu komprimieren wären. Das Kategoriensystem, nach dem die einzelnen Rubriken zu erstellen sind, entwickelt—in anderem Zusammenhang—wiederum die Apodemik, wobei nicht selten dieses System in Form eines ausgeklügelten Fragenkatalogs präsentiert wird.

Auch anhand der apodemischen Literatur kann man ablesen, wie sich der Blickwinkel der Reisenden zunehmend von einem Interesse an Kuriositäten über eine eher gelehrte Bestandsaufnahme des rein Faktischen hin zu politischen Fragen, denen nunmehr eine praktische Bedeutung zukommt, verschiebt. Martin Zeiller (1589–1661) bietet in *Fidus Achates, oder getreuer Raeßgefert* 1651 in seiner Liste von Punkten, auf die man auf einer Reise achten sollte (darunter Topographie, Flüsse, Städte, Klima), als einen unter vielen Punkten, „wie er [der betreffende Ort] jederzeit regiert worden".[24] Ähnlich äußerte sich Rantzau 1587 in *Methodus describendi regiones, urbes et arces* (165).

---

[24] Vgl. bes. 3(.)5r.

Es gehört zu den Topoi der apodemischen Literatur, dass sich der Reisende gut vorbereitet haben müsse. Die Apodemik formulierte, katalogisierte und systematisierte ja gerade die aus dieser Forderung resultierenden Regeln.

Die Apodemik erschien nicht nur in Form selbständiger Schriften, sondern auch—in diesem Punkt der Historik vergleichbar—in Form programmatischer Vorreden zu Reisebeschreibungen: So enthalten zahlreiche Reisebeschreibungen programmatische Vorreden oder Einleitungskapitel.[25] In der Vorrede zu Birkens (1626–1681) Darstellung einer klassischen Kavaliersreise wird die Bildungsabsicht dieses Reise-Typs aus der Sicht des absolutistischen Herrschers wie folgt formuliert: „Viel Regirungen beschauen / machet wohl und löblich regiren." [2):(1^{v}].

Die Übergänge zwischen apodemischer Literatur und den Fragekatalogen sind fließend. Blotius' (1534–1608) Katalog ist der wohl erste sozialwissenschaftliche Fragebogen überhaupt (2Y7^{r}–2Z6^{v}):[26] Blotius' *Tabula* beinhaltet 117 Fragen, die in Richtung einer „empirisch-vergleichenden Staats-, Rechts- und Religionswissenschaft" gehen.[27] Der Fragebogen, der 1645 der *Beschreibung des mächtigen Königreichs Japan* von Caron zugrunde liegt, ist dagegen sehr viel weniger systematisch, ja: oberflächlicher; er umfasst 31 Fragen mit den jeweils folgenden Antworten.[28]

[25] Hier sei exemplarisch nur hingewiesen auf Adam Olearius, *Vermehrte Newe Beschreibung der Muscowitischen und Persischen Reyse*, Schleswig 1656. Reprint: Dieter Lohmeier, Hg., Tübingen 1971, 1–5; Johann Sigmund Wurffbain, *Vierzehen Jährige Ost-Indianische Krieg- und Ober-Kauffmanns-Dienste*, Hg. Johann Paul Wurffbain, Sulzbach Nürnberg 1686. Nachdruck als: *Reise nach den Molukken und Vorder-Indien 1632–1646*, Hg. R. Posthumus Meyjes, in: S.P. L'Honoré Naber, Hg. *Reisebeschreibungen von deutschen Beamten und Kriegsleuten im Dienst der Niederländischen West- und Ost-Indischen Kompagnien 1602–1797*, Bde. 8 u. 9, Den Haag 1931, 5–10.

[26] Hugo Blotius, *Tabula Peregrinationis continens capita politica* [ca. 1570], in: Paul Hentzer, *Itinerarium Germaniae, Galliae, Angliae, Italiae*, Nürnberg 1629.

[27] Justin Stagl, „Vom Dialog zum Fragebogen: Miszellen zur Geschichte der Umfrage," *Kölner Zeitschrift für Soziologie und Sozialpsychologie* 31 (1979): 615.

[28] Der Vollständigkeit halber sei der Fragenkatalog hier abgedruckt (Caron 1645 hier und im Folgenden nach der Ausgabe von Haberland 2000: 5f): (1) Wie groß das Land Japan und ob es eine Insel ist; (2) Wie viele Provinzen dieses Land hat; (3) Welche Eigenschaften und welches Ansehen der oberste Herr in Japan hat; (4) Welchen Hofstaat, welche Pracht und welches Gefolge seine Majestät hat; (5) Von der Anzahl seiner Soldaten und ihren Waffen; (6) Von der Autorität und dem Ansehen der Reichsräte und Lehnsleute; (7) Von den Qualitäten der Fürsten und Landesherrn samt ihrer Macht; (8) Welches ihre Einkommen sind und worin sie bestehen; (9) Wie und auf welche Weise Gerechtigkeit geübt wird; (10) Welche Missetaten am schwersten bestraft werden; (11) Was für einen Gottesdienst die Einwohner pflegen; (12) Was für

Vergleicht man die Texte von Caron und Varenius miteinander, so zeigt sich, dass Varenius sich offensichtlich in seiner Stoffeinteilung an Caron orientiert hat, dass er allerdings seiner Vorlage keineswegs sklavisch folgt: Für einige seiner Kapitel benutzt er Carons Fragenkatalog als Einteilungsschema. Es gibt viele Gemeinsamkeiten zwischen beiden Autoren: die Ausführungen über Herrscher und Herrschaft, die sozialen Hierarchien, die Geschlechterbeziehungen, die Kindererziehung, die Rechtsprechung und Wehrverfassung entsprechen einander, ebenso die Kapitel über Münzen, Maße und Gewichte.[29] An anderer Stelle geht Varenius über Caron hinaus: er präsentiert eine andere Stoffaufteilung und gewichtet anders. So fehlen bei Caron ein Kapitel über die landesüblichen Speisen und die Kleidung.[30] Dafür findet sich bei ihm jeweils ein Kapitel über die verschiedenen Provinzen, eine detaillierte Schilderung des prunkvollen Hofstaates und Informationen über die Thermalbäder.[31] Das Thema 'Religion', bei Varenius ausgeklammert, wird von Caron in insgesamt fünf Kapiteln (129–138) behandelt. Der Handel, der bei Varenius vergleichsweise kurz vorkommt,[32] wird von Caron (145–149) in immerhin vier Kapiteln recht ausführlich abgehandelt.

Die Kapitelgliederung des Vareniusschen Japan-Buches ist nicht systematisch, aber sie umfasst ein breites Spektrum, wobei sie zahlreiche Kuriosa, interessante Merkwürdigkeiten, enthält.[33]

Tempel sie haben; (13) Was für Priester sie halten; (14) Was für Sekten sie betreiben; (15) Von der Verfolgung der katholischen Christen; (16) Wie diese Nation in ihren Häusern und Hausrat beschaffen ist; (17) Wie und womit sie einander empfangen; (18) Ob sie den Ehestand halten; (19) Von der Erziehung ihrer Kinder; (20) Wie bei ihnen die Erbfolge ohne Testament beschaffen ist; (21) Ob diese Nation treu oder untreu ist; (22) Welcher Handel in diesem Reich getrieben wird und von welchen Völkern; (23) Worin der inländische Handel besteht und welche Fahrten sie in andere Länder unternommen haben; (24) Vom Profit des Handels; (25) Von ihrem Kontakt mit oder Wissen von anderen Ländern; (26) Von des Landes bequemlichen Nutzbarkeiten; (27) Von Münze, Maß und Gewicht; (28) Was sie für Vieh und Geflügel im Land haben; (29) Welche warmen Bäder sie haben; (30) Wie Könige, Fürsten, Herren und Edelleute Audienz bei ihrer Majestät bekommen und welches Geleit sie haben; (31) Von ihrer Sprache, Schreib- und Rechenkunst, und ob sie für die Nachwelt ihre Geschichtsbücher herausgeben.

[29] Vgl. etwa: Caron 1645 (Haberland 2000): 150; Varenius 1649: 199–202, dt 188f.

[30] Varenius 1649: 156–65, dt 160–66.

[31] Caron 2000: 88f, 90–103 u. 151f.

[32] 1649: 184–199, dt 172–184.

[33] Blanke 2001: 41f, 45.

Fassen wir die Befunde zusammen, so zeigt sich, dass—so weit überschaubar—es relativ wenige direkte Übereinstimmungen zwischen Varenius und anderen Autoren gibt: Das trifft, wie wir gesehen haben, auf Caron, die Elzevirischen *Republiken* und auch auf die apodemische Literatur zu. Varenius hat die ihm vorliegende Literatur in großem Stil rezipiert, aber das Wenigste intensiv genutzt (d.h. ausgeschrieben). Sein Werk macht den Eindruck, dass er viele Informationen aus dem Gedächtnis anführt, er also auf eine länger zurückliegende Lektüre zurückgreift und nur selten einzelne Belege überprüft und nochmals nachschlägt. Eine solide Kenntnis der apodemischen Literatur lässt sich zwar für die *Descriptio* nicht nachweisen, darf aber unterstellt werden: sie gehörte zum Bildungskanon der damaligen Zeit, die zumindest vermittelt, in der Form abgesunkenen Kulturgutes, bekannt gewesen sein dürfte. Ist die Stoffeinteilung der *Descriptio Regni Iaponiae* noch relativ unstrukturiert (was auch daher resultiert, dass sich Varenius nicht explizit mit dem Theorieangebot der Apodemiken auseinandergesetzt hat), so ändert sich das in der *Geographia Generalis*, in der auch apodemische Literatur systematisch herangezogen wurde.

Ein intellektueller Durchbruch: Die Geographia Generalis *und die Systematisierung und Reflexion der Themengliederung*

Oben wurde bereits kurz auf Varenius' Konzeption einer Geographia specialis, die er beiläufig entwickelt hat, hingewiesen. Varenius galt lange Zeit als Begründer der wissenschaftlichen Geographie; zum einen wurde die enge Verbindung von Geographia generalis und Geographia specialis, die sich wechselseitig ergänzten und zum ersten Male eine „Vollgeographie" konzipierten, besonders hervorgehoben;[34] zum anderen wurde die thematische Vollständigkeit und reflexive Strukturierung der *Geographia Generalis* betont.[35] In letzter Zeit wurden dagegen erhebliche kritische Vorbehalte angemeldet.[36]

Diese Kritik ist nicht gänzlich unbegründet. Varenius war niemals eine zentrale Gestalt der Geographiegeschichte; er ist erst mit Beginn

[34] Günther 60f; Kastrop 1982: 89.

[35] Günther 61ff.

[36] So bes. Büttner 2001: 244ff (ein eindrucksvoller Versuch einer Demontage des großen Geographen), 238–42, 251f speziell zu Günther.

des 20. Jahrhunderts als großer Geograph wiederentdeckt worden. Um die These von der Vorwegnahme moderner Standards und vor allem deren systematischer Vollständigkeit aufrechterhalten zu können, musste man den Stoff der einzelnen Kapitel der *Geographia Generalis* neu arrangieren—und auch großzügig darüber hinwegsehen, dass sich die „nautischen Kapitel" mit Gegenständen beschäftigen, die nach modernem Verständnis nicht zur Geographie (sondern in ein Handbuch zur Navigation oder Schiffbautechnik) gehören.[37]

Die *Geographia Generalis* ist in drei Bücher und insgesamt 40 Kapitel unterschiedlicher Länge gegliedert. Die Inhaltsübersicht, die sich aus dem den inhaltlichen Ausführungen vorweggestellten detaillierten 32seitigen Inhaltsverzeichnis, den Falttafeln am Ende des ersten Kapitels und nicht zuletzt den Überschriften im laufenden Band gewinnen lässt,[38] sieht wie folgt aus:

Buch I: Absoluter Teil:

Sektion 1: Vorkenntnisse:
• Kap.1. Einleitung: Begriff, Einteilung, Gegenstand, Methode usw.;
• Kap.2. Geometrische Voraussetzungen;

Sektion 2 erklärt in 5 Kapiteln die allgemeinen Eigenschaften:
• Kap.3. Von der Gestalt der Erde;
• Kap.4. Von ihrer Abmessung und Größe;
• Kap.5. Von ihrer Bewegung;
• Kap.6. Von ihrem Ort im Weltsysteme;
• Kap.7. Von der stofflichen Beschaffenheit;

[37] Günther hat die einzelnen Kapitel der *Geographia Generalis* neu arrangiert und sechs Großabschnitte unterschieden [die erste Literaturangabe bezieht sich auf Varenius, die zweite auf Günther]: (1) die Einleitungskapitel, in denen der Gesamtplan des Werkes entwickelt wird (1–19 [Kap. 1f]; 53–61 u. 159–62), (2) die Kapitel der mathematischen Geographie (20–57, 437–667 [Kap. 3–5, 22–31]; 61–73, 162–71), (3) die kartographischen Kapitel (688–744 [Kap. 32–34]; 74–79, 171–74), (4) die nautischen Kapitel (744–786 [Kap. 35–40]; 79–82, 174–77), (5) die auf Meteorologie und Klimatologie bezüglichen Kapitel (338–436 [Kap. 19–21]; 83–92, 177–84), (6) die ozeanographischen Kapitel (118–216 [Kap. 12–14]; 93–103, 184–90) und schließlich (7) Kapitel zur allgemeinen physikalischen Geographie und der terrestrischen Morphologie (62–118, 216–338 [Kap. 7–11, 15–18]; 103–25, 190–212).—Zur Kritik vgl. Büttner 2001: 240f mit 251f (Anm.12).

[38] Varenius 1650: *7^{r}–2*10^{v}, Faltblatt gegenüber S. 8 (siehe Fig./Abb. 14), *passim*. Es gibt kleinere Abweichungen zwischen den drei Versionen, auch in der Groß- und Kleinschreibung. In der folgenden Anmerkung sind abweichend von den jeweiligen Vorlagen die Textpassagen konsequent in Kleinschreibung gehalten.

Sektion 3, in welcher 4 Kapitel die Konstitution der Erde und ihrer Teile schildern:
- Kap.8. Von der Verteilung des Landes und des Wassers:
- Kap.9. Von den Bergen überhaupt, und ihren Ausmaßen;
- Kap.10. Von der Verschiedenheit der Berge;
- Kap.11. Von Wäldern, Wüsten und Bergwerken;

Sektion 4, Hydrographie, deren Inhalt in 6 Kapiteln die Beschaffenheit und Eigenart der Gewässer begreift:
- Kap.12. Von der Einteilung der Gewässer:
- Kap.13. Von Ozean und Meer;
- Kap.14. Von den Meeresbewegungen und zumal von den Gezeiten;
- Kap.15. Von den Flüssen;
- Kap.16. Von den Seen, Mooren und Sümpfen;
- Kap.17. Von den Mineralwässern;

Sektion 5:
- Kap.18. Von der wechselseitigen Umwandlung trockener in die von Wasser bedeckte Örtlichkeiten;

Sektion 6, die Atmosphäre behandelnd:
- Kap.19. Von Atmosphäre und Luft;
- Kap.20. Von den Winden überhaupt:
- Kap.21. Von der Verschiedenheit der Winde und von besonderen Winden;

Buch II: Respektiver Teil, welcher die Kapitel enthält, in der die Erscheinungen vom Himmel zur Sprache kommen:
- Kap.22. Von den Beziehungen der Erde zum Himmel überhaupt;
- Kap.23. Von der geographischer Breite und Polhöhe eines Ortes;
- Kap.24. Von der Einteilung der Erde in Zonen;
- Kap.25. Von der Tageslänge und der Einteilung der Erdkunde in Klimate;
- Kap.26. Von Licht, Wärme und Jahreszeiten;
- Kap.27. Von den Schattenverhältnissen und der auf sie begründeten Einteilung der Erdbewohner;
- Kap.28. Von der verschiedenartigen Beeinflussung der Erdbewohner durch die Himmelkörper, also auch von Gegenwohnern, Nebenwohnern und Gegenfüßlern;
- Kap.29. Von der Verschiedenheit der Zeitverhältnisse verschiedener Orte;

- Kap.30. Von der Verschiedenheit des Aufganges der Sonne und des Mondes, sowie von einigen anderen, hier in Betracht kommenden himmlischen Erscheinungen;

Buch III: Komparativer Teil, jene Eigenschaften untersuchend, die aus der Vergleichung der Lage zweier Erdorte sich ergeben:

- Kap.31. Von der geographischen Länge der Orte;
- Kap.32. Von der gegenseitigen Lage zweier Orte;
- Kap.33. Von der Entfernung eines Ortes von einem anderen;
- Kap.34. Vom sichtbaren Horizonte;
- Kap.35. Von der Seefahrtskunst im allgemeinen und vom Schiffbau;
- Kap.36. Von der Last, die ein Schiff tragen kann;
- Kap.37. Vom ersten Teile der Schiffslenkung, von der Distanz;
- Kap.38. Vom zweiten Teile der Schiffslenkung, vom Kompassstriche;
- Kap.39. Vom dritten Teile der Schiffslenkung; die Histiodromie, Bestimmung des Schiffsweges;
- Kap.40. Vom vierten Teile der Schiffslenkung; Ermittlung des augenblicklichen Ortes des Schiffes.[39]

---

[39] Vgl. Günther 57f.—Im einzelnen: I. *Pars absoluta* (1–436); sectio 1: *De præcognitis geographiæ*; 1. De geographiæ definitione, divisione, methodo & aliis prolegomenis, 1–9; 2. quædam ex geometria & trigonometria, quæ studiosis geographiæ scire oportet, 9–19; sectio 2 *continens affectiones generales & absolutas totius telluris, quinque capitibus*: 3. De figura telluris, 19–30; 4. de telluris dimensione & magnitudine, 30–47; 5. de motu telluris, 48–57; 6. de situ seu loco telluris respectu planetarum & stellarum, 57–62; 7. de telluris substantia & constitutione, 62–70; sectio 3, *in qua capitibus quatuor explicatur constitutio terræ, sive siccæ partis telluris*: 8. de divisione partium terræ naturali in partes integrantes, facta a mari, 70–83; 9. de montibus in genere, & dimensione altitudinis, 83–96; 10. de montium differentiis, & tractu, & in specie de montibis ardentibus, 96–112; 11. de fodinis, sylvis & desertis, 112–18; sectio 4 *continens hydrographiam sex capitibus explicatam*: 12. de oceani divisione per terras, 118–32; 13. de oceani atque ejus partium proprietatibus quisbusdam, 132–74; 14. de motibus oceani in genere, & in specie, de fluxu & refluxu, 174–216; 15. de lacubus, stagnis & paludibus, 216–30; 16. de fluviis in genere, 230–86; 17. de aquis mineralibus, thermis, acidulis & aliis, quas vulgus admiratur, 287–313; sectio 5: 18. de mutatione & generatione locorum aridorum & aquarum telluris, 313–38; sectio 6 *continens explicationem athmosphæræ & venterum capitibus tribus*: 19. de athmospæra & aëre, 338–87; 20. de motu aeris & ventis in genere; de plagis mundi, 387–404; 21. de ventis in specie, differentiis & de tempestatibus, 403–36; II*: Pars respectiva: De affectionibus locorum telluris cælestibus* (437–616): 22. definitiones & præcognita necessaria, 437–45; 23. de locorum, latitudine & elevatione poli, 445–56; 24. de divisione telluris zonas, & apparentiis cælestibus in diversis zonis, 456–69; 25. de longitudine diversa dierum in diversis locis, & inde ortâ divisione superficiei in climata, 469–91; 26. de luce, calore, frigore, anni tempestatibus in diversis telluris locis & zonis, 491–557; 27. de umbris & denominatione locorum inde orta, 557–75; 28. de comparatione affectionum cælestium in diversis locis & inde orta denominatione, antœcorum, &c., 575–85; 29. de comparatione temporis in diversis telluris locis, 586–96; 30. de diverso ortu atque

Varenius hat am Anfang seiner *Geographia Generalis* einen kurzen wissenschaftsgeschichtlichen Abriss gegeben (7–9). Im Laufe seiner Arbeit setzt er sich meist implizit, an vergleichsweise wenigen Stellen explizit mit seinen Vorgängern auseinander. Er konnte an Werke wie die von Keckermann (1573–1609) oder Goellnitz anknüpfen;[40] er trug mehrfach kritische Vorbehalte gegenüber Aristoteles vor, der häufig zitiert wird (22). Es überrascht nicht, dass apodemische Literatur nicht im großen Stil herangezogen worden ist; anders als im Zusammenhang mit der Geographia specialis gibt es bei den Themen der *Geographia Generalis* nur relativ wenig Berührungspunkte. Ohnehin werden mehr antike als frühneuzeitliche Autoren zitiert—an geographischen Werken besonders das von Goellnitz (2). So handelt Varenius ausführlich über die Nilüberschwemmung, in Auswertung und Auseinandersetzung mit Seneca, der auch auf den Seiten 261–271 des längeren zitiert wird.[41] Aber es gibt durchaus Bezugnahmen auf apodemische Literatur, allerdings eher in der Form indirekter Zitate und Anspielungen denn als wörtliche Übernahme.

Varenius erläutert die Kugelgestalt der Erde, wobei er—dies ist nur eines von mehreren schlagkräftigen Argumenten—auf die vollzogenen nautischen Umkreisungen (etwa durch Magellan) verweist (26). An anderer Stelle räsoniert er, ebenfalls mit Bezug auf Magellan, über das Gewinnen oder Verlieren eines Kalendertages, je nachdem in welcher Richtung man die Erde umkreist (592ff). Seine Ausführungen über die

---

occasu altitudine solis & stellarum in diversis locis, 596–616; III. *Pars comparativa terrestris* (617–786): 31. de longitudine locorum, 617–67; 32. de situ locorum mutuo, de compositione globi terrestris & mapparum, 668–725; 33. de distantia locorum, 725–39; 34. de horizonte sensibili sive visibili, 739–44; 35. de artis nauticæ in genere & in specie de navium structura, 744–47; 36. de onere navibus imponendo, sive secunda artis nauticæ parte, 747–50; 37. de limeneuretica sive arte dirigendæ navis, & primus ejus parte, 751–53; 38. de secunda artis directoriæ parte/De cognitione plagarum in singulis locis, 753–63; 39. de histiodromia sive linea cursus navis, 763–79; 40. de cognitione loci, ad quem navis pervenit ad quodlibet navigationis tempus, & ejus signatione in mappis, 779–86.—Die Problematik der deutschen Übersetzung kann man gut an der Übertragung des Begriffes affectiones festmachen; s. Richter und Staffhorst in diesem Band sowie Büttner 2001: 240, 243, 247f.

[40] Dazu Büttner 1973; Kastrop 1982: 88ff.

[41] Varenius 1650: 261–71, im Anschluss an Senecas *Naturalium quaestionum libri viii* (Hg. Alfred Gercke, Stuttgart 1970) Buch iva, 137ff. Er war relativ nah an der Wirklichkeit. Vgl. Adolf Bauer, „Antike Ansichten über das jährliche Steigen des Nil," *Historische Untersuchungen*, Festschrift für Arnold [Dietrich] Schäfer, Bonn 1882, 70–97; Albert Rehm, „Nilschwemme", *Paulys Real-Encyclopädie der Classischen Altertumswissenschaft. Neue Bearbeitung*, Hg. Wilhelm Kroll, Bd. 17, Stuttgart 1937, 571–90.

Gezeitenlehre zeigen ihn als großen Systematiker (174ff, bes. 186ff). Gegen Fabulieren hat er dagegen eine starke Abneigung (79).

Es überrascht nicht, dass er für seine Beispiele nur einige wenige Male auf Spezifika Kanadas eingegangen ist: nämlich an zwei bzw. drei Stellen auf die Insel Neufundland und den St.-Lorenz-Strom (76, 233, 285). Es überrascht aber, dass er für seine Beispiele nicht auf Japan, über das er doch kurz zuvor ein kleines Bändchen publiziert hatte, eingegangen ist: Japan wird an drei Stellen kurz erwähnt, einmal im Zusammenhang mit dem Insel-Dasein, das andere Mal im Zusammenhang mit heißen Quellen, und beim dritten Mal wird ein namenloser Berg (Vulkan) erwähnt (75, 309, 108). Im Kontext der Beschäftigung mit Breiten- und Längengraden von Städten wird eine Tabelle mit knapp 150 Orten präsentiert (658–665); japanische Städte tauchen indes nicht auf—wahrscheinlich fehlten Varenius verlässliche Daten (und so hat er denn konsequent auf jegliche Bezugnahme verzichtet).

Bei allen Vorbehalten gegenüber der Stringenz von Varenius' Einteilungskriterien im Einzelnen bleibt festzuhalten, dass sein Kategorien-System und die Gliederung durchdacht ist. Der Schritt von der *Descriptio* zur *Geographia Generalis* kommt einem Quantensprung, einem Paradigmenwechsel gleich.[42]

An vielen Stellen tappen wir noch im Dunkeln. Weiter helfen würde eine zweisprachige historisch-kritische Ausgabe der *Geographia Generalis* des Varenius, die ein dringendes Desiderat bleibt.

Das Eigenleben der bildlichen Tradition durch Eingriffe des Verlegers: Die Staatsform 'absolute Monarchie' und der Kotau

Die *Descriptio Regni Japoniae* besteht nahezu vollständig aus geschriebenem Text; sie verzichtet—von einer Ausnahme abgesehen—auf die Integration von Karten und Bildmaterial. Diese Ausnahme ist (wie bei allen Bänden der Elzevirischen *Republiken*) das Titelkupfer (siehe Fig./Abb. 1).[43] Das Titelkupfer ist zweigeteilt. Der obere Bereich,

[42] Vgl. Thomas S. Kuhn, *Die Struktur wissenschaftlicher Revolutionen*, 5. Aufl. Frankfurt (Main) 1981.

[43] Die Titelkupfer nehmen in einer meist nur losen Form Bezug auf die Inhalte der betreffenden Bände. In der *Respublica Moscoviæ* 1630 werden der Fürst auf dem

der etwa ein Viertel der Seite ausmacht, zeigt den Kotau von neun Personen vor dem japanischen Shogun: er gibt also in bildlicher Form jenes symbolische Zeichen bedingungsloser Unterwerfung wieder, das für die Regierungsform der 'absoluten Monarchie', über die im Text so ausführlich gehandelt worden war, charakteristisch ist. Der größere untere Teil zeigt—neben der Titelei—drei Japanerinnen, die eine Landkarte Japans präsentieren, und eine vierte, europäische Gestalt mit breitkrempigen Hut, die an einem Buch schreibt—sehr wahrscheinlich ist dies, wie Schwindt vermutet,[44] ein Porträt des Varenius.

Es werden durch das Titelkupfer Assoziationen hergestellt, die quer zur Varenius' Interpretation stehen: Der Kotau war keine genuin japanische Form der Ehrfurchtserweisung, sondern stammte aus dem Chinesischen. Aus diesem Grunde taucht er in Varenius' Schilderungen nicht auf. Mehr noch wird die Spezifik der japanischen Herrschaftsform—die 'absolute Monarchie'—von ihm recht anschaulich beschrieben, aber keineswegs kritisiert. Die Integration des Bildes evoziert nicht nur Fremdheit, sondern eine Fremdheit, die für europäische Zeitgenossen inakzeptabel war.[45]

Zur Problematik der Wirkungsgeschichte des Japan-Buches

Varenius' Japan-Bericht beruht nicht einmal zu Teilen auf Autopsie; aber er stützt sich auf Autopsie-Berichte. Damit nicht genug. Er gibt selbst seine Gewährsleute an: und zwar in einem Quellenverzeichnis, zum Teil auch im Text, manchmal in Form allgemein gehaltener Hinweise, mitunter auch mit Angabe einzelner Kapitel oder sogar genauer Seitenzahlen.[46] Das Quellenverzeichnis „Autores et scriptores,

---

Thron und ihm zur Seite ein orthodoxer Geistlicher und ein Soldat als Garanten und Repräsentanten der Staatsmacht dargestellt.

[44] Schwind, „Die wissenschaftliche Stellung der *Descriptio Regni Japoniae*," *Descriptio regni Japoniae: Beschreibung des japanischen Reiches*, Hg. Martin Schwind u. Horst Hammitzsch, Übers. Ernst-Christian Volkmann, Darmstadt 1974, xii.

[45] Das Titelkupfer wurde auch für die deutsche Caron-Ausgabe von 1663 übernommen (Caron 2000: 80). Siehe dazu Reichert in diesem Band.—Ein derartiges Vorgehen ist nicht einzigartig, sondern lässt sich an weiteren Beispielen verfolgen.

[46] Siehe z.B. die Hinweise auf Linschoten und Caron bei Varenius 1649: 10, 51 (dt 41, 73) u.ö. Vgl. Francisus Xaverius u. Orazio Torsellini, *Apostolisches Leben und Thaten des heiligen Francisci Xaveri*, München 1674, sowie *Gesammelte Briefe des heiligen Franciscus Xaverius*, Hg. Anton Eglauer, 3 Bde., Augsburg 1794; Giovan Pietro Maffeus, "De

ex quibus haec desumpta sunt“[47] enthält eine Liste von Namen, darunter u.a. den Bericht von François Caron und den Reisebericht von Linschoten; es wird hier in der Übersetzung von Volkmann[48] vollständig zitiert:

1. Maffeus Geschichte der Begebenheiten in Indien.
2. Aus Indien gesandte Briefe in Auswahl, herausgegeben ebenfalls von Maffeus.
3. Briefe des Xaverius, herausgegeben von Tursellinus.
4. Leben des Xaverius, beschrieben ebenfalls von Tursellinus.
5. M. Paulus Venetus.
6. Belgische Beschreibung des Kaiserreichs Japan von dem Holländer François Caron. Er war noch als Knabe nach Japan verschlagen worden. Als er nämlich Schiffsjunge war (Kocksjung) und ohne Schuld vom Kapitän recht häufig geprügelt wurde, versteckte er sich, als das Schiff wieder abfahren sollte. Daraufhin wurde er durch die wohlwollende Förderung der dort weilenden belgischen Prokuratoren im Lesen, Schreiben und Rechnen unterrichtet. Diese Fähigkeit eignete er sich mit gutem Glück an und erlernte obendrein die japanische Sprache so, dass er sich wendig in ihr ausdrücken konnte, was vor ihm noch kein Belgier vermocht hatte. Aus diesem Grunde wurde er allmählich zu immer größeren handelspolitischen Aufgaben hinzugezogen und wurde schließlich im Auftrage der Belgier Leiter für den Handel in Japan und Mitglied des Indischen Senats, der im Auftrag der föderierten Belgier begründet wurde und De Raet van Indien genannt wird.
7. Anmerkungen zu Caronius von dem Belgier Hagenerus, der gleichfalls lange in Japan lebte.

Japonicis rebus epistolarum" (1571), *Opera omnia latine scripta*, Bd.2, Dillingen 1747, 91–316; John Hay, *De rebus Iaponicis, Indicis, et Pervanis epistolæ recentiores*, Antwerpen 1605; Caron geschrieben 1636 (holländ. 1645/46) 1648; Jan Huygen van Linschoten, *Itinerario: Voyage ofte Schipvaert / van Jan Huygen van Linschoten naer Oost ofte Portugaels Indien 1579–1592*, 4 Bde., Amsterdam 1596. Nachdruck Hg. Hendrik Kern et al., 5 Bde., Den Haag 1910–1957. Vgl. auch die Texte und Abbildungen in Kapitza 1990 Bd.1: 47a–573a (dort Auszüge aus Xaver und Caron: 73b–95a, 232a–42b bzw. 533a–63b).

[47] Vgl. Varenius 1649: 3*7rf (dt 33f) und Schwinds Kommentar zu Varenius' Gewährsleuten in 1974: 197–202.

[48] Varenius 1974: 33f.

8. Belgische Beschreibung der Verfolgung, von Rejerus Gisbert, der sich zu jener Zeit dort aufhielt.
9. Belgische Beschreibung der Erdumsegelung des Oliverius.
10. Belgische Beschreibung der Seefahrt unter der Leitung des Führers oder Admirals Matelief.
11. Itinerar des Linschotenius.

Xaverius erwähnt in den Briefen auf S.149 eine Beschreibung Japans und der Sitten des japanischen Volkes, verfasst von Paulus v. Japan, einem Mann von höchster Religiosität und Treue. Diese aber ist, soweit ich in Erfahrung bringen konnte, nicht herausgegeben worden. Gleichfalls berichtet Xaverius auf S. 126, er habe im Auftrag der römischen Gesellschaft (Societas Romana) sorgfältig geschriebene Kommentare zu einer volks- und landeskundlichen Beschreibung eines portugiesischen Kaufmanns geschickt. Aber auch diese Kommentare sind nicht herausgegeben worden. Ich glaube aber, sie werden bis heute in Rom verwahrt und verbergen sich in irgendeiner Bibliothek der Jesuiten. Auch eine Grammatik der japanischen Sprache haben die Jesuiten herausgegeben, die viele Angaben über den Zustand des Kaiserreichs enthält, wie ich von Herrn Golius, einer hochangesehenen Persönlichkeit, erfuhr. Es war aber nicht erlaubt, sie zu sehen. Der gleiche Gewährsmann zeigte mir zwar ein Lexikon mit portugiesischen und lateinischen Ausdrücken, doch waren darin weder historische noch volkskundliche Angaben enthalten.

Ein derartiges Quellen- und Literaturverzeichnis, das heute für jede historische Proseminararbeit obligatorisch ist, war zu Varenius' Zeiten ungewöhnlich. (In der *Geographia Generalis* fehlt ein derartiges Verzeichnis, was aber möglicherweise auf Varenius' frühen Tod zurückzuführen ist.) Dabei ist zu berücksichtigen, dass Varenius nicht nur die von ihm benutzten Schriften aufgeführt hat, sondern auch auf die ihm nicht zugänglichen Arbeiten eingegangen ist, vor allem auf unveröffentlichtes Archivmaterial.

Varenius listet insgesamt elf Titel auf, aus denen er seine *Descriptio* kompiliert hat. Grundsätzlich muss man ihm bescheinigen, dass er die wichtigsten Texte, die zu seiner Zeit gedruckt vorlagen, für sein Überblickswerk herangezogen hat.[49] Hier sind zuerst die maßgeblichen

[49] Kapitza 1990 Bd. 1: 47a–573a.

Jesuiten-Relationen (meist in Briefform) über die Missionsarbeit in Japan (Franz Xaver, 1506–1552; und Giovanni Pietro Maffei, 1533–1603) zu nennen, ferner Orazio Torsellinis (1544–1599) stark hagiographisch zugespitzte Biographie *Apostolisches Leben und Thaten des heiligen Francisci Xaverii*, die sich ihrerseits auf Xavers Schriften stützt, aber—in der deutschen Übersetzung der lateinischen Ausgabe—auch Materialien aus dem Familienarchiv ausgewertet hat. Diese Texte waren unzweifelhaft wichtige Quellen zur Japan-Kenntnis; sie beruhten zumindest teilweise auf Autopsie. Was die Briefe Xavers und seiner Ordensgenossen für das 16. Jahrhundert waren, das ist François Carons *Beschrijvinghe van het machtigh Coninckrijck Japan* (1645) für die Zeit nach der Abschließung des Landes (1637): Seine alles in allem vorzügliche *Beschreibung* beruht zu großen Teilen auf Autopsie (immerhin sprach Caron so gut japanisch, dass er als Dolmetscher eingesetzt werden konnte); er hatte bereits über zwanzig Jahre in Japan verbracht, als er sein Werk druckfertig machte. (Bei den von Varenius erwähnten „Anmerkungen zu Caronius" des „Belgiers Hagenerus" handelt es sich um die von Hendrik Hagenaer besorgte Fassung des Caronschen Berichtes in der Sammlung von Isaac Commelin (1598–1676) *Begin ende Voortgangh van de Vereenighde Nederlantsche Geoctroyeerde Oost-Indische Compagnie* aus dem Jahre 1646, also einem Werk, das erst kurze Zeit, bevor sich Varenius an seine *Descriptio* machte, erschienen war.) Schließlich hat Varenius auch noch verschiedene Reiseberichte herangezogen, die Japan am Rande streifen, so die Berichte von Jan Huyghen van Linschoten (1563–1611) aus dem Jahre 1595 und von Cornelis Matelief (1569–1632) aus dem Jahre 1608. Somit stellt seine *Descriptio* in ihrem Versuch, die verschiedenen Informationen in systematischer Form zu bündeln, eine beachtliche intellektuelle Leistung dar: zwar gewiss nicht die Wissenschaftsgeschichte revolutionierend, aber doch herausragend.

Doch Varenius beschränkte sich nicht darauf, ein allgemeines Quellen- und Literaturverzeichnis seiner Darstellung vorwegzustellen, sondern im Laufe der Darstellung verwies er häufig auf seine Gewährsmänner.

Was nun die Rezeptionsgeschichte betrifft, so ist festzustellen, dass Varenius' *Geographia Generalis* zu Recht berühmt ist, dass dagegen die *Descriptio Regni Japoniæ* für lange Zeit kaum bekannt gewesen ist.[50]

---

[50] Dazu Schwind 1974.

Während die *Geographia Generalis* wiederholt aufgelegt wurde und auch in verschiedene europäische Sprachen übersetzt wurde—nämlich ins Englische (1682 u.ö.), Französische (1755), Holländische (1750), Russische (1718), aber, merkwürdig genug, niemals ins Deutsche,[51] geriet die *Descriptio* schon bald nach Erscheinen in Vergessenheit. Die *Geographia Generalis* ist von so bedeutenden Wissenschaftlern wie Isaac Newton, Alexander von Humboldt und Friedrich Ratzel enthusiastisch gewürdigt worden, doch bereits der nur dreißig Jahre jüngere Kaempfer kannte die *Descriptio* nicht mehr. (Dabei hätte Kaempfer, wenn er denn Varenius gekannt hätte, sich kaum entgehen lassen, auf ihn zu verweisen, d.h. sich kritisch von ihm abzusetzen.[52]) Eine neuere Einschätzung hat zwar dagegen betont, dass Varenius den Kenntnisstand „für mindestens hundert Jahre gültig zusammen"gefasst habe;[53] tatsächlich aber ist diese Zusammenfassung unbeachtet geblieben. Erst Mitte der 1970er Jahre ist die *Descriptio Regni Japoniæ* wiederentdeckt worden.

Es stellt sich die Frage nach den Gründen, warum Varenius in Vergessenheit geraten ist. Lag es etwa an der lateinischen Sprache? Das ist kein zwingendes Argument. Auch Kaempfer hat seine *Aemonitates* 1712 in lateinischer Sprache veröffentlicht, der klassischen Gelehrten-Sprache. Andererseits, um bei diesem Beispiel zu bleiben, ist Kaempfers Japan-Buch nicht in Latein, sondern in den Nationalsprachen moderner europäischer Länder erschienen: in englischer, französischer und deutscher Sprache.[54] Oder, um ein weiteres Beispiel anzuführen:

[51] Nachweis der Ausgaben bei Lange, „Das Werk des Varenius," 1961: 9–15.—Nur das erste Kapitel der *Geographia Generalis* erschien in deutscher Sprache: Lange, „Varenius über die Grundfragen," 1961: 275–78, Neudruck in Hanno Beck, Hg., *Geographie: Europäische Entwicklung in Texten und Erläuterungen* (Orbis Academicus), Freiburg (Br.) 1973, 116–22.

[52] Vgl. bereits Engelbert Kaempfer, *Geschichte und Beschreibung von Japan*, Hg. Christian Wilhelm Dohm, 2 Bde., Lemgo 1777/79. Nachdrucke: Stuttgart 1964; Berlin 1980, 97f. Das Kapitel „Über den Ursprung der Japaner" leitete Kaempfer (um sie später widerlegen zu können) mit der überlieferten Auffassung ein, die bei Varenius im 3. Kapitel der *Descriptio* enthalten ist. Als Quelle nannte Kaempfer die „unsrigen", d.h. die Holländer, die sich diese „Geschichte" erzählen, die zuerst von Linschoten unter die Schriftsteller gebracht worden sei. Hier wäre der Ort für einen Hinweis auf Varenius gewesen, wenn Kaempfer dessen Schriften gekannt hätte (Bd. 1, Buch 1, Kap. 6). Vgl. auch bereits Schwind 1973: 81f; in Varenius 1974: xviii. Vgl. historisch-kritische Ausgabe: Engelbert Kaempfer, *Heutiges Japan*, Hg. Wolfgang Michel u. Barend J. Terwiel, 2 Bde. (*Werke* 1/1 u. 1/2), München 2001.

[53] Josef Kreiner, „Das Bild Japans in der europäischen Geistesgeschichte," *Japanstudien* 1 (1990): 16.

[54] Dazu Kapitza 1980; Blanke 1997 Bd. 2: 28–33.

Nieuhofs China-Darstellung ist nahezu gleichzeitig in holländischer (1665), französischer (1665), deutscher (1666), lateinischer (1668) und englischer (1669) Sprache erschienen.

Ein nicht unbedeutender Strang der Rezeption der Vareniusschen *Geographia Generalis* lief über die in die modernen Nationalsprachen übersetzten Ausgaben (insbesondere im englischsprachigen Raum); für die *Descriptio* fehlen solche Übersetzungen. Und dies hat sicherlich einerseits etwas mit der Qualität des Werkes zu tun (der *Descriptio* fehlt die Genialität der *Geographia Generalis*) und andererseits mit der Tatsache, dass es durch eine Neuerscheinung hoffnungslos überholt wurde: Kaempfers Japan-Werk stellte alles vorher Publizierte in den Schatten.

Ausblick: Fragen, die bleiben

Es bleibt, eine Bilanz zu ziehen. Auf einige der offenen Fragen haben wir Antworten gefunden; oft haben sie den Status von Vermutungen und sind zu vorläufig.

Darüber hinaus bleibt einiges zu tun:

(1) Es ist dringend notwendig, einmal den Charakter der Textgattung 'Kompilation' genauer zu untersuchen. Am Beispiel der *Allgemeinen Historie der Reisen* wird das für die frühe Aufklärung geschehen. Am Beispiel von Morhof könnte dies für die Polyhistorie versucht werden. Wie auch immer, Varenius muss wissenschaftsgeschichtlich präzise eingeordnet werden. Jedenfalls besser, als das bislang geschehen ist.

(2) Dazu gehört auch die Rezeptionsgeschichte, die in diesem Beitrag nur angerissen werden konnte. Es ist dabei auch zu unterscheiden zwischen bildlicher und literarischer Tradition. Man muss in Rechnung stellen, dass der Autor für Dinge herhalten muss, an die er gar nicht gedacht hat, dass sich der Verleger ins Spiel bringt—bzw. der Kupferstecher. Etwaige Abweichungen von den Intentionen des Autors können entweder aus Inkompetenz oder auch aus kommerziellen Gründen erfolgen.

(3) Es ist die innere Entwicklung einer Einzelperson, in diesem Falle von Varenius, genauer nachzuzeichnen. Varenius freilich ist meines Erachtens nicht so sehr personengeschichtlich interessant, sondern weil

sich vor dem Hintergrund seiner Werke die Produktionsbedingungen von Wissenschaftlichkeit nachvollziehen lassen. Genauer: Es geht um die Rekonstruktion der Entstehung von Wissen und von Wissenstransfer.

III

GEOGRAPHIA GENERALIS AND ITS LEGACY

DIE *GEOGRAPHIA GENERALIS* UND IHRE NACHWIRKUNGEN

VARENIUS—EIN GEOGRAPH ZWISCHEN ALLEN STÜHLEN?

Frank Richter †

Abstract

Varenius between All Philosophical Stools?

Can we establish Varenius' position in relation to philosophical models of his time, did he follow Aristotle or Descartes? Could he be appropriated to Newtonianism in the eighteenth century? These questions will be discussed in five sections:

1) Isaac Newton edited Varenius' *Geographia Generalis* twice (1672, 1681); further additions and corrections were added by Jurin and Shaw, and according to Warntz the work acquired more and more Newtonian characteristics while moving away from the predominantly Cartesian approach taken by Varenius. Why Newton chose Varenius' rather than someone else's book is not known. He may not have been familiar with Keckermann, a possible candidate. Perhaps Newton liked Varenius' attempt to find a synthesis between general and special geography and make geography practicable. The combination of mathematics and physics within the study of geography may have been attractive to him. Even though we do not know what philosophical or theological opinions Varenius held (no explicit statements by him have yet been found), it appears to be useful for an analysis of the influence of the *Geographia Generalis* on later generations to locate his position (regardless of chronology) 'between' Descartes and Newton.

2) Warntz considers Varenius' work in natural philosophy Cartesian and says that his *Geographia Generalis* departs from the Aristotelian-scholastic approach in geography. Newton, however, rejected a number of Descartes' (and Cartesianists') fundamental hypotheses and therefore modified Varenius in certain aspects, conformable with the Cambridge Platonists and Henry More. It is not easy to follow Warntz's argument, let alone to find evidence for it, since Newton's hand as editor is only apparent in the geometrical matters added. There are too many possible lines of transmission for any certainty. Such lines might be:

a) Aristotle—16th century scholasticism—Descartes
b) Aristotle—16th century scholasticism—Bacon
c) Platonism—Kepler—Galilei—Descartes (as regards a mathematical model of nature)
d) Democritus/Epicure—Gassendi—Descartes (as regards a mechanist model of nature)

e) Gassendi—Jungius—Varenius
f) Bacon—Jungius—Varenius
g) Cartesianism—Platonism—Cambridge Platonists—Newton
h) Aristotelian science as it still existed

Whether and to what degree Varenius was a Cartesian, is not clear at all, and the fact that he still relied on Aristotle to a considerable degree in his physical geography is normal and to be expected.

3) Cartesianism is a rationalist philosophy of science, which starts from innate ideas, intuitively grasped, and does not deny its Platonic-Aristotelian background; it is opposed to empiricism. The idea of God is rationally deduced. Cartesian natural philosophy combines the idea of mathematical structures and harmonies in nature (Kepler) with Galilei's method of experimenting and leads to a consistently mechanist view of the world. Aristotle's explanation of motion is no longer accepted, neither is his concept of entelechy. In trying to apply this new departure to Varenius, an analysis of his use of the term *affectiones* (see the book's title) might be helpful. It would be anachronistic to see it in the light of causal-mechanistic processes. Many different translations for *affectiones* have been suggested. Varenius occasionally uses *proprietates* as its synonym (and Dugdale translates it as 'properties'). Perhaps 'Eigenschaften' would be best in German. It is strange that Varenius does not use Keckermann's term *accidentia*, which would combine Aristotelian and Cartesian meanings.

Zur Wirkungssphäre des Bernhard Varenius[1]

Das Thema „Varenius" erweist sich auch für den Nicht-Geographie-Historiker, der eher in der zeitgenössischen Wissenschaftsphilosophie zu Hause ist, als ein in vielfältiger Weise interessantes Objekt. Bei den komplexen Zusammenhängen wie der Einordnung eines Wissenschaftlers in den Entwicklungsgang seiner Wissenschaft unter den Bedingungen seiner Epoche gibt es immer mehrere Entscheidungs- oder Vorschlagsmöglichkeiten. Im Falle des Varenius geht es dabei z.B. um solche Fragen wie: War Varenius Cartesianer und wenn ja, wieso hat

[1] Das umfangreiche Manuskript des Autors war vor seinem Tod abgeschlossen, aber für diese Publikation noch nicht überarbeitet worden. Vgl. das vollständige MS. unter: http://www.thur.de/philo/gast/richter/Varenius.pdf. Hier wird es in einer von Margret Schuchard und Ulrich Staffhorst gekürzten Version vorgelegt. Die ursprünglichen Zwischenüberschriften lauteten: 1. Vorbemerkungen: Macht und Ohnmacht der Begriffe, 2. Das Jahr 2000 als Wende in der Varenius-Forschung?, 3. Zur Wissenschaftsphilosophie des 17. u. 18. Jahrhunderts, 4. Empirismus und Rationalismus: Der Begriff affectio bei Varenius und in der Wissenschaftsphilosophie seiner Zeit, 5. Die Bewertung der Leistungen von Varenius.

Newton dann sein Hauptwerk zweimal herausgegeben? Kann Varenius als Begründer der Geographie als Vollgeographie gelten oder wurde und wird seine Bedeutung zum Beispiel bei Alexander von Humboldt oder bei Siegfried Günther weit überschätzt?

Die Arbeit von Manfred Büttner von 2001 macht deutlich,[2] wie man im Verlauf einer Forschungsarbeit zu einem bestimmten Objekt von den anfangs eingenommenen Positionen immer mehr abrücken und am Ende zu einem ziemlich überraschenden Ergebnis gelangen kann: Varenius sei gar kein Cartesianer und sein Beitrag zur Entwicklung der Geographie sei sehr gering.

Büttners Begründung für seine Thesen scheint schlagend zu sein: Die Wenigsten, die Varenius' Leistung so priesen, hätten ihn wirklich gelesen, und wenn doch, dann hätten sie ihn falsch übersetzt oder in falschen Übersetzungen studiert. Niemand habe so richtig wahrgenommen, wie wenig Varenius selber seinen (in Inhaltsverzeichnissen wiedergegebenen) Intentionen in der praktischen Durchführung hat entsprechen können. Letztlich liege die Wurzel im verkehrten Verständnis eines Schlüsselbegriffs, der im Titel von Varenius' Hauptwerk *Geographia Generalis* vorkommt: dem Terminus ‚affectiones'. Die Sachlage kompliziert sich durch das Problem adäquater Übersetzungen von einer Sprache in die andere. Die Wissenschaftssprache des 16. und auch noch vielfach des 17. Jahrhunderts ist im Allgemeinen Latein, wenn es auch schon Ausnahmen gibt. Die Wissenschaftsphilosophie des 17. Jahrhunderts war ja nicht allein durch den Streit um Geo- oder Heliozentrismus (Kopernikus, Bruno, Kepler, Galilei) bestimmt, sondern wenigstens genau so (freilich nicht so spektakulär und öffentlichkeitswirksam) durch begriffliche Entwicklungen, die Problemkreise begleiteten und realisierten wie den Universalienstreit, die Ablösung des Aristotelismus in der Wissenschaftsphilosophie und Naturwissenschaft, die Kritik am Empirismus durch den Descartesschen Rationalismus, die Kritik Newtons an Descartes, die Beziehungen von Wissenschaft, Ethik und Theologie, u.ä.

Wenn man Varenius nicht nur als einen geographischen Dilettanten versteht, der seine geographischen Arbeiten, die ja im Jahre 1650 abrupt abbrachen, allein aus Gründen des Broterwerbs und der Erlangung

[2] Manfred Büttner, „Begründung der wissenschaftlichen Geographie durch Bernhard Varenius (1622–1650)?: Gedanken zu den Defiziten der Varenius-Forschung aus Anlaß des Varenius-Jahres 2000," *Berichte zur Wissenschaftsgeschichte* 24 (2001), 237–254.

beruflicher Reputation verfasst, sondern wenn dieser einen Markstein in der Geschichte geographischen Denkens gesetzt hat, dann kann man von einem solchen Großen der Wissenschaften mit einiger Berechtigung und Wahrscheinlichkeit erwarten, dass er auch zu den wichtigen wissenschaftsphilosophischen Streitfragen seiner Zeit Stellung bezogen hat und zu ihrer Lösung etwas beizutragen suchte. Wir werden sehen, dass es nicht einfach ist, dafür Belege zu finden und ein verbindliches Urteil über Varenius zu fällen. Was ja wohl auch in Ordnung ist.

Man kann die neuere Position Büttners natürlich nicht so einfach widerlegen. Selbst wenn es gelingt zu zeigen, dass Varenius nicht einfach vergessen worden ist, so bleibt der Nachweis seiner Wirkung auf Deutschland bzw. Kontinentaleuropa tatsächlich schwierig, wenn nicht unmöglich. Es gibt aber auch andere Wertungen: Margarita Bowen[3] siedelt Varenius nicht nur einfach in der Epoche von Francis Bacon bis zu Alexander von Humboldt an, sondern spricht ihm dabei eine ganz besondere Rolle zu. Freilich sei dann der Versuch des Varenius, Geographie in hoher Komplexität aufzufassen, zunächst nicht fortgeführt, sondern vor allem unter dem Eindruck des Siegeszuges der Methodik des Isaac Newton (!) abgebrochen worden. Es gibt schon einige bedeutende Gelehrte nach Varenius, die ihn zur Kenntnis genommen und sich auch auf ihn bezogen haben. Dazu sollen im Folgenden einige Beispiele angeführt werden.

Christian Wolff, der in der Nachfolge von Leibniz als entscheidender Mitbegründer der deutschen philosophischen Sprache gilt, wegen seiner natürlichen Theologie in der Gestalt der Physikotheologie schon damals in Halle wie auch später vielen Anfeindungen ausgesetzt war, von dort nach Marburg ging und da interessanterweise Lehrer des Begründers der russischen Wissenschaft, M.W. Lomonossow, wurde,[4] hatte sich in seiner Lehr- wie Publikationstätigkeit auch der Geographie zugewandt. Eine Durchsicht der entsprechenden Werke hat ergeben, dass er Varenius des öfteren nennt und ihn mit J.B. Riccioli vergleicht. Riccioli habe das umfangreichere Werk verfasst, Varenius habe dagegen auch zugleich den natürlichen Zustand der Erde beschrieben. In

---

[3] Margarita Bowen, *Empiricism and Geographical Thought: From Francis Bacon to Alexander von Humboldt*, Cambridge 1981.

[4] Dazu Frank Richter, „Michail Wassiljewitsch Lomonossow: Der Einfluß Christian Wolffs auf die russische Aufklärung," *Forschungen zur Physikotheologie im Aufbruch*, Bd. 1, *Naturwissenschaft, Theologie und Musik in der Aufklärung*, Hg. Frank Richter u. Manfred Büttner, Referate des Symposiums in Halle 1994 (Physikotheologie im historischen Kontext Bd. 1), Münster 1995, 225–252.

seinem *Mathematischen Lexicon*[5] von 1716 bezieht er sich elfmal auf Varenius, u.a. im Zusammenhang mit der Verfertigung von Landkarten und der Verwendung von Erdgloben, mit den Jahreszeiten auf der Erde, mit der Mittagshöhe der Sonne und der Ortslänge. Wolff nennt die Jurinsche Ausgabe von 1712 als die beste Auflage, offensichtlich deshalb, weil sie einige neue, den Erkenntnisstand nach dem Tod von Varenius ausdrückende Anmerkungen enthält. Wolff nennt auch die Cambridger Ausgabe von 1672 sowie die Ausgabe von 1693 in Jena, welche ein Nachdruck der ersteren sei.

Für Wolff ist die Geographie immer noch ein Teil der Mathematik; die mathematische Geographie untersuche alle Dinge (Figuren, Größen), die sich an der Erde ausmessen und ausrechnen lassen. Die daraus ableitbaren Eigenschaften seien dann als Grund für den Bau der Erdkugel, für die Landkarten sowie Witterung, Tages- und Nachtlänge u.a. anzusehen. Die Geographie baue auf der Astronomie auf (so in *Anfangsgründe*). Insofern verwendet Wolff konsequenterweise in einer entsprechenden Arbeit aus dem Jahre 1731 den Terminus *Cosmologia generalis*.[6] Diese befasst sich mit der Welt bzw. dem Universum und existiere als theoretische und experimentelle Wissenschaft. Die *Cosmologia generalis* (S. 2, 7) erkläre die allgemeinen Prinzipien der Modifikation der materiellen Dinge und behandele in drei Sektionen die Bewegung der Welt bzw. des Universums, die Bewegung der Körper sowie die Natur des Universums und ihre Vollkommenheit. Physik und Astronomie sind für Wolff (1723) nicht identisch, auch wenn viele physikalische Entdeckungen von Astronomen gemacht worden seien. Astronomie wird in diesem Zusammenhang als Beobachtung und Berechnung von Sternenbewegungen verstanden. Er bezieht sich auf Descartes sowie auf die Hamburger Logik von Jungius.[7]

Auch Immanuel Kant hat die Arbeiten des Varenius gekannt, wie er selbst in seinem „Entwurf der physischen Geographie" schreibt.[8] Er habe die Werke von Varenius, Buffon und Lulof ausgewertet. In diesem Zusammenhang betont er auf der ersten Seite, dass es sowohl für die

---

[5] Christian Wolff, *Mathematisches Lexicon* (1716), *Gesammelte Werke*, Abt. 1, Bd. 11, 2. Hildesheim 1978.

[6] Wolff, *Cosmologia generalis*, *Gesammelte Werke*, Abt. 2, Bd. 4, Hildesheim 1964.

[7] Wolff, *Vernünftige Gedanken von den Würckungen der Natur* [1723], *Gesammelte Werke*, Abt. 1, Bd. 6. Hildesheim 1981, 150, 12.

[8] Immanuel Kant, „Entwurf und Ankündigung eines Collegii der physischen Geographie (1757)", *Gesammelte Werke*, Hg. Ernst Cassirer, Bd. 2, Hg. Artur Buchenau. Berlin 1912 [Edition 1922, 1–13].

mathematische wie für die politische Geographie genügend Hilfsmittel gebe, für die physische Betrachtung der Erde jedoch noch ein Lehrbuch fehle. Das bedeutet hier, dass Kant die *Geographia Generalis* des Varenius nicht für eine physische Geographie gehalten hat.

Die Tatsache allerdings, dass es zunächst der große Newton war, der in seiner Eigenschaft als Inhaber des Lucasian Chair im Trinity College zu Cambridge zweimal die Vareniussche *Geographia Generalis* mit bestimmten, moderaten Bearbeitungen neu herausgegeben hat, der dann weitere Auflagen durch Bentley bzw. Jurin und im Jahre 1733 eine Übersetzung ins Englische durch Dugdale[9] folgten, hat auch Alexander von Humboldt beeindruckt und zur positiven Wertung der Arbeit von Varenius beigetragen.

Humboldt[10] hat sich offensichtlich stark für die tatsächliche Herkunft des Varenius interessiert. In seinem Nachlass findet sich kein Werk des Varenius, was wahrscheinlich durch den teilweisen Verkauf der Humboldtschen Bibliothek nach seinem Tode durch die Erben bedingt ist. Er hat aber zumindest die Newton-Ausgabe von 1681 besessen, denn er schickte sie an Spiker mit einer Anfrage zur *dedication* und zur Herkunft des Varenius.[11] Ob er sie zurückerhalten hat, ist mir nicht bekannt. In einem Brief an Galle vom 13. März 1844 drückt er seine Verwunderung darüber aus, dass in der doch von Newton herausgegebenen Ausgabe nichts von einer Polabplattung der Erde zu lesen ist, und er vermutet eine Geheimniskrämerei des großen Mathematikers und Physikers im Zusammenhang mit dessen Streit mit Hooke.[12] An anderer Stelle dieses Briefes bedauert Humboldt, dass Newton in diesem Band keine Ergänzungen aus seiner Sicht vorgenommen habe.

Humboldt sagt 1845 zu Varenius,[13] indem er die Beziehung des planetarisch-terrestrischen zum Sonnensystem diskutiert, dass die richtige systematische Unterordnung des uranologischen Teils der Geographie unter den planetarisch-terrestrischen schon bei Varenius zu finden sei. Dieser habe scharfsinnig allgemeine und spezielle Erdbeschreibungen unterschieden, und es sei das bleibende Verdienst von Varenius, dass die Ausführung seines Entwurfes der Erdkunde Newtons Aufmerksamkeit

[9] Siehe dazu den Beitrag von Mayhew in diesem Band.

[10] Dazu Rebok in diesem Band.

[11] Abschrift eines Briefes im Bestand der Humboldt-Forschungsstelle Berlin (189 H. an Spiker, 1830 IX 28).

[12] Literaturarchiv Berlin Tegel, Varia II/7, Ka 107.

[13] Alexander von Humboldt, *Kosmos: Entwurf einer physischen Weltbeschreibung*, Bd. 1, Stuttgart 1845; Seitenangaben fortan nach dieser Ausgabe.

auf sich gezogen habe. Freilich habe bei dem mangelhaften Zustande der Hilfswissenschaften, aus denen Varenius schöpfte, die Bearbeitung nicht der Größe des Unternehmens, des Vorhabens, entsprechen können.[14] In einer sehr langen Anmerkung erörtert Humboldt die mögliche Herkunft und Nationalität des Varenius (er sei ein Deutscher und zwar ein Lüneburger)[15] und sagt im einzelnen:

> Das überaus wichtige Werk des Varenius ist im eigentlichen Sinne des Wortes eine physische Erdbeschreibung. Seit der vortrefflichen Naturbeschreibung des Neuen Continents, die der Jesuit Joseph de Acosta (1590) entwarf, waren die tellurischen Phänomene nie in solcher Allgemeinheit aufgefaßt worden. Acosta ist reicher an eigenen Beobachtungen; Varenius umfaßt einen größeren Ideenkreis, da ihn sein Aufenthalt in Holland, als dem Mittelpunkt eines großen Welthandels, in Berührung mit vielen wohlunterrichtenden Reisenden gesetzt hatte. [...] Die allgemeine Erdbeschreibung des Varenius (Pars absoluta 1–22) ist in ihrem ganzen Umfange eine vergleichende, wenn gleich der Verfasser das Wort Geographia comparativa (cap. 33–40) in einer viel eingeschränkteren Bedeutung gebraucht.[16]

In der Anmerkung geht er speziell auch auf Newton ein:

> Die 1681 von Newton veranstaltete Ausgabe enthält leider keine Zusätze des großen Mannes. Der sphäroidalen Gestalt und Abplattung der Erde geschieht nirgends Erwähnung, obgleich Richer's Pendelversuche um 9 Jahre älter als die Ausgabe von Cambridge sind: aber Newtons Principia mathematica Philosophiae naturalis wurden erst im April 1686 der königlichen Societät zu London im Manuscripte mitgeteilt.[17]

In einer weiteren Arbeit, *Kritische Untersuchungen über die historische Entwicklung der geographischen Erkenntnis von der Neuen Welt*, konstatiert er „merkwürdige" Erkenntnisse bei Varenius, was hier als Lob zu interpretieren ist, obwohl er auch mit Kritik nicht spart.

Auch in der Gegenwart spielt der Name des Varenius eine Rolle, wie aus der Ankündigung des US National Center for Geographic Information and Analysis (NCGIA) von 1997 hervorgeht:

> [...] The objective of NCGIA's new research plan, entitled Project Varenius, is to advance geographic information science through basic research, education, and outreach. The research is motivated by scientific,

[14] *Kosmos*, 60.
[15] *Kosmos*, 75.
[16] *Kosmos*, 74.
[17] *Kosmos*, 75.

> technical, and societal concerns. First, the research serves science and scientists in two ways, focusing on areas in which our knowledge of formalizable geographic concepts is currently incomplete, and contributing to the development and refinement of tools and methods that scientists can use to study geographically distributed phenomena. Second, the research provides basic understanding of geographic concepts, which is required for the production of new technologies. Third, the research examines the impacts that these technologies have on individuals, organizations, and society, and that other digital technologies have in the context provided by geographic space. [...]

Ausdrücklich wird darin auf die *Geographia Generalis* Bezug genommen und erläutert:

> To Varenius, geography was a field of mixed (or applied) mathematics which considered the quantitative states of the earth including its shape, size and motion and the distribution and characteristics of land, water, mountains, woods, deserts and the atmosphere. [...] Geography, geometry and graphics make up an important element of this textbook, published in a number of annotated and revised versions, including some edited by Sir Isaac Newton. This work was very much a part of the debate between the Cartesian and the Newtonian scientific systems and thus provides a philosophical foundation for research to advance the science of geographic information.[18]

Damit wird sichtbar, wie und warum sich ein solch anspruchsvolles, interdisziplinäres und komplexes Forschungsgebiet wie die Entwicklung eines globalen geographischen Informationssystems auf den Namen Varenius bezieht: dieser wird als früher Repräsentant einer qualitativ neuen Art komplexen geographischen Denkens angesehen.

Varenius und die Wissenschaftsphilosophie des 17. und 18. Jahrhunderts

Wir wollen die unmittelbare Wirkungssphäre des Bernardus Varenius bis in die erste Hälfte des 18. Jahrhunderts hinein ausdehnen, da noch in den Jahren 1712 bis 1765 Auflagen der *Geographia Generalis* in England erschienen sind. In der Wissenschaftsphilosophie dieser Zeit dominieren mit ihren verschiedenen Konzepten Kepler, Galilei, Bacon, Descartes, Leibniz und schließlich Isaac Newton selbst; inhaltlich geht es—wie bereits weiter oben—vorrangig um die Ablösung des Aristotelismus, also im wesentlichen um die Zurückdrängung des

[18] Vgl.: http://www.ncgia.ucsb.edu/varenius/jec.html.

teleologischen Moments in den Wissenschaftsgegenständen und des Aristotelischen Formbegriffs, also um die Beziehung von Empirie (Beobachtung und Experiment) und Theorie (physikalische Gesetze in ihren mathematischen Strukturen). Überlagert wurden diese Debatten durch Auseinandersetzungen zum Für und Wider einer mechanistischen Naturanschauung, um den Atomismus und um das Verhältnis von Wissenschaft und Glauben. Letzteres kann wieder unterschiedliche Inhalte haben: Geht es bei Keckermann (*Systema Geographicum*, 1611), einem der Vorläufer von Varenius, noch um die Möglichkeit einer systematischen Herauslösung der Geographie aus der Theologie, so debattieren nachfolgende Generationen von Wissenschaftlern und Theologen über die Konsequenzen, die sich aus solchen Differenzierungsvorgängen ergeben. Hier kommen dann natürliche Theologie und Physikotheologie zu ihrem Recht, aber auch Versuche der Reaktivierung des Gedankengutes des Görlitzer Schuhmachers Jakob Böhme sind in England zu beobachten. Lässt sich Varenius hier irgendwie einordnen? Leider hat sich dieser selber zu einer solchen Parteinahme nicht geäußert, so wie er auch sonst nicht so sehr gerne zitiert hat—auch die vorangegangenen oder zeitgenössischen Geographen nicht.

Wie ich schon andernorts erörtert habe,[19] ist es ein kompliziertes Problem, einem Naturforscher die Attribute des Aristotelismus, Cartesianismus oder Newtonianismus beilegen zu wollen, so wie man streng genommen selbst einem Aristoteles, Descartes oder Newton mit diesen Ismen nur bedingt gerecht wird: Solche entstehen in der Regel erst durch die Schüler der betreffenden Persönlichkeiten im Gefolge der Verabsolutierung eines gewählten Ansatzes, und ihre verkürzende Erläuterung folgt hier mit allem Vorbehalt.

Während der Aristoteliker zu dieser Zeit noch davon überzeugt ist, dass ein betrachteter Körper, in seiner Bewegung durch immanente, zielsetzende Kräfte (Entelechie) bestimmt, schließlich seinem natürlichen Ort zustrebt, wobei es immer bestimmte Qualitäten sind, die sich in den Erscheinungen, Akzidentien, ausdrücken, so hat sich Galilei von solchen Vorstellungen zu trennen begonnen: Gestalt, Größe und Masse als die primären, nun prinzipiell als quantitativ fassbar vorgestellten ‚Qualitäten' der physikalischen Körper lassen sich untereinander durch mathematische Prinzipien zu physikalischen Gesetzen verbinden, wobei es nun ausschließlich die äußeren Beziehungen der

[19] Richter, „Gibt es eine antipositivistische Strömung in der Entwicklung der bürgerlichen Wissenschaftstheorie?" *Deutsche Zeitschrift für Philosophie* 5 (1980): 624–628.

Körper zueinander sind, die solche Gesetze ‚hervorbringen'. Wichtige Vorarbeiten hinsichtlich eines generellen Verständnisses der Wirklichkeit als durch mathematische Formen und Strukturen bestimmte hatte Kepler geleistet, der damit zugleich neuplatonistische Aspekte in die neue Wissenschaftsphilosophie hineintrug. Dehnt man diese Verbindung von Erfahrungswirklichkeit und mathematischer Strukturbildung auf die gesamte Realität aus (einschließlich der des menschlichen Körpers), gelangt man zu einer mechanistischen Naturauffassung, die ohne jegliche außerweltliche oder innerweltliche separate geistige Kraft auskommt. Das kann man als Cartesianismus bezeichnen, der sich als rationalistisch versteht und insofern vom Baconismus unterschieden werden kann, der die hohe Wertschätzung des Mathematischen nicht teilt und dafür das Empirische besonders betont. Ihre letzte Konsequenz findet jene mechanistische Naturauffassung im Atomismus des Gassendi. Freilich erfordert die Existenz von Atomen die Existenz von leeren Räumen zwischen ihnen und damit die Widerlegung des mittelalterlichen Grundsatzes des horror vacui. Boyle zeigte dann bald, dass dieser Gedanke des Atomismus auch und gerade im Verständnis chemischer Prozesse zum Tragen kommt. Naturphilosophisch bleibt das Thema jedoch noch lange Zeit umstritten. Auch Immanuel Kant schlug sich damit noch herum, und schließlich lehnt er einen naturphilosophisch begründbaren Atomismus ab. In der Naturforschung selber könne man natürlich von kleinsten Teilen sprechen und nach ihnen suchen, wie er in *Metaphysische Anfangsgründe*[20] darlegte.

Die Aufgabe der Naturforschung nach Descartes besteht in der Ableitung der Naturerscheinungen aus den durch die erkenntnistheoretische Grundlegung gewonnenen Bestimmungen der körperlichen Substanz. Das Wesen der körperlichen Substanz besteht in ihrer Ausdehnung in Länge, Breite, Tiefe. Daher fällt die Körperwelt geradezu mit dem Raum zusammen; somit gibt es für Descartes keinen leeren Raum, seine ‚letzten' Teile, die Atome im Sinne der physikalischen Atomistik, sind immer noch durch Gott weiter teilbar. Für Descartes gibt es auch keine stofflichen Unterschiede zwischen den verschiedenen Teilen der einen sich unendlich ausdehnenden Substanz. Es gibt nur eine überall identische Materie und nur eine Welt. Die Materie ist teilbar und

---

[20] Kant, „Metaphysische Anfangsgründe der Naturwissenschaft (Riga 1786)". *Gesammelte Werke.* Hg. Ernst Cassirer, Bd. 4, Hg. E. Cassirer u. A. Buchenau. Berlin 1913. [Edition 1922: 444].

in ihren Teilen beweglich. Alle Variation oder alle Verschiedenheiten ihrer Formen hängen von der Bewegung ab. Bewegung und Ruhe sind als zwei verschiedene modi des Körpers zu verstehen, erstere als Ortsveränderung, als Wechsel der Nachbarschaft.

Die Leibnizsche Monadenlehre kann in diesem Zusammenhang als ein Versuch der Versöhnung mit dem Aristotelismus betrachtet werden, und sie erscheint uns heute unter den Bedingungen jener Zeit als eine Sackgasse.[21]

Newton schließlich grenzt sich in bestimmten Fragen von Descartes ab: Er verweigert sich einem unbedingten Natur-Mechanismus, obwohl er derjenige ist, der die Mechanik zu einem allumfassenden Erklärungsmuster der Natur entwickelt, und er will ihn mit dem realen kosmologischen Geschehen durch gelegentliche göttliche Eingriffe in Übereinstimmung bringen und somit relativieren. Ansonsten ist er jedoch naturphilosophischer Spekulation abhold. Descartes und seine Anhänger waren da schnell mit der Mechanik, der Attraktion der Atome, zur Hand. Dabei musste Descartes, um sein Konzept halten zu können, ziemlich weitreichende und phantastische Annahmen zur Gestalt der Atome machen, die er mit Auswüchsen, Zweigen und Ecken versah. Weiterhin hat Descartes die Äthertheorie vertreten, was dann über Huygens zur Wellentheorie des Lichtes führte. Newton favorisierte jedoch bekanntlich die Korpuskulartheorie (wohl ohne sich dabei endgültig festzulegen, da er von der Komplexität des Phänomens wusste), und er nahm an, der Raum sei leer. Also primär gegen den Cartesianismus, weniger gegen Descartes selbst gerichtet war seine berühmt gewordene Forderung *hypotheses non fingo*; die Physik hat es nicht mehr mit den aristotelischen und scholastischen ‚Substanzen' zu tun, und solange wir nicht wissen, was Gravitation in physikalischer Hinsicht ist, müssen wir uns mit ihrer mathematischen Darstellung in Form des Gravitationsgesetzes begnügen. Descartes vertrat (wie auch Leibniz) die Auffassung, das Maß der Bewegung im Kosmos sei konstant, insofern die Bewegung selber unzerstörbar ist. Newton ist dagegen Deist, für den die Bewegung (streng genommen nur die Tangentialbewegung)

[21] Wenn Krafft die moderne Theorie der Selbstorganisation mit dem Entelechie-Prinzip des Aristoteles in eine neuartige Beziehung bringt, so passt erstere wahrscheinlich genau so oder gar noch besser und eher mit der Leibnizschen Monadenlehre zusammen. Vgl. Krafft, „. . . *denn Gott schafft nichts umsonst!*": *Das Bild der Naturwissenschaft vom Kosmos im historischen Kontext des Spannungsfeldes Gott—Mensch—Natur*, Münster 1999, 11.

geschaffen worden ist. Außerdem trennte er Bewegung (Kraft) und Materie vollständig voneinander.

Zu bedenken ist freilich auch, dass Galilei, Descartes, Gassendi und erst recht Newton die quantitativ-mechanistische Naturerklärung nur als methodischen Grundansatz für die naturwissenschaftliche Erklärung der Wirklichkeit ansahen. Eine allgemeine Weltanschauung aus diesem Ansatz machten erst die längst nicht so bedeutenden Nachfolger des praktischen Wissenschaftsbetriebes der nachfolgenden Jahrzehnte und Jahrhunderte; für die Großen blieb immer noch die eigentliche Welterklärung zu leisten, und da waren sie allesamt tief religiös und alles andere als Materialisten oder ‚Mechanisten'.

Will man Varenius hier einordnen, so steht ein solcher Versuch unter dem Anspruch, einige auf Varenius bezogene und noch weitgehend offene Fragen mit zu beantworten. Diese Fragen sind:

1. Warum hat Newton die *Geographia Generalis* des Varenius gewählt, um seinen Aufgaben als Nachfolger von Bentley auf dem Lucasian Chair an der Universität Cambridge nachkommen zu können?
2. Welche Rolle spielt Varenius in der damaligen Auseinandersetzung zwischen Aristotelismus und Cartesianismus in der Wissenschaftsphilosophie, oder besser gefragt: welche Position nimmt er in dieser Auseinandersetzung ein?
3. Spiegeln sich diese Debatten im Gegenstand der *Geographia Generalis* des Varenius (speziell in deren Gliederung und dem Stellenwert sowohl einer physischen als auch einer Humangeographie) wider?

Warntz[22] versuchte, diese Fragen zu beantworten, und brachte dazu Varenius wie Newton in einen Zusammenhang zu dem sogenannten Cambridger Platonismus. Dieser ist eine philosophische Strömung des 17. Jahrhunderts und unter seinen Hauptvertretern sind Cudworth sowie More. Ihre philosophische Herkunft ist der Renaissance-Platonismus, der bekanntlich gegen den in scholastische Erstarrung versetzten Aristoteles gerichtet war, natürlich in neoplatonistischer Sichtweise. Die *Encyclopaedia Britannica* (1999) unter „Cambridge Platonists" charakterisiert sie so:

> Their metaphysics derives from Renaissance Platonism, which interpreted Plato in a Neoplatonic light. They learned much from Descartes's critique

[22] William Warntz, "Newton, the Newtonians, and the *Geographia Generalis Varenii*," *Annals of the Association of American Geographers* 79:2 (1989): 165–191.

> of Empiricism; but, fearing that the new "mechanical" theories might undermine the religious world view, they supported (against Descartes) a teleological interpretation of natural processes.

Die Sachlage ist in der Tat ziemlich verworren und komplex. Mystizismus, Religiosität und Rationalismus bilden mit der Ablehnung von Hobbes' materialistischem Gesellschaftskonzept eine schwierige Gemengelage. Eine einfache Konfrontation Aristotelismus gegen Cartesianismus anzunehmen, würde eine unzulässige Einengung und Idealisierung der Wissenschaftsphilosophie in der damaligen Zeit bedeuten, und da sich Varenius zu diesen Fragen selber nicht bzw. kaum äußert, muss vieles spekulativ bleiben, wenn man hier überhaupt über den eigentlichen Text der *Geographia Generalis* des Varenius hinaus weiterdenken will.

Der (im Jahr 1988 verstorbene) William Warntz vertrat in diesem Zusammenhang folgende Positionen:[23] Varenius widerlegte mit seinem Werk 1650 eindeutig die aristotelisch-scholastische Konzeption der Geographie und Wissenschaft. Gefördert durch die Cambridger Platonisten, wurde das Werk durch Newton zweimal herausgeben und ergänzt (1672 und 1681), dabei in vorsichtiger und konservativer Weise mit Descartes umgehend. Bentley machte aus der cartesianischen *Geographia Generalis* des Varenius eine newtonische, wobei aber durchaus auch die (zunächst noch dominierenden) Cartesianer in Cambridge die Neueditionen nutzten, freilich selektiv.

Die *Geographia Generalis* erwies sich als attraktiv für die Cambridger Platonisten hinsichtlich ihrer Suche nach einer geeigneten Form einer Anpassung von Religion und Naturphilosophie. Gleichzeitig stellen diese den Zusammenhang zum Universalienstreit der damaligen Zeit her. Die Kritik des Descartes am Empirismus ließ sich nicht zuletzt gegen Hobbes richten, dessen eher materialistische Auffassung vom Menschen ethisch problematisch schien—wie freilich auch die ganze Cartesische Konzeption von einer Welt, die in ihrer Substantialität völlig ohne Gott auskommen sollte, eine theologische Entgegnung nötig zu haben schien.

Es stellt sich die Frage, ob hier möglicherweise von Warntz unabsichtlich eine falsche Spur gelegt worden ist. Es erscheint aus meiner Sicht mindestens zweifelhaft, ob ein ‚waschechter' und konsequenter Cartesianer, wenn Varenius denn einer gewesen wäre, sich für eine solche Mittlerrolle zwischen rationalistischer Wissenschaftsphilosophie und

[23] Vgl. Warntz 1989.

(natürlicher) Theologie tatsächlich geeignet hätte. Auch bleibt es bisher unbestimmt, worin denn ein Cartesianismus bei einem Geographen zu suchen wäre, der im wesentlichen noch ohne Physik und Geologie auskommen musste—Disziplinen, die sich erst später als ‚harte Kerne' für die Geographie darstellen sollten. Insofern scheint ein Konzept wie die mechanistische Weltbetrachtung des Descartes auf die Geographie überhaupt nicht zu passen—damals nicht, und auch heute wohl immer noch nicht so richtig. Ob man die Zweiteilung der Geographie durch Varenius in eine Geographia generalis und eine Geographia specialis als Reaktion auf den Cartesianismus betrachten darf, ist auch sehr zweifelhaft: diese Einteilung ist nicht originäres Verdienst von Varenius, sondern ist auch bei Keckermann und Göllnitz zu finden.

Da sich Varenius in der Regel nicht auf Quellen bezieht, ist die Beantwortung der oben gestellten drei Fragen ziemlich schwierig. Es wäre freilich denkbar, über die Person von Joachim Jungius, bei dem Varenius studiert hat und der in der Wissenschafts- und Philosophiegeschichte den Namen eines aufgeklärten Mannes besitzt, auch nähere Aufschlüsse über eben einen seiner Studenten zu erlangen. Aber auch das gelingt kaum: Jungius, der die sich stürmisch entwikkelnden Naturwissenschaften frei von theologischen Konzeptionen und Auseinandersetzungen halten wollte, vertrat einen konsequenten Atomismus im Rahmen seiner sogenannten synodialkritischen Philosophie, was nun wiederum auf Descartes nicht zutrifft. Ob nun atomistisch angelegt oder nicht—die mechanistische Naturauffassung der damaligen Zeit eignete sich eher für solche Disziplinen, die ihren Erkenntnisfortschritt von einem Fortschreiten in das Innere und Mikrostrukturelle der jeweiligen Gegenstände erwarten durften, nicht aber für Disziplinen, die wie die Geographie eher in globalen und regionalen Zusammenhängen zu denken gezwungen waren. Bei Günther[24] findet man einige Aussagen zu Jungius und Varenius und auch einige Passagen aus Briefen, die Varenius von Königsberg und Leiden aus an seinen Lehrer gerichtet hat und in denen er auch die ‚philosophische' Situation speziell in Leiden schildert. Leider findet sich im erhaltenen Briefwechsel mit Jungius nichts Näheres, zumal Jungius auf wichtige Briefe des Varenius nicht geantwortet hat. Gleiches gilt für Tassius, an

[24] S. Günther, *Varenius*, Leipzig 1905. Nachdruck Amsterdam 1970.

den Varenius Briefe und Bücher geschickt hatte.[25] Da geht es wohl um den Streit zwischen Aristotelikern und Cartesianern, um die Atomistik des Gassendi, aber eindeutige Positionierungen findet man nicht. Auch scheint Jungius in der Geographie auf dem ‚klassischen' Standpunkt einer mathematischen Geographie stehengeblieben zu sein.

Bei Margarita Bowen (1981), die die Entwicklung der Geographie bis hin zu Alexander von Humboldt analysiert, findet sich eine relativ ausführliche Darstellung der Leistungen des Varenius—eingeordnet sowohl in die Geschichte des geographischen Denkens wie der Wissenschaftsphilosophie der damaligen Zeit.

Im Werk des Varenius gebe es eine aktive Reaktion auf die Ideen von Descartes, Bacon und Galilei, obwohl er sich nicht explizit auf sie beziehe. Aber es sei klar, dass er die Geographie als Wissenschaft in deren Terminologie entwickeln wollte. Mit Descartes unterstreiche Varenius die Bedeutung der Mathematik bei der Suche nach allgemeinen Prinzipien, die eine Wissenschaft überhaupt erst ausmachen, Bacon folgend charakterisierte er die Geographie als einen Zweig der angewandten (*mixed*) Mathematik und betonte die Abhängigkeit von der Erfahrung bei der Suche nach Erklärungen.[26] Keckermann in Bezug auf eine allgemeine Geographie folgend, habe sich Varenius um die wissenschaftlichen Grundlagen einer modernen Geographie bemüht. Dabei habe er es mit zwei Herausforderungen zu tun gehabt: die Einbeziehung der speziellen Geographie, speziell also der Humangeographie, gegen den Rationalismus in der damaligen Wissenschaft zu rechtfertigen und andererseits den Unwillen der Geographen zu überwinden, die theoretische Basis für ihre eigene Wissenschaft zu entwickeln.[27] Dabei sei die spezielle Geographie für Varenius ein integrierter Bestandteil der Geographie, und das Nützlichkeitsargument wäre für ihn hinreichend gewesen, um eine wissenschaftliche Theorie zu modifizieren, die sich von praktischen Erfordernissen fernhalten wollte.[28] Damit sei er freilich in Konflikt mit dem strengen Cartesianischen Wissenschaftsverständnis gekommen. Varenius löse das Problem für sich so: die allgemeinen

---

[25] Vgl. Robert C.B. Avé-Lallemant, *Des Dr. Joachim Jungius aus Lübeck Briefwechsel mit seinen Schülern und Freunden [. . .], aus den Briefen der Hamburger Stadtbibliothek zusammengestellt*, Lübeck 1863.

[26] Bowen 80.

[27] Bowen 81.

[28] Bowen 82.

geographischen Kategorien sind in der speziellen Geographie anzuwenden.

Da es schwierig war, in der Geographie mit mathematischen Ableitungen zu arbeiten, kommt nun der Gedanke Bacons zum Tragen: Also seien neben den Sätzen der Geometrie und Astronomie drittens auch die gesammelten Beobachtungen in Bezug auf die einzelnen Regionen der Erde als wissenschaftlich begründet anzuerkennen. Im Abschnitt „Methodus" verknüpfe Varenius insofern das neuplatonische Konzept einer mathematisch begründeten Wissenschaft mit dem Baconschen Sensualismus-Empirismus-Schema.[29]

Er musste also den Wissenschaftsbegriff ausweiten. Er unterscheidet drei Weisen von Wissenschaft: nur wahrscheinliches Wissen; sicheres Wissen, begründet durch strenge Ableitung oder Beobachtung; streng mathematische Ableitungen. Chronologie, Astrologie und Geographie genügen dem zweiten Kriterium. Dabei, referiert Bowen,[30] verwende Varenius den Terminus ‚empirisch' selber nicht. Varenius sei es beim ‚Empirischen' aber auch nicht primär um Nützlichkeit o.ä. gegangen, sondern primär um die Prüfung allgemeiner theoretischer Sätze. Damit stelle sich Varenius eindeutig als ein Vorgänger von Humboldt dar. Beide hätten eine Wissenschaftsphilosophie abgelehnt, die auf Geographie nicht anwendbar sein sollte.

Aber eine solche praktische Anwendbarkeit verlangt die (vorherige) Bereitstellung von erforderlichem allgemein-geometrischem, mathematischem Wissen. Das erfolgt zumeist durch Mathematiker, eigentlich nie durch jene, die dieses Wissen benötigen und anwenden. Diese beschränken sich zumeist (auch in der Lehrtätigkeit) auf ein minimales Wissen. Das ist ein Hindernis für die Weiterentwicklung der Geographie. Insofern sollten die theoretischen Denker hier eine größere Rolle spielen—was als ein Appell für einen stärker rational orientierten Empirismus und einen weiteren Begriff von Nützlichkeit gewertet werden müsse.[31]

Der Hauptgrund für den Niedergang der Geographie in jener Zeit ist nach Bowen gerade in Versäumnissen in dieser Hinsicht begründet: Negierung allgemeiner Prinzipien und Reduzierung des Gegenstandes auf eine Beschreibung individueller Regionen. Insofern verdiente die

[29] Bowen 83.
[30] Bowen 85.
[31] Bowen 86.

Geographie eigentlich den Namen einer Wissenschaft nicht. Deshalb habe Varenius eine allgemeine Geographie zu schreiben begonnen. Er beginne seine *Geographia Generalis* deshalb mit einer Einführung, in der er Theorie, Methode und Geschichte der Disziplin als Ganzes behandelt. Dabei gehe er weiter als jeder andere Autor vor ihm, um eine konzeptionelle Basis entwickeln zu können. Diese Vorgehensweise vergleicht Bowen mit dem allgemeinen Teil von Humboldts *Kosmos*.

Freilich konnte Varenius dieses Programm noch nicht streng durchführen. Der Versuch seiner Synthese von Descartes und Bacon bleibe ein Versuch, der von seinen Nachfolgern nicht anhaltend weitergeführt wurde. Dagegen dominierte ein Empirismus bzw. Empirizismus. Der synthetische Ansatz bei Varenius hatte zunächst also keine Folgen. Die Humangeographie blieb ‚draußen', die regionale Geographie verharrte in einer zwiespältigen und defensiven Position. Der Niedergang der Geographie währte bis hin zu Kant.[32]

Versucht man, voranstehende Erörterungen zusammenzufassen, so drängt sich einem schon der Eindruck auf, dass Varenius zwischen allen wissenschaftsphilosophischen Stühlen seiner Zeit sitzt, wobei man aber auch zu der Überzeugung gelangen kann, dass er sich einfach nirgendwo hingesetzt hat, also praktisch diese Bewegungen nicht oder nur sehr partiell zur Kenntnis genommen hat. Wenn man davon ausgeht, dass Varenius im Prinzip zwischen die nachstehenden Fronten, hier als Wirkungslinien bezeichnet, geraten war, so kann man seine Situation wirklich als nicht beneidenswert ansehen: Zu viele, zum Teil widersprüchliche Tendenzen und Aspekte würden sich in Person und Werk des Varenius verknüpfen—wenn sie denn hier tatsächlich ihren Widerhall gefunden haben, als dass man schlüssig bestimmte Wirkungslinien aufzeigen könnte. Solche Wirkungslinien könnten sein:

1. Aristoteles—Neuscholastik des 16. Jh.—Descartes
2. Aristoteles—Neuscholastik des 16. Jh.—Bacon
3. Platonismus—Kepler—Galilei—Descartes hinsichtlich der mathematischen Naturauffassung
4. Demokrit/Epikur—Gassendi—Descartes hinsichtlich der mechanistischen Naturauffassung
5. Gassendi—Jungius—Varenius

[32] Bowen 90.

6. Bacon—Jungius—Varenius
7. Cartesianismus—Platonismus—Cambridger Platonisten—Newton.

Außerdem existierte ja immer noch der Aristotelismus in den Naturwissenschaften.

Zunächst muss man sich von der Vorstellung lösen, als hätten Bacon oder auch Descartes jeweils einen absoluten Bruch gegenüber der aristotelischen und neuscholastischen Wissenschaftsphilosophie vollzogen. Descartes suchte weiter nach Gottesbeweisen, glaubte an angeborene Wahrheiten und war kein Atomist. Bacon war zwar Induktivist und Atomist, aber die mathematisch-mechanistische Naturauffassung (wie auch die Lehre des Kopernikus) war ihm fremd, und er blieb demzufolge weitgehend bei der aristotelischen Formenlehre stehen.

Ob und inwiefern Varenius dann Cartesianer gewesen sein kann, ist unklar. Dass er sich in seiner physischen Geographie weitgehend von der Faktenlage her auch auf Aristoteles stützt (was ihm Büttner ankreidet), ist zunächst normal und hat auch nicht unbedingt etwas mit dem Form- und Entelechie-Konzept des Aristoteles zu tun. Auch gibt es ja bei Descartes keine (physische) Geographie. Erst recht ist dann unklar, inwiefern Newton die *Geographia Generalis* des Varenius in Richtung einer Abschwächung cartesianischer Elemente hat bearbeiten können, wie Warntz meinte. Die Cambridger Platonisten waren für Descartes und zugleich gegen ihn—letzteres insbesondere dort, wo Hobbes die cartesianische Trennung von Denken und Sein in Richtung eines materialistischen Menschen- und Gesellschaftsbildes weiterführte. Bei Dugdales Übersetzung des Varenius ins Englische gibt es den Hinweis, er habe die Stellen, an denen sich Varenius auf Descartes beruft, durch solche ersetzt, in denen er die weitaus bessere Theorie des Newton anwendet.

Empirismus und Rationalismus: Der Begriff affectio bei Varenius und in der Wissenschaftsphilosophie seiner Zeit

Wenn die wichtigsten Philosophen der hier interessierenden Zeit Bacon, Descartes und Leibniz sind, so wird deutlich, dass damit zugleich auch verschiedene Wissenschaftsphilosophien aufeinandertreffen, die in der Literatur zumeist als Empirismus und Rationalismus einander gegenübergestellt werden. Natürlich gibt es eigentlich keinen wirklich bedeutenden Philosophen, der entweder nur die Sinnesorgane oder nur den logisch arbeitenden Verstand bzw. die Vernunft als Kriterium für Wissenschaftlichkeit und Wahrheit des Wissens anerkennen würde. Normalerweise versucht jeder, beide Kriterien miteinander zu vermitteln, eine Synthese o.ä. zu finden. Der Titel der *Geographia Generalis* des Bernhardus Varenius lautet präzise *Geographia Generalis, In qua affectiones generales Telluris explicantur.* Die geographiegeschichtliche Wertung und Einordnung des Varenius hat sich dabei stark auf die Interpretation des Begriffes der affectiones[33] konzentriert. Varenius hat den Begriff selber nicht definiert, sondern gleich dreifach angewendet und auf kosmologische, tellurische, also irdische, und schließlich auf humane Aspekte der Geographie bezogen. Affectiones können für Varenius vieles und Verschiedenes sein: mathematisch fassbare Eigenschaften der Erdoberfläche (Gestalt, Größe, die Lage bestimmter Punkte auf der Erde relativ zum Äquator o.ä.), erfahrungsmäßig gegebene natürliche Eigenschaften (wie z.B. die Verteilung von Land und Wasser in bestimmten Regionen der Erde, klimatische Verhältnisse o.ä.) wie auch Eigenschaften des Menschen (Hautfarbe, Sitten, Gebräuche, Religionen) in den verschiedenen Regionen der Erde. Insofern sind affectiones also auch nicht einfach das, was wir heute als Fakten, als Erfahrungstatsachen oder Erscheinungen im Sinne von empirisch fassbaren Eigenschaften verstehen. Es ist also bereits der Charakter der Geographie als einer sehr komplexen Wissenschaft, der verhindert, dass Varenius etwa Gefahr laufen könnte, die Geographie empiristisch, etwa im Sinne einer Naturbeschreibung und -geschichte misszuverstehen.

Interessant ist, wie Bowen affectiones interpretiert. Sie übersetzt nicht jedesmal, sondern lässt den Terminus als ‚affections' stehen und gibt nur vorher eine Legende: ‚Affections' kann bei Varenius bedeuten: Eigenschaft, Status, Bedingung, Beziehung, Einfluss, Bedingungsänderung.[34] Sie übersetzt die Stelle „triplicis generis esse videntur illa, quae in singulis regionibus considerari merentur", wo Varenius den Terminus affectiones einführt, mit "there are, it appears, three kinds of things [...], which in individual regions deserve consideration". Günther[35] übersetzt einfach mit ‚Eigenschaften', wobei er dann drei grundlegende Gruppen von Eigenschaften unterscheidet.

[33] Dazu auch Staffhorst im Anschluss an diesen Beitrag.
[34] Bowen 277.
[35] Günther 53 u. Anm. 232.

Varenius hatte es mit zwei extremen Standpunkten zu tun, zwischen denen er offensichtlich zu vermitteln sucht: die Reduzierung der Geographie auf eine mathematische Disziplin einerseits, die Simplifizierung der Geographie auf Reisebeschreibungen o.ä. andererseits. In diesem Sinne versteht er die Geographie als eine mathematica mixta—möglicherweise abgeleitet vom entsprechenden Wissenschaftsverständnis des Francis Bacon. Den hatte Varenius wahrscheinlich durch Jungius kennengelernt. Die Passage, in der er sich beim Leser dafür entschuldigt, dass er doch Beschreibungen spezieller geographischer Regionen einfügt (quasi nur, um dem Geschmack der damaligen Zeit wenigstens etwas zu entsprechen, und wohl auch, um Leser anzulocken), macht deutlich, wie unausgereift der Entwicklungsstand der ‚Vollgeographie' (im Sinne Büttners) damals noch war und wie unsicher sich Varenius dementsprechend fühlte.

Während Varenius mit ganz wenigen Ausnahmen von affectiones spricht, operiert Keckermann mit dem Terminus accidentia, das Zufällige, Unwesentliche, Veränderliche an einem Gegenstand, die lateinische Übersetzung des griechischen ta symbebēkóta, das von Aristoteles eingeführt und in der *Metaphysik* 1025 a 14 bestimmt wurde als „das, was einem Gegenstand zukommt und was man von ihm aussagen kann, aber nur das, was ihm nicht notwendig und nicht meistenteils zukommt", d.h. die zufälligen, wechselnden, unwesentlichen Eigenschaften eines Gegenstandes, etwas ‚Hinzugekommenes'.[36] Der Gegensatz dazu ist essentiell, substantiell. Die Frage ist, warum Keckermann und Varenius hier nicht den gleichen Terminus verwendet haben und ob man etwas aus dieser verschiedenen Terminologie für das jeweilige Verständnis von Geographie bei den beiden herleiten kann.

Die Verwendung von accidentia ähnelt doch sehr dem, was wir in der Regel mit der Beziehung von Wesen und Erscheinung bezeichnen, überlagert weiterhin durch die Beziehungen von Notwendigkeit und Zufall sowie von Allgemeinem und Einzelnem. Essenz/Akzidens hat von all dem etwas, bzw. es ist eine Synthese von allen drei Beziehungen. Koppelt man diese Relation an die Form/Materie-Problematik bei Aristoteles und Thomas, dann ist Akzidens materiell, dagegen Essenz Form, ideell. Form wird bei Galilei und Descartes jedoch nicht mehr als qualitative Bestimmung, sondern als mathematische Struktur verstanden. Diese mathematische Struktur muss immer Struktur von etwas sein, also

[36] Johannes Hoffmeister, Hg., *Wörterbuch der philosophischen Begriffe*, Hamburg 1955, 23.

Beziehung bestimmter physikalischer ‚Akzidentien'. Die mathematische Naturauffassung ist damit an eine mechanistische Naturauffassung geknüpft. Das ist nun seit Galilei die Masse, die man dann als von atomarer Struktur auffassen kann, dies aber nicht muss. Varenius setzt an wenigstens einer Stelle accidentia mit affectiones gleich[37] (im 1. Kapitel im Punkt divisio, Einteilung der Geographie, wo es um den Gegenstand des relativen Teils der Geographie geht, s.u. Appendix), bevorzugt jedoch ansonsten eindeutig affectiones. Wir wissen nicht, warum er das tut, und können darüber nur Mutmaßungen anstellen. In Caput I variiert Varenius an wenigstens zwei Stellen affectiones mit proprietates,[38] und der Zusammenhang ist derart, dass man mit gutem Grund annehmen kann, er meint hier jeweils dasselbe. Nun ist der Eigenschaftsbegriff aber auch wieder vieldeutig, so dass wir damit immer noch keine ausreichende Bestimmung von affectiones haben. Ein Stück weiter kommen wir, wenn wir uns ansehen, wie Varenius die Voraussetzungen (praecognitis) der Geographie bestimmt und dabei die Unterscheidung von definitio, divisio, objectum, affectiones, principia, ordo, methodus, origo und praestantia geographiae vornimmt. Die affectiones werden also von Prinzipien abgehoben und erhalten damit eine Bedeutung, die nun von Bestimmungen, Beziehungen, Typen, Verhältnissen, Erscheinungen bis hin zu Dingen und Eigenschaften reichen kann.

Wäre Varenius als eigentlicher Nachfolger von Keckermann anzusehen, bekäme seine Verwendung von accidentia anstatt von affectiones einen Sinn, den man dann mit Günther als Abgrenzung auch von Aristoteles interpretieren könnte. Bei diesem ist die Gegenüberstellung von Substanz und Akzidens immanent, und man könnte versucht sein, affectiones und accidentia einfach gleichzusetzen. Es könnte aber auch sein, dass sich Varenius von der Gegenüberstellung von Substanz und Akzidens frei machen wollte. Dafür spricht zum einen, dass er eben nicht accidentia, sondern affectiones sagt. Darüberhinaus gibt es jedoch bestenfalls einige Indizien, streng genommen eigentlich nur eins—dort, wo er in der Überschrift zum 7. Kapitel die hier zu erörternden affectiones

[37] Dugdale (Varenius 1733) übersetzt dabei *affectiones* mit *appearances* und lässt *accidents* für *accidentia* stehen. Hanno Beck (*Geographie: Europäische Entwicklung in Texten und Erläuterungen*, Freiburg, Br. 1973, 117) übersetzt hier: „Erscheinungen und Einzeltatsachen", Bowen: *affections* und *occurrences*.

[38] Im zweiten Buch, Caput XXII, wo er sagt „...affectiones seu proprietates..." und schon in Caput I „In Generali vero Geographia...considerantur primo affectiones absolutae Telluris, & partium constitutio. Deinde coelestes proprietates in genere contemplabimur...".

als „substantia et constitutio telluris“ wiedergibt. Hier ist der Substanzbegriff—wenn man so viel hineindeuten will—nicht mehr aristotelisch, sondern cartesianisch gefasst.

Die von Warntz angenommene Verbindung zwischen Varenius und Descartes, seine Auffassung, Varenius habe den Aristotelismus in der Geographie bekämpft und dafür dem Cartesianismus zum Durchbruch verholfen, würde nahe legen, dass Descartes auch affectiones benutzt hat. Es gibt bei Ueberweg/Heinze eine Stelle, wo der Terminus im Zusammenhang mit der Darstellung von Descartes' Philosophie auftaucht: „Was wir perzipieren, sind teils res und rerum affectiones (sive modi), teils aeternae veritates, nullam existentiam extra cogitationem non habent.“[39] Zu den aeternae veritates zähle Descartes u.a. Sätze wie „ex nihilo nihil fit, impossible est, idem simul esse et non esse“; die res gliedert Descartes in zwei oberste genera: rerum intellectualium, rerum materialium sive quae pertinent ad substantiam extensam, hoc est ad corpus.

Wenn ich es richtig sehe, versteht Descartes hier unter affectiones Modifikationen von Dingen, die als intellektuale oder materiell-körperliche existieren können. Für Descartes ist die Natursubstanz im Körper zu suchen. Völlig abwegig ist eine solche Interpretation nicht. Von Descartes wird Varenius den Terminus nicht haben, denn er hatte ja noch keine lateinische Übersetzung von dessen Werken gesehen.

Allerdings diskutiert Varenius die Existenz verschiedener Wissenschaftsbegriffe und deren Verknüpfung in der einen Wissenschaft Geographie und verleiht damit einer meines Erachtens neuartigen Perspektive der geographischen Wissenschaft Ausdruck. Wenn man versucht, über den Terminus affectiones etwas mehr über das Wissenschafts- und Geographiekonzept von Varenius zu erfahren, bleibt auch hier ein zwiespältiger und letztlich unbefriedigender Zustand. Wahrscheinlich meint er eben wirklich nur Eigenschaft.

Zusammenfassung: Die Bewertung der Leistungen des Varenius

Die Bedeutung des Varenius und seines Wissenschaftsverständnisses für die Entwicklung der Geographie ist nur schwer einschätzbar. Ganz

---

[39] Friedrich Überweg, *Grundriss der Geschichte der Philosophie der Neuzeit: Vom Aufblühen der Altertumsstudien bis auf die Gegenwart*, Hg. Max Heinze, Berlin 1880, 58.

anders als Büttner findet Bowen, wie schon oben gezeigt, außerordentlich lobende Worte für Varenius als Schaltstelle in der Entwicklung der Geographie zwischen Bacon und Alexander von Humboldt. Besonders positiv schätzt Warntz Varenius ein.

Varenius' Differenzierung in Geographia generalis und Geographia specialis sowie die Hereinnahme von physischen und sozialen Aspekten in die Geographie lässt sich (entsprechend der von Büttner aufgeworfenen Fragestellung) verschiedenartig bewerten:

(a) Varenius als Abschluss einer Etappe (mit dem wichtigen Vorläufer Keckermann),
(b) Varenius als Beginn einer neuen Etappe (über Keckermann weit hinausgehend),
(c) oder aber Varenius als ein Zwischenglied in der Kette von Bacon bis Humboldt, die bis zur modernen Geographie führt (so wie Bowen es sieht).

Die Umbewertung des Varenius, wie gegenwärtig von Büttner vorgenommen, erscheint mir nicht zwingend, zumal sie unnötig akzentuiert die Leistungen des Varenius vom heutigen Gesichtspunkt aus zu werten versucht. Ursprünglich hatte Büttner angenommen, Varenius habe die Vollgeographie begründet. Aus wissenschaftshistorischer wie wissenschaftsphilosophischer Sicht scheint die Vollgeographie des Varenius, die auch humangeographische Betrachtungen in den Blick nimmt, dem damaligen Zeitgeist, speziell auch gerade dem Newtonschen Wissenschaftsverständnis, nicht zu entsprechen und viel zu früh zu kommen. Da Varenius den entsprechenden Teil jedoch nicht mehr ausführen konnte, war eine Rezeption des Vareniusschen Hauptwerkes *Geographia Generalis* auch von Gegnern einer solchen Ausweitung immer möglich.

Nach meiner Einschätzung können die Untersuchungen noch nicht als abgeschlossen gelten. Wahrscheinlich ist es doch so, dass sich Varenius—sofern er sich wirklich auf irgendeine wissenschaftsphilosophische Position festgelegt hat—zwischen allen oben genannten ‚Stühlen' gesessen hat. Da Naturwissenschaftler aber in der Regel weit nach ihrem 28. Geburtstag zu philosophieren beginnen, sollte es uns nicht wundern, wenn bei Varenius dazu kein Befund ausgemacht werden kann.

MISZELLEN ZUR *GEOGRAPHIA GENERALIS* DES BERNHARD VARENIUS

Ulrich Staffhorst

Abstract

Bernhard Varenius' Geographia Generalis*: Miscellania*
The following topics are briefly discussed:

1 The influence of Varenius' *Geographia Generalis* (reprints, revised editions, translations) resulted from the systematic general account of the standard of geographical knowledge at that time, useful for students.
2 The term 'affectiones' already appears in the title; it is often equated with 'proprietates', which means 'properties, typical features'.
3 Although Varenius explains equally the arguments of the Ptolemaics (Aristoteleans) and Copernicans, there can be no doubt about his supporting the heliocentric system.
4 Varenius' work exhibits signs of reasoning by means of causal connections.

Im Jahr 2000 versammelte sich in Hitzacker, der Heimat des Bernhard Varenius, ein kleiner Kreis von Menschen, um seines Todes vor 350 Jahren zu gedenken. Im ,Hitzacker-Kreis' entwickelte sich in den darauf folgenden Monaten eine lebhafte briefliche Diskussion über wichtige Fragen, die sich auf das Hauptwerk des Varenius, die *Geographia Generalis*, beziehen. Besonders intensiv gestaltete sich der Gedankenaustausch mit Manfred Büttner, dem hoch verdienten Nestor der wissenschaftsgeschichtlichen Erforschung des Faches Geographie. Seine Überlegungen fanden schließlich ihren Niederschlag in einer Abhandlung von 2001. Da der Aufsatz „Begründung der wissenschaftlichen Geographie durch Bernhard Varenius (1622–1650)? Gedanken zu den Defiziten der Varenius-Forschung aus Anlass des Varenius-Jahres 2000" sozusagen den letzten Stand seiner Forschung zu Varenius darstellt, sollen im Folgenden einige wesentliche Punkte daraus aufgegriffen und der Versuch unternommen werden, sie zu präzisieren und ggf. in ein neues Licht zu rücken.

1. *Zur Nachwirkung der Geographia Generalis des Varenius*

Büttner (250) meint, das Buch sei nur „als Reiselektüre, als Unterhaltungslektüre oder auch als Lektüre in unterhaltsamer und nicht sehr wissenschaftlicher Form für Nautiker und Seefahrer" wahrgenommen worden. Es habe „keine Nachwirkungen gehabt". Dieses Urteil verwundert angesichts der europaweiten Nachdrucke, Bearbeitungen und Übersetzungen sehr. Wir können wohl kaum annehmen, dass Newton beim Nachdruck der inhaltlich so trockenen *Geographia Generalis* an Reisen und Unterhaltung seiner Studenten oder an die Unterhaltung von Nautikern und Seefahrern gedacht hat, die zum großen Teil wohl gar kein Latein beherrschten (außerdem umfassen die spezifisch nautischen Kapitel nur die Seiten 744–786).

War es dann die Hereinnahme physiogeographischer und humangeographischer Aspekte, die sein Werk so attraktiv erscheinen ließen? Diese Erklärung ist nicht abwegig, reicht aber wohl nicht aus. Varenius unterscheidet säuberlich zwischen Geographia generalis und specialis. Er hat erstere—wenn offenbar auch unter großem Zeitdruck—zu einem alles in allem passablen Abschluss gebracht. Für letztere, eine Art Länderkunde, plante er eine eigene Publikation: Nos [...] adjungemus Speciali Geographiae multa capita de Usu Geographiae (5). Dass er dennoch im 1. Kapitel der *Geographia Generalis* relativ ausführlich von der speziellen Geographie spricht und auf zwei beigegebenen Falttafeln[1] den Kapitelüberschriften der *Geographia Generalis* 28 Punkte (nicht Kapitel!) gegenüberstellt, die bei der Behandlung der speziellen Geographie beachtet werden müssen, hat nicht nur mit der Abgrenzung der beiden Geographien gegeneinander zu tun, sondern ist auch deshalb sinnvoll, weil diese Gesichtspunkte ja generell bei Ländern und Städten ins Auge gefasst werden sollen.

Wenn wir uns die beiden Falttafeln freilich näher anschauen, werden wir schnell feststellen, dass die Kategorien der Geographia specialis weitgehend denen der Geographia generalis entsprechen—mit einer Ausnahme: den zehn Punkten der Humangeographie. Der Verfasser selbst begegnet ihnen mit einer gewissen Reserve ([...], etsi [...] minus recte ad Geographiam referantur: Sed dandum est aliquid consuetudini & utilitati discentium, [...] 4f), aber sie sind ohne Zweifel anregend und reichhaltig. Nur: völlig neu sind sie keineswegs! Das kann

[1] Siehe unten Appendix und Fig./Abb. 14 u. 15.

man schnell sehen z.B. bei einem Blick in die Rede *De Creichgoia* des David Chytraeus,[2] des Urgroßvaters des Varenius, in die *Introductio in universam Geographiam tam veterem quam novam* des Philipp Clüver, der im Geburtsjahr des Varenius starb, in die schwer zu entziffernden Blätter der „Geographia Generalis" des Lehrers des Varenius, Joachim Jungius,[3] die u.a. einen Sprachenvergleich an Hand des Vaterunsers bieten, und schließlich in die *Descriptio regni Japoniae* des Varenius selbst. Man kann da im Einzelfall streiten, ob es sich überhaupt um Geographie bzw. ein Exemplar der Geographia specialis handelt—Martin Schwind z.B. bestreitet das für die *Descriptio*;[4] das sind Fragen der Ponderierung im Verhältnis zu den Terrestria und Caelestia; aber jedenfalls waren Kategorien wie „menschliche Kultur, Sprache, Verfassung, Religion, Geschichte" usw. in geographischem Zusammenhang nicht ganz neu.

Offensichtlich besteht das Verdienst des 28jährigen Varenius nicht in lauter bahnbrechenden Neuigkeiten, sondern mehr in der umfassenden und zugleich handlichen systematisierenden Zusammenschau des damaligen Wissensstandes, die sich gut als Lehrbuch einsetzen ließ.[5] Man muss auch aus heutiger Sicht jemanden bewundern, der in einem knappen Jahr eine solche intellektuelle Aufgabe bewältigte, durch die er (laut „Epistola dedicatoria") hoffte, die Geographie erst zu einer anerkannten Wissenschaft zu machen.

Zum Abschluss des Themas ‚Nachwirkung' noch ein Streiflicht auf das Karlsruher Bismarck-Gymnasium, das einstige Gymnasium Illustre. Die Schule ist älter als die Stadt; sie feierte 1986 ihr 400jähriges Bestehen. Unter den aus diesem Anlass damals ausgestellten Büchern der alten Gymnasialbibliothek befand sich auch die *Geographia Generalis* des Bernhard Varenius in der Jenaer Ausgabe des Jahres 1693. Das Buch war einst im Besitz des Theologen Johannes Wasmuth gewesen, der laut einer Eintragung das vorliegende Exemplar 1697 in Rinteln, wo er sieben Jahre lang die Universität besuchte, erworben hatte. Wasmuth war von 1708 bis 1750 am Gymnasium in Karlsruhe tätig. Nun findet

[2] Dazu auch Düchting in diesem Band.

[3] Dazu auch Lehmann in diesem Band.

[4] Schwind, „Die wissenschaftliche Stellung der *Descriptio Regni Japoniae*," *Descriptio regni Japoniae: Beschreibung des japanischen Reiches*, Hg. Martin Schwind u. Horst Hammitzsch, Übers. Ernst-Christian Volkmann, Darmstadt 1974, XXIVff.

[5] Ähnlich Rainer Kastrop, *Ideen über die Geographie und Ansatzpunkte für die moderne Geographie bei Varenius unter Berücksichtigung der Abhängigkeit des Varenius von den Vorstellungen seiner Zeit.* Diss. Saarbrücken 1972, 12.

sich in einer alten Geschichte dieser Schule[6] der explizite Hinweis auf eine Einführung der *Geographia Generalis* des Varenius in den Unterricht der Schule in der zweiten Hälfte des 17. Jahrhunderts. Das Werk blieb Grundlage des Unterrichts bis ins erste Jahrzehnt des 18. Jahrhunderts (1705). Wir haben hier also punktuell 1. den Nachweis vom Einsatz des Buches im deutschen Schulunterricht (für Schüler war das sicher eine schwer verdauliche Kost); 2. davon unabhängig sehen wir das Buch im Besitz und Gebrauch eines Lehrers, der es während seines Studiums kennen gelernt hatte und erst drei Jahre nach dem Ersatz dieses Lehrwerkes durch ein anderes (einfacheres)[7] an die Schule kam; es finden sich darin handschriftliche Bemerkungen in vorwiegend lateinischer, aber auch deutscher und französischer Sprache, die sich mit bestimmten Kapiteln auseinandersetzen und etliche Ergänzungen und Verweise auf andere einschlägige Werke enthalten.

2. *Affectiones generales*

Eine der umstrittensten Fragen ist die nach der Interpretation des im Titel erscheinenden affectiones-Begriffs.[8] Man kann dabei von dem lateinischen Wort sive bzw. seu ausgehen. Wenn es zwischen zwei Wörtern steht (x sive y), bedeutet es im Gegensatz zu aut (vel) in der Regel, dass die beiden Wörter inhaltlich nicht recht zu unterscheiden sind, z.B. generalis sive universalis geographia / specialis sive particularis geographia / de situ sive loco Telluris respectu Planetarum & Stellarum / liquorem seu aquas. Wenn wir also die Formel quatuor generales Telluris affectiones seu proprietates (zu Beginn von Caput 7) oder proprietates seu affectiones (zu Beginn von Caput 22) finden und dazu schon im 1. Kapitel die beiden Wörter dauernd synonym nebeneinander gebraucht werden (1ff), so dürfen wir affectiones ganz sicher mit „Eigenschaften,[9] Eigentümlichkeiten, Merkmale" übersetzen und affectiones coelestes, terrestres, humanae wiedergeben mit „vom Himmel, von der Erde, von Menschen bedingte Eigenschaften (einer

---

[6] Karl Friedrich Vierordt, „Geschichte der im Jahre 1724 aus Durlach nach Karlsruhe verpflanzten Mittelschule." Beilage zum Programme des Karlsruher Lyceums. Karlsruhe 1858/1859, 74.

[7] Julius Reichelt, *Elementa Geographica*. Straßburg 1705.

[8] Vgl. dazu auch Richter in diesem Band.

[9] So schon S. Günther, *Varenius* (Klassiker der Naturwissenschaften.4), Leipzig 1905, 53 u. Anm. 232 (ohne nähere Begründung).

Landschaft)". Natürlich entspricht das nicht mehr unserer heutigen Begrifflichkeit; wir würden differenzieren zwischen „Abhängigkeiten / Einwirkungen" (vom Himmel), „Eigenschaften / Erscheinungsformen" (der Erdoberfläche), „Prägung" (durch den Menschen). Aber es ist wohl nicht statthaft, Varenius unsere Vorstellungen überzustülpen.

Dabei ist von Interesse, dass im Titel ausdrücklich von affectiones generales die Rede ist; gelegentlich heißt es auch affectiones propriae (z.B. 1. Kap., Divisio). Wenn diese Attribute einen Sinn haben sollen, so muss man sich „allgemein bzw. weithin anzutreffende, generelle bzw. generelle Erörterung erfordernde, charakteristische (propriae), wesentliche" Merkmale vorstellen. Man trifft sie, wohlgemerkt, in der Geographia specialis—observans regulas generales (Epistola dedicatoria)—wieder an:

> Nos in Generali Geographia quasdam affectiones generaliter explicavimus, quas in speciali Geographia ad singularum regionum explicationem accommodabimus (1. Kap., Ordo) / Deinde caelestes proprietates in genere contemplabimur, quae deinde in Speciali Geographia ad singulas regiones applicanda sunt (1. Kap., Affectiones).

Natürlich sieht die Welt in den einzelnen Regionen dann sehr verschieden aus; in diesem Sinne spricht 1645 David Christiani in *Systema Geographiae Generalis* von affectiones speciales (610), die in der Geographia specialis behandelt werden: das an einer bestimmten Örtlichkeit so und nicht anders anzutreffende Bündel bestimmter Merkmale.—Die Kategorie des eher Ungewöhnlichen, Untypischen, nicht Berechenbaren, ganz Individuellen dagegen wird man des öfteren in der Humangeographie vorfinden, die von Varenius zwar ebenfalls der Geographia specialis zugewiesen, aber in der *Geographia Generalis* nicht behandelt ist.

Büttner hat seinerzeit brieflich darauf hingewiesen, dass Bartholomäus Keckermann, dessen *Systema geographicum* 1616 erschien und der erstmals die Geographia generalis und specialis unterscheidet, im zweiten Kapitel von De Globi Terreni generalibus affectionibus seu Accidentibus spricht, also affectiones mit Accidentia gleichsetzt. Er unterteilt dann in § 4: Tres affectiones globi terreni [...]: Quod globus Terrae exacte sit in medio mundi. Quod immobilis quiescat [...] Quod sit absolutissimum centrum universi. Man sieht sofort, dass die übliche Übersetzung von accidentia mit „das Zufällige, Unwesentliche, Veränderliche" hier nicht zutrifft. Aristoteles definiert in der *Metaphysik* (1025a30) das Akzidens (τὸ συμβεβηκός) noch in einer zweiten Weise: „In einer anderen Bedeutung aber nennt man Akzidens das, was sich an jedem Einzelnen an sich findet, ohne aber in seinem Wesen enthalten zu sein, so etwa am Dreieck die Winkelsumme von 2 Rechten. Das Akzidens in dieser Bedeutung kann ewig sein, in jener (ersteren) aber nicht."[10] Ross bemerkt dazu in seinem *Metaphysics*-Kommentar: „Aristotle now proceeds to what he elsewhere calls the καθ' αὑτὸ συμβεβηκός, that which, since it is not included in the definition of the subject, is a συμβεβηκός but which yet flows from the nature of the subject,—in other words, the property."[11]

Da haben wir also genau die Umschreibung 'property'. In diesem Sinne wären die Keckermannschen Vermutungen, dass die Erdkugel unbeweglich in der Mitte des Weltalls ruht, nicht zur Definition der Erde gehörig, wohl aber als affectiones anzusehen, d.h. als Eigenschaften, die nur von der Erde ausgesagt werden können, so wie die Winkelsumme vom Dreieck. Dass Keckermann, inhaltlich gesehen, ganz andere ‚Eigenschaften' im Auge hat als Varenius, steht auf einem anderen Blatt. Wir dürfen jedenfalls annehmen, dass accidentia (im zweiten aristotelischen Sinne), proprietates und affectiones ganz eng beieinander angesiedelt sind.

Übrigens geht Varenius dem accidens-Begriff und dem Verbum accidere nicht völlig aus dem Weg. So heißt es im 1. Kapitel (unter: Divisio): In Parte Respectiva illas affectiones atque accidentia considerabimus, quae Telluri a caelestibus causis accidunt; ob wir hier affectiones und accidentia gleichsetzen dürfen, ist nicht gesichert, da als Bindeglied atque, nicht sive/seu fungiert—ähnlich Christiani: Geographia Theoretica (Generalis) ex Arithmeticis, Geometricis, Opticis [. . .] varia accidentia et affectiones [. . .] explicat et inquirit.[12] Hinzufügen lässt sich aus Varenius z.B. (Kap. 9, Prop. 10): Montibus accidunt (widerfahren) ruina, ruptura, transpositio. Rara quidem horum accidentium exempla sunt; sowie (Kap. 22, 437): [. . .] proprietates seu affectiones illas considerabimus, quae Telluri a motu Solis atque stellarum apparente accidunt. Diese (seltenen) Stellen erwecken alle nicht den Eindruck, als ob sie mit philosophischem Hintersinn aufgeladen wären. Es ist von einer Einwirkung die Rede, die der Erde a caelestibus causis oder a motu Solis widerfährt. Das ist normales kausalgenetisches Denken mit Hilfe eines im Normalsinn verwendeten Verbs.

---

[10] Aristoteles, *Metaphysik*, Übers. Franz F. Schwarz, Stuttgart 1970, 154.

[11] *Aristotle's Metaphysics*: A Revised Text with Introd. and Commentary by William D. Ross, vol. 1, Oxford 1924, 349.

[12] Kastrop 1972, Anm. 240.

Vor allem denkt man hier nicht an einen Gegensatz accidentia—substantia im Sinne des Descartes (*Meditationes de prima philosophia*, Synopsis 4). Es gibt aber eine diesbezügliche interessante Stelle am Beginn des 7. Kapitels der *Geographia Generalis*, wo den generales Telluris affectiones seu proprietates die ipsa substantia seu essentia gegenübergestellt wird. Unter ersterem—also den affectiones, die übrigens in den *Meditationes* nicht auftauchen—versteht Varenius Telluris figura, dimensio & magnitudo, motus, locus in mundi systemate, unter letzterem—also der Substanz oder Essenz—die klassischen vier Elemente oder die Atome oder die drei Prinzipien der Alchemisten oder, seiner eigenen, hier sehr spekulativen Ansicht entsprechend, die fünf prima simplicia corpora, nämlich Aqua, Oleum seu sulphur, Sal, Terra & Spiritus quidam, quem acidum nonnulli appellant, sive is sit Mercurius Chymicorum („das Quecksilber der Alchemisten"). Alle Körper und Teile der Erde ließen sich in jene fünf elementares substantias auflösen; doch wolle er nicht leugnen, dass jene selbst sich nicht so sehr durch ihr inneres Wesen als durch die einzigartige Verschiedenheit der Figuren und Größen unterschieden. Da sind wir freilich ganz nah bei der antiken Atomlehre und bei Descartes, auf den sich Varenius ausdrücklich beruft.

3. *Varenius—ein Vertreter des heliozentrischen Weltbildes*

Wenden wir uns nun der Frage zu, ob Varenius Kopernikaner war oder nicht. Man meint, das müsse sich leicht entscheiden lassen und sei längst geklärt. Aber dem ist nicht so; das liegt vor allem daran, dass Varenius hier wie sonst alte und neue Meinungen und Argumentationen nebeneinander zu Wort kommen lässt, wie schon an der oben erörterten Substanzfrage deutlich wurde. So meint Büttner: „Nicht einmal das copernicanische Weltbild hat er konsequent vertreten! Varenius schaut nach ‚rückwärts' und nicht nach ‚vorwärts' [...]".[13] Doch das ist ein Irrtum.

a) Noch am unbestimmtesten äußert sich Varenius in der „Epistola dedicatoria" der *Geographia Generalis*. Es mag sein, dass er in seiner Widmung nicht sofort irgendwelche Kreise vor den Kopf stoßen wollte. Immerhin sagt er, es sei „bis jetzt nicht mit schlüssigen Argumenten gezeigt, dass die Erde den Mittelpunkt des Himmels einnimmt."

[13] Büttner 2001: 250.

b) Was die Umdrehung der Erde um ihre eigene Achse (und um die Sonne)—motus primus (und motus secundus)—betrifft, so waren sich bekanntlich die Pythagoreer (Aristarch von Samos) und Kopernikaner (Kopernikus, Kepler, Galilei usw.) auf der einen Seite, die Ptolemaiker auf der anderen Seite nicht einig. Um diesen Streit geht es auf den Seiten 48–57. Die Argumente der Kopernikaner werden in acht Punkten breit und mit unverhohlener Zustimmung auf den Seiten 49–52 abgehandelt. Abschließend sagt Varenius (52) folgendes:

> Das sind die ziemlich leicht zu führenden Beweise, mit denen die Kopernikaner die Erdbewegung erfolgreich dartun wollen; wenn diese auch nicht zwingend (Apodictica) sein sollten, so machen sie diese Hypothese doch äußerst wahrscheinlich (admodum probabilem) im Vergleich zu jener, die eine Bewegung des Himmels annimmt.

Es folgen zwar in aller Kürze acht Argumente der Ptolemaiker (52f); sie werden jedoch durch die einleitende Bemerkung des Varenius von vornherein entwertet: „Jene Gründe aber, die einige (Ptolemaiker) dagegen anführen, lassen sich mit leichter Mühe entkräften“ (facili negotio dissolvuntur). In der Tat folgt (53f) eine eingehende Widerlegung. Auf den Seiten 55f wird angedeutet, was sich aus dem Kopernikanischen Weltbild ergibt.

c) Im Kapitel 6 (57–62) geht es um „die Lage bzw. den Ort der Erde im Hinblick auf Planeten und Sterne“. Nach dem Ptolemäischen Weltbild nimmt die Erde das Zentrum des Universums ein, nach den Kopernikanern dagegen die Sonne, wogegen die Erde zwischen Mars und Venus angesiedelt ist und in jährlichem Lauf die Sonne umkreist. Für die Kopernikanische Weltsicht führt Varenius ziemlich ausführlich und mit Zustimmung sieben Punkte auf den Seiten 59f an (vgl. z. B. zu Punkt 6: Est vero hoc argumentum meo quidem judicio praecipuum.—Punkt 7: Sic quoque commode explicatur variatio distantiae planetarum a Tellure). Viel kürzer lässt er mit acht Punkten die Aristoteliker mit den Ptolemaikern auf S. 60 zu Wort kommen, nur um sie gleich darauf (61) zu widerlegen: „Diese Gründe der Aristoteliker entkräften die Kopernikaner mit leichter Mühe“ (facili opera diluunt).

d) In der zweiten Falttafel, die die Gesichtspunkte zur Geographia specialis enthält,[14] steht bei den Coelestia octo unter Nr. 8: Quantitatem vel celeritatem motus juxta Copernicanam hypothesin (ebenso S. 3: quantitas vel celeritas motus, quo juxta Copernicanam hypothesin sin-

[14] Siehe unten Appendix und Fig./Abb. 15.

gulis horis locus quilibet rotatur). Von Aristotelikern und Ptolemaikern ist im Zusammenhang mit der Geographia specialis auf der Tafel also gar nicht mehr die Rede.

Abschließend können wir daher die heliozentrische Sicht des Varenius umso höher bewerten, als seine Vorgänger sich noch nicht zu dieser Erkenntnis durchgerungen haben: weder Merula noch Keckermann noch Alstedt noch Göllnitz noch Christiani. Der Letztgenannte ist zwar durchaus mit den Gedanken des Kopernikus, Kepler, Galilei vertraut und führt ihre Argumente in seinem Werk *Systema Geographiae Generalis* von 1645 (!) an, entscheidet sich dann aber dennoch (36) für das geozentrische System (Terra in medio Universo immota quiescit).

4. *Zum Wissenschaftsverständnis des Varenius*

Zuletzt noch ein paar Worte in ganz rudimentärer Form zum Wissenschaftsverständnis des Varenius. Ausgelöst sind sie wieder durch eine Bemerkung Büttners: „Er [V.] hat auch nicht die Beobachtung oder gar kausalmechanisches Denken in Verbindung mit Experimenten in die Geographie eingeführt.“[15]

Nach heutiger Auffassung kann sich Wissenschaft nicht allein mit einer anschaulichen Beschreibung begnügen, sondern muss sich auch um eine kausale Erklärung mittels deduktiv oder induktiv gewonnener Argumente bemühen. Varenius selbst hat sich im ersten Kapitel zur „Methode“, d.h. zur „Art, die Wahrheit geographischer Sätze zu prüfen“, geäußert. In der Geographia generalis kann sehr viel bewiesen werden (Demonstrationibus confirmentur), vor allem caelestes affectiones, deren Verständnis sich auf Astronomie und Mathematik stützt (wir erinnern uns daran, dass Varenius von der Mathematik herkommt); anders sieht es in der Geographia specialis aus, wo es der Erfahrung und Beobachtung (experientia & observatio) bedarf, d.h. des Zeugnisses der Sinne (sensuum testimonium).

Das Wunder schätzt Varenius nicht: [...] neque miraculo opus est, sagt er (67) und fügt hinzu: [Geographia] nec veterum opiniones curat, nec in explicandis Telluris proprietatibus confugere debet ad miracula. Wie Kastrop resümiert: „Zwar berichtet auch er zuweilen von wunderlichen Dingen [...], aber dann nennt er seinen Gewährsmann, um

[15] Büttner 2001: 250.

so anzudeuten, dass er für diese Nachrichten nicht verantwortlich ist. [...] Ausdrücke wie mihi videtur [...], verisimile est [...] oder puto [...] zeugen dagegen von einem kritischen Selbstbeschränken, von der klaren Erkenntnis, daß sowohl das Wissen der Zeit als auch das eigene noch nicht ausreichen, um alle strittigen Fragen zu klären."[16]

Wir haben schon beim heliozentrischen Weltbild gesehen, dass er weder gegenüber der Kirche noch gegenüber der Antike autoritätsgläubig ist: [...] scripturam sacram in rebus physicis loqui secundum apparentiam et vulgi captum („...dass die Heilige Schrift in naturwissenschaftlichen Fragen gemäß der Erscheinung und der Fassungskraft der Menge spricht"), sagt Varenius und erklärt dazu: Namque sacrae litterae nobis non ad philosophandum, sed ad pietatem colendam concessae sunt (54).

Bezüglich der Antike ist sich Varenius bewusst, dass deren Wissen noch recht beschränkt war, dass damals die Erde noch nicht umsegelt werden konnte, dass Aristoteles noch nichts von Ebbe und Flut wusste usw. (1. Kap., Origo Geographiae). So stellt er dem antiken Gedankengut, wo immer möglich, neuzeitliches gegenüber. Es war freilich nicht immer möglich. Denn bei dem Wissensstand seiner Zeit konnten viele Phänomene nicht befriedigend erklärt werden.

Immerhin ist eines der in der *Geographia Generalis* am häufigsten gebrauchten Wörter ‚causa', und es sollte uns doch wundernehmen, wenn sich bei Varenius nicht wenigstens Ansätze zu einer kausalgenetischen bzw. kausalmechanischen Betrachtungsweise fänden. Die richtige Methode garantiert freilich noch keine richtigen Ergebnisse. Varenius, dem sein kurzes Leben keine Zeit zu eigenständigen geographischen Forschungen ließ, musste sich auf den Stand des Wissens seiner Epoche und auf seine Gewährsleute verlassen. Bei falschen physikalischen Prämissen, bei mangelhaften und unvollständigen Beobachtungen und fehlerhaften Beschreibungen fremder Länder konnte er leicht falsche Schlüsse ziehen, was den heutigen Leser so frustriert. Gleichwohl müssen wir versuchen, ihm gerecht zu werden.

Ein Beispiel: In Kapitel 15 (Prop. 8) spricht Varenius davon, dass so viele Ströme ins Kaspische Meer, ein Binnenmeer, münden. Wo bleibt dieses Wasser? Wir würden natürlich antworten: es verdunstet. Aber das hielt man bei solchen Wassermassen damals nicht für möglich. So schloss man auf „unterirdische Höhlen und Gänge", durch die das

[16] Kastrop 102f.

Wasser in den Ozean abfließe. So führt eine fehlende physikalische Erkenntnis zu einem zwar falschen, aber nicht unlogischen Schluss.

Ein zweites Beispiel (aus Kap. 19) bezieht sich auf die Entstehung von Winden: „Die Atmosphäre bzw. Luft, die sich erwärmt, nimmt größeren Raum ein als zuvor. Je mehr sie aber der Wärme beraubt ist, desto mehr zieht sie sich zusammen und beansprucht geringeren Raum. [. . .] Deshalb ist die Höhe der Luft bzw. Atmosphäre nicht konstant, sondern sie nimmt ab und zu. Natürlich muss sie am Mittag am größten sein, bei Mitternacht am kleinsten [. . .]." Solche Unterschiede haben dann eine Bewegung der Luft zur Folge: „Wenn auch die sommerliche Wärme in unserer Gegend auch unsere Luft mehr verdünnt und aufsteigen lässt als im Winter, so wird sich die Luft dennoch, weil zur gleichen Zeit an einem anderen Ort der Erde Winter herrscht, an jenem Ort weniger erheben. Deshalb wird sich ein Teil unserer Luft in Richtung der Luft jener Gegenden bewegen, wo die Luft weniger hoch steht" (350). Auf denselben Gedanken kommt er später (393ff) unter dem Stichwort causae ventorum variae zurück:

> Der besondere und allgemeine Grund ist die Sonne selbst, die durch ihren feurigen Glanz die Luft auflockert und verdünnt, insbesondere jene, auf die sie ihre senkrechten Strahlen fallen lässt bzw. über der sie steht. Denn die verdünnte Luft erfordert einen viel größeren Raum. Daher kommt es, dass die von der Sonne in Bewegung gesetzte Luft andere benachbarte Luft mit großer Heftigkeit fortstößt, und da die Sonne von Osten nach Westen wandert, wird sie die Luft vornehmlich nach Westen treiben (394).

Gleichviel ob man hier an den Passat denkt oder nicht, so könnte man diese Ausführungen zumindest im Ansatz als kausalmechanisch bezeichnen.—Noch ein letztes Beispiel: Varenius meint (163ff), das Wasser des Ozeans sei an den Polen weniger salzhaltig als am Äquator, und führt dafür sechs Gründe an: davon spricht z.B. der erste von größerer Verdunstung in der heißen Zone, was den Salzgehalt steigere.

Mit Sicherheit lassen sich weitere derartige Belege finden. Forschungen auf diesem Gebiet würden wesentlich erleichtert, wenn der Wissenschaft eine—möglichst mit kommentierenden Anmerkungen versehene—deutsche Übersetzung des ‚Urvarenius' (von 1650) zur Verfügung stünde.

NOTES ON *GEOGRAPHIA GENERALIS* AND ITS INTRODUCTION TO ENGLAND AND NORTH AMERICA

Margret Schuchard

Abstract

Zur Geographia Generalis *und seiner Verbreitung in England und Nordamerika* Mit der *Geographia Generalis* verband Varenius auch eine didaktische Zielrichtung, denn die Fundierung der Geographie als einer mathematischen Wissenschaft war systematisch zu entwickeln und sollte zielgerichtet lehr- und lernbar werden. Dass er sich bei seinen Vorgängern möglichst umfassend informierte, gehörte zu seiner Arbeitsweise. Die Publikationsgeschichte in England zeigt, dass Newtons Ausgabe, hier besonders die zweite Auflage von 1681, eine besonders große Verbreitung fand, die über England hinaus wirkte (auch auf den Nachdruck von Cröker in Jena) und nicht zuletzt in den Colleges Nordamerikas, und auf indirekten Wegen bei Jefferson, nachwirkte. Während Richard Blomes ambitioniertes kosmographisches und geographisches Projekt in englischer Sprache, ausgehend von Sanson und Varenius, nur zum Teil realisiert wurde (er imitierte John Ogilbys Publikationsstrategie) und keinen originalen Ansatz aufwies, zeigte James Jurins Edition wissenschaftlichen Ehrgeiz, ermuntert von Richard Bentley. Von Newton über Jurin bis Shaw standen die Editoren von Varenius im Laufe ihres Lebens in enger Verbindung zur Royal Society.

Varenius explained his aim in the dedication of his *Geographia Generalis*. In geography he saw a need for establishing a theoretical foundation based on mathematics. Once geography was based on reliable proven foundations, it could be easily taught and learned; Varenius' didactic purpose is explicitly stated here. Again he begins his book with a theoretical part. His often quoted phrase of definition that geography is a "scientia Mathematica mixta",[1] i.e. mixed mathematics, makes use of a well established terminology (which in the early eighteenth century was replaced by the term 'applied mathematics'). Bacon had first distinguished between pure and mixed mathematics in 1605 and again in

[1] If "Mathematica" were an ablative case (the context explains that it is not), the translation would be "a science mixed with mathematics".

1623, and others followed (among them Goclenius in 1613 and afterwards Snellius).[2] Varenius, while still a student in Königsberg, referred to the distinction between "mathematica pura" and "mathematica mixta" as a matter of course in 1644. (He explained in a letter to Jungius that after having studied pure mathematics—Euclid's elements, demonstrations in trigonometry and logarithms—he now wanted to learn more about mixed mathematics).[3] Although it was regarded inferior to pure mathematics within the hierarchical organisation of knowledge, because in dealing with quantity it was not abstracted from matter, Varenius set out to establish geography on a foundation within a respectable and respected position. The design of his *Geographia Generalis* unfolds before the reader both in the listed chapter headings and the inserted tables (see appendix below) that illustrate his classifications and systematise the subject matter by moving from the general to the increasingly specific. In his division of geography into general and special geography Varenius intends to devote himself to general geography, but that does not mean degrading 'special' geography to a less important second rank. As always, he does not belittle the practical use of scholarship; students, merchants and voyagers would profit from it for their trade world wide. Varenius' concept of 'general geography' has been often discussed and need not be repeated here.[4]

Varenius' work superseded what in particular Bartholomäus Keckermann (1572–1609) in his *Systema Geographicum* of 1611 (reprinted twice in Varenius' life time) had done before him; Keckermann had already

---

[2] In Bacon's tree of knowledge geography is not named but it would be placed alongside with perspective, music, astronomy, cosmography, architecture and engineering (Gary I. Brown, "The Evolution of the Term 'Mixed Mathematics'," *Journal of the History of Ideas* 52 (1991): 82). See also Daniel Garber, "Descartes and the Scientific Revolution: Some Kuhnian Reflections," *Perspectives on Science* 9 (2001): 405–422; and Christoph Meinel, „Reine und angewandte Wissenschaft," *Das Magazin* (2000: no. 1) (publ. by Wissenschaftszentrum Nordrhein-Westfalen).

[3] G.E. Guhrauer, *Joachim Jungius und sein Zeitalter*, Stuttgart und Tübingen 1850, 371.

[4] See the work of William Warntz; J.N.L. Baker, "Nathanael Carpenter and English Geography in the Seventeenth Century," *The History of Geography*, Oxford 1963, 1–13 [reprint from *Geographical Journal* 71 (1928)]; Manfred Büttner, *Die Geographia Generalis vor Varenius*, Wiesbaden 1973; and Büttner's „Begründung der wissenschaftlichen Geographie durch Bernhard Varenius (1622–1650)?: Gedanken zu den Defiziten der Varenius-Forschung aus Anlaß des Varenius-Jahres 2000," *Berichte zur Wissenschaftsgeschichte* 24 (2001), 237–254; Elizabeth T. Roberts, *The* Geographia Generalis *of Bernhard Varenius*, MA Thesis in History, U. of New Hampshire 1994, unpubl. Typescript; Margarita Bowen, *Empiricism and Geographical Thought: From Francis Bacon to Alexander von Humboldt*, Cambridge 1981.

classified geography into the two categories of general and special. Varenius used Keckermann, as Büttner has shown, and in Büttner's view it is Keckermann rather than Varenius who must be regarded as the innovator in geographical thought; Büttner (2001) now places Varenius (he calls him an 'amateur' geographer) at the end of a development, a last Aristotelian.

Among other predecessors of Varenius were Paul Merula (*Cosmographiae Generalis*, Amsterdam 1636), David Christiani (*Systema Geographiae Generalis*, Marburg 1645) and Philipp Clüver.[5] Nathanael Carpenter (who had made use of Keckermann, "a late German writer", without naming him) may have been known to Varenius, but there is no conclusive evidence of this, since similarities may be due to the use of the same sources (Göllnitz [Golnitzius], Keckermann).[6] Among the authors Varenius quoted by name (he cited about 300 different classical and modern authors according to Warntz)[7] more than half of them were from classical antiquity. Aristotle's works stand out for the most part, but Strabo (and through him Eratosthenes and others, as Kießling[8] pointed out), Ptolemy and Plinius, Seneca, Vitruvius and Pomponius Mela are also important sources for certain subjects. Varenius' division of geography into the two parts is already in Ptolemy (which he called 'geography' and 'chorography'), and Merula, Christiani and Göllnitz appear to have taken up the bipartite classification.[9] Varenius is firmly rooted in the tradition, which he has critically surveyed, yet his purpose

[5] For Merula and Christiani see A. Philippson, „Zwei Vorläufer des Varenius," *Ausland* 52 (1892); and the critical remarks by Josef Schwerdfeger, „Bernhard Varenius und die morphologischen Capitel seiner *Geographia generalis* (Amsterdam 1650): Ein Beitrag zur Geschichte der Geographie", Offprint from *Jahresberichte des K.K. Staatsgymnasium in Troppau*, 1897/1898. Clüver's *Introductionis in universam geographiam* of 1624 was widely distributed, even though its general part was not up to date on the level of scholarship. Clüver's lasting achievement was in what Varenius would call special geography, the description of a country (but in Clüver's case including a special interest in ancient history). See Gottfried Lange, "Philipp Clüver (1580–1622)," *Geographisches Taschenbuch* 1962/1963, 251–258.

[6] See Lange 1962/63, Baker 1963: 11.

[7] William Warntz, "*Geographia Generalis* and the Early Development of American Academic Geography," *Origins of American Academic Geography*, ed. Brian Blouet, Hamden CT 1981, 245–263.

[8] M. Kießling, "Varenius und Eratosthenes," *Geographische Zeitschrift* 15 (1909): 12–28.

[9] Abraham Göl(l)nitz had his *Compendium geographicum* published by Louis Elzevier in 1643 and 1649. See Dietrich zum Felde, „Die Abhängigkeit der Kenntnisse und Auffassungen des Varen vom klassischen Altertum," *Geographische Zeitschrift* 38 (1932): 13–22 [based on his dissertation Heidelberg 1921].

is to make use of his materials, old and new, in laying reliable foundations for future application and in presenting a systematised concept of geography. In his study of 1981 Warntz (258f) gave an estimate of the number of sources that were used for the *Geographia Generalis*. From Varenius' original citations of about 300 authors, the figures in Newton's edition of 1712 have increased to almost 400, and by 1733 Dugdale and Shaw list about 450 authors (they added eighteenth-century authorities in geographical studies).

Varenius' concept of geography was far ahead of his time and could therefore dominate the discourse for over a hundred years, not least because, as Elizabeth Roberts has pointed out, the didactic design of his book made it easy to add to it and update it. How its influence spread across Europe and beyond is a subject we cannot possibly exhaust. My remarks will be restricted to what happened in the Anglo-Saxon world.

Daniel Elzevier, Louis' successor, reprinted the *Geographia Generalis* in 1664 without any reference to the author (personal notes about the author were rarely given by the printers anyway), and in 1671 another edition based on the version of 1664 followed in which nothing but the engraved title was changed. After 1671 the Elzeviers were no longer involved in the distribution of Varenius' works. (A fourth Elzevier edition of 1672, mentioned by Baker and others following him, appears to be a ghost.)

The Geographia Generalis *in England*

At the time of Daniel Elzevier's final activity on Varenius' behalf, the *Geographia Generalis* was seriously studied in Cambridge by Isaac Newton who in 1672 edited it as a textbook for students with the original author's unchanged dedication; nowhere did Newton show his hand in the text. When he did make a few corrections or additions, he did so as unobtrusively as possible. Richard Westfall and other biographers of Newton have pointed out again and again how shy Newton was of publicity; he was a perfectionist who had to be begged and sometimes tricked into parting with his manuscripts for printing. Completely unselfish himself, he discreetly put himself at the service of another author. Newton was Barrow's successor as Lucas professor and would have been required by the statutes to teach geography accordingly, but whether he actually did so is uncertain. The fellows and students of Cambridge and Oxford were not very studious at the time and not many students would attend lectures, yet Varenius' book was available for teaching and studying purposes. One of the students who remembered his study of Varenius all his life was the antiquary and Newtonian physician William Stukeley (1687–1765), who was admitted to Cambridge University as an undergraduate in 1704 when he was taught geography with Newton's edition of Varenius,[10] and in his *Memoir of Sir Isaac Newton's Life* of 1752 he noted that "he [Newton] put out a new edition of that curious piece, Varenius's geography. the method of that author pleasd him; it is like his own.",[11] or, as he phrased it elsewhere in 1727, "whose method he has some measure imitated in his own works".[12] Stukeley's impression of Newton's high regard of Varenius for his clarity may partially explain why Newton did not interfere much with the text and only added the schemes.[13]

This Latin version must also have had readers outside the universities, because a second edition was published in 1681 and again by the Printer to Cambridge University. It was widely distributed, also in Germany where it was reprinted in 1683 by Heinrich Christoph Cröker (Crokerus) in Jena.[14] But no German translation was ever undertaken in Germany, and to this day a full version of the *Geographia Generalis* in German is unfortunately still lacking. In England, however, Latin as the lingua franca began to be less common when books were no longer primarily aimed at an exclusive readership. English translations would reach a wider audience, which is why Richard Blome had long been thinking about an ambitious project in English. He explained it in the preface to *A Geographical Description of the Four Parts of the World Taken from the Notes and Works of the Famous Monsieur Sanson* (1670):

> The next Vollume to be finished will contain the Arts of COSMOGRAPHY and GEOGRAPHY, being (in Part) a Translation from the Works of the eminent and much esteem'd Geographer VARENIUS, wherein are at large handled all such Arts as are necessary to be understood in the true

[10] David Boyd Haycock, *William Stukeley: Science, Religion and Archaeology in Eighteenth-Century England*, Woodbridge 2002, 37f.

[11] Royal Society Ms. 142.

[12] King's College Cambridge, Keynes MS 136.

[13] Cf. also John Conduitt's remarks (Keynes Ms. 130.5, 1v), quoted by Mayhew below. (These Newton references are available online, published by The Newton Project, Imperial College London.)—Cf. Mayhew's assessment of this issue below.

[14] His German name is actually Cröker (not Cröck, as Lange assumed). See his title pages in Bircher, *Deutsche Drucke des Barock 1600–1720.*

> knowledge thereof: To which shall be added the much wanted Schemes omitted by the Author: And for the further compleating the same, shall be added about 100 GEOGRAPHICAL and HYDROGRAPHICALL TABLES of all the Remarkable and known-Kingdoms, Countreys, and Isles in the World, with their Chief Citties and Sea-Port-Towns...

He planned a hydrographic description of the world with sea maps and finally a volume on the British Isles.[15] The combination of Sanson with Varenius was actually realized in 1682, and the title page reads *Cosmography and Geography in Two Parts*, but his further plans are no longer mentioned in the otherwise unaltered preface. (*Britannia* did appear in 1673.) Blome's dedication indicates how well known Varenius was by that time: "Here are the United Labours of Varenius and Sanson, Two so well known Authors / to the Learned and Ingenious World, that they need no other Encomium, than to name them." Other editions of Blome's *Cosmography and Geography* were published in 1683 (dedicated to Charles II) and 1693 ("The Third Edition", which retained the visual tribute to the king and the "CR" mentioned by Mayhew below), but the title page to Sanson's second part remained unchanged as it was in 1680. Richard Blome was interested in geographical works and in travel literature and used to respond to what was in the air. (For instance, he imitated John Ogilby's publishing concepts in this respect and tried to surpass him by realizing them more quickly.)

Newton had nothing to do with Blome's editions in London, and they would not have satisfied his standards of perfection and accuracy anyway. James Jurin's version of Newton's version is a different matter, for Jurin, an ardent Newtonian, was a young scholar and a scientist. But as far as we know, Newton did not have a hand in Jurin's version of his text either (which would have been easy for him, had he wanted to). Jurin had been a student at Trinity College Cambridge (when Newton had already left for London), then spent some time in Leiden and afterwards became master of Newcastle grammar school where he lectured on mathematics and mechanics; later he became an esteemed member of the Royal Society and eventually president of the

[15] In justifying the latter project he states that Peter Heylin, Johnson (probably Robert Johnston) and Blaeu either gave incomplete maps and tables or did not use 'our' language. Here as in the projects of the Royal Society we see that the systematic survey of England is undertaken and documented in the context of 'Englishness', a very topical subject today. On their agenda were the country's topography and all human traces in history and prehistory (cf. the ambitious projects of John Ogilby and John Aubrey).

Royal College of Physicians. Jurin died exactly 100 years after Varenius. When still at Newcastle, he wrote a letter to Richard Bentley, Master of Trinity College, Cambridge. Bentley replied from Westminster on 7 June 1711:

> ...I shall here give you my advice, that by all means you would publish your supplements to Varenius; [...] and when your papers are sent down to Cambridge I shall either read them over my self [...or Dr. Syke...] & correct ye few Escapes of ye Stile and for ye Matter you will take sufficient care.[16]

Bentley was another keen supporter of Newton's "natural philosophy"; he was all-powerful in Cambridge and completely controlled the scene. Although he was intelligent and demanding, he was also dogmatic in his opinions; his colleagues found him unbearable. So luckily for Jurin, who was 22 years younger and at the beginning of his career, he enjoyed Bentley's blessing and could go ahead, and his appendix to Newton's edition of Varenius appeared in 1712. It was Jurin's first publication, dedicated, of course, to Bentley. Both his dedication as well as his preface to the reader that introduces his appendix were translated into English by K.D. Hargin.[17] As Warntz points out,[18] Bentley was keen on having Varenius' *Geographia* presented in the light of Newton, not of Descartes, but Jurin continued Newton's work on Varenius and added new findings without being critical of Descartes.[19]

Jurin's edition was not reprinted in England, and for more than twenty years no other edition of Varenius was published there. Only after Newton's death was there another initiative to present his *Geographia* to the public when Stephen Austen published it in London in 1733, but by then the time for a Latin edition was past and it appeared in English under the title of *A Compleat System of General Geography.* Two men were responsible for this new version and its additions and corrections, the physician Dr. Peter Shaw and one Mr. Dugdale who did the translation. So far nobody has been able to discover who the translator was. On the title page he is respectfully called by his last name only, which may suggest that this Mr. Dugdale was a member of the well-known Dugdales who in the seventeenth century were heralds at

[16] *The Correspondence of James Jurin (1684–1750): Physician and Secretary to the Royal Society*, ed. Andrea A. Rusnock, Amsterdam 1996, 62f.

[17] Reprinted in Warntz 1989: 184f.

[18] Warntz 1989: 183.

[19] Warntz 1989: 186.

the Herald's College in London. Peter Shaw (1694–1763) was one of those scientists, like Varenius, who were both natural philosophers and physicians. In his later life he became physician in ordinary to George II and George III, became a member of the Royal Society and made a name for himself through his editions of Francis Bacon and Robert Boyle. This English version was, as Mill saw it in 1905, "largely spoilt by stupid notes and interpolations".[20]

As in the case of Newton and Jurin, Shaw's study of Varenius marks a point of departure for him in his scientific work, since these men at the beginning of their careers evidently felt the need to assure themselves of a foundation of knowledge, which they then intended to impart to the public. Stephen Austen, the publisher, made a success of it, because the first edition of 1733 was succeeded by a second one in 1734 and its reissue in 1736. The last issue of this version by Newton/Jurin/Dugdale-Shaw was published in 1765, and it ended the transmission of Varenius' work in an independent edition.

Thanks to Newton and his successors, Varenius became an export article in the eighteenth century, because many translations were based on their versions. The Dutch translation of 1750 is based on the Dugdale-Shaw version and critically consults Jurin's Latin edition to make it as accurate as possible. In France, where Varenius was mainly transmitted through Sanson,[21] a French translation in four volumes was made by Philipp-Florent des Puisieux in 1755. In Naples Varenius was reprinted in Jurin's version in 1715, and in Moscow a Russian translation was published in 1718 by command of Tsar Peter the Great.

The Geographia Generalis *in North America*

William Warntz pointed out in 1981 that Varenius was used in the colleges of the American colonies long before the War of Independence and held an important place in their curriculum for a century and a half. Although Varenius was studied there first in Latin as edited by Newton and afterwards by Jurin, the English translation by Dugdale

[20] Hugh Robert Mill, "The Present Problems of Geography," *The Geographical Journal* 25 (1905): 3.

[21] Lange, "Das Werk des Varenius" (1961), 13.

and Shaw became the 'modernised' standard text book in the colleges.[22] Astronomy and geography as "mixed mathematics" could be taught by (pure) mathematicians as well as by natural philosophers in all their diversifications.[23] In the colonial period the colleges continued the European heritage of the enlightenment, based on natural law and the natural rights of the individual.[24] When in 1756 William Smith set up the curriculum for the University of Pennsylvania, he made Varenius required reading for first year students of theology, and Smith's influence helped to spread it to Columbia and elsewhere.[25] In the universities of New England, where Varenius had become the standard text book for geography, Jedidiah Morse (1761–1826), whose achievements between 1784 and 1802 made him the "father of American geography", was an important transmitter. In *The American Universal Geography* (1793) he took up Varenius' ideas and repeated many of his words and phrases, even though Morse never made his borrowings explicit, as Wright explained in 1955.[26] Morse developed Varenius' ideas, especially those on special geography which rely on experience rather than mathematical models; they were to be put to use in describing his own country.[27]

Thomas Jefferson, himself a knowledgeable geographer, doubtless came across Varenius' ideas by direct or indirect channels. As John Allen, the expert on the Lewis and Clark Expedition, wrote in an email on 10 Dec. 2001, "Varenius is mentioned in several items of correspondence between Jefferson and other scientists in Europe and America between 1787 and 1808 (source: microfilm of Jefferson's entire correspondence, 1784–1824)." Varenius is, however, not listed among his books of 1783; Jefferson may have owned a copy before the disaster when his first library was destroyed by fire in 1770. He then built up another library with great care, which he sold to the Library of Congress in 1815. Among the long list of geography books he made in his own hand (catalogue of

---

[22] Warntz 1981: 255.

[23] Warntz 1981: 251.

[24] Warntz 1981: 253.

[25] Warntz 1981: 66f.

[26] J.K. Wright, "Some British 'Grandfathers' of American Geography," *Geographical Essays in Memory of Alan G. Ogilvie*, London 1959, 155f.

[27] Geoffrey J. Martin, *All Possible Worlds*, New York 1993; and his "The Emergence and Development of Geographic Thought in New England," *Economic Geography* 74, special issue (1998): 1–13.

1783) we find Ptolemy, Strabo, Cluverius, Pomponius Mela, and many others; Jedidiah Morse's *American Geography* is also listed.[28]

Jefferson began his *Notes on the State of Virginia* in 1781, and in the process of his work they were continually enlarged and eventually published in London in 1787, in Philadelphia in 1788. The survey follows the methodology of question and answer as refined by the Royal Society (among them John Aubrey, Robert Hooke).[29] The list of topics Jefferson discusses shows much the same scheme as Varenius had outlined it (in *Geographia Generalis*) in his classification of what special geography should deal with, starting with the boundaries and moving on to rivers, sea-ports, mountains, cascades, productions (mineral, vegetable, animal), climate, population, and on to political aspects, religion and manners. In the context of Jefferson's ideology of nation building and expansion, the exploration of the west to the source of the Missouri River and the Pacific became a task that he had on his agenda for a long time. After years of planning such an expedition he sent Meriwether Lewis to Philadelphia in 1803 to study natural sciences under distinguished professors and to learn the use of instruments for determining positions. When Jefferson had finally brought the Louisiana Purchase to a close (it was signed in Paris on 2 May 1803), the plan was made public and the Lewis and Clark Expedition was set up formally in Jefferson's letter to Captain Meriwether Lewis of 20 June 1803, which instructed him in great detail as to his mission and what information he was expected to collect and bring home. The range of topics is wide, and particular emphasis is in human geography in the interest of trading with the natives. Exploring the unknown west, taxonomy, mensuration and naming became political instruments of conquest and taking possession. Direct evidence that Varenius was one of the sources consulted by Jefferson in preparing the mission has not

[28] James Gilreath and Douglas L. Wilson, eds., *Thomas Jefferson's Library: A Catalog with the Entries in His Own Order*, Washington DC 1989 [full-text version also online through Library of Congress].

[29] See, for instance, John Ogilby's twenty-two "Queries in Order to the Description of *Britannia*" of 1673, Bodleian Library Oxford, MS Aubrey 4, f. 244 (reproduced in J.B. Harley's "Bibliographical Note" to the facs. edition of Ogilby's *Britannia, 1675*, xiii). They were drawn up and discussed by members of the Royal Society after the Cambridge editions of Varenius had already become available. (Newton was not involved in the discussion of these queries, but he had become a member of the Royal Society on 11 January 1672.) Hooke's diary informs us that he was personally acquainted with Pell and Isaac Vossius, and that he "Received Maginus, Varenius Japan" on 18 September 1678.

been found so far.[30] As it turned out, the Lewis and Clark Expedition became a huge success, and the explorers brought back a wealth of information on their return in 1806.[31]

Another explorer, Alexander von Humboldt, visited Jefferson in Philadelphia in May 1804, and they became close friends; Humboldt had access to the president's rooms at any time. Geography was a favourite subject with them. Humboldt's fine tribute to Varenius, expressed long after his visit to America, is well known.

It is interesting to see how the meaning of geography changed. In colonial times geography was important in the curriculum and for thinking on an international scale, but when America in the period before the Civil War receded into isolation, it was no longer important as a subject in the colleges and led to specialised descriptions of the huge country in an encyclopaedic manner.[32] In the romantic period scientific thinking and the romantic imagination merged, and explorers like Humboldt, Lewis and Clark were also seen as questers, epic heroes whose heroic struggles roused the enthusiasm of an admiring public.

---

[30] John Allen in a preliminary search found no index entries to Varenius in the collected works of Lewis and Clark papers and related documents, done by the American documentary editor Donald Jackson, nor in the voluminous Lewis and Clark Journals, edited by Gary Moulton.

[31] On Jefferson and the Lewis and Clark Expedition see int. al.: Jefferson, *Writings*, New York 1984 for "Notes on the State of Virginia" and Jefferson's Letters; Malcolm Kelsall, *Jefferson and the Iconography of Romanticism: Folk, Land, Culture and the Romantic Nation*, London & New York 1999, ch. 3; *The Journals of...Lewis and Clark*, ed. Biddle, 2 vols., New York 1962, with Jefferson's relevant letters of 20 June 1803 and 18 August 1813 reprinted in the introduction.

[32] Warntz 1981: 261.

FROM HACKWORK TO CLASSIC: THE ENGLISH EDITING OF THE *GEOGRAPHIA GENERALIS*

Robert J. Mayhew

Abstract

Von der Lohnschreiberei zum Klassiker: Die englische Herausgabe der Geographia Generalis

Seit der Mitte des 18. Jahrhunderts zieht Bernhard Varenius' *Geographia Generalis* bis heute das breit gefächerte Interesse von Historikern der Geographie auf sich. Alexander von Humboldt lobte Varenius in seinem *Cosmos*, Fred Lukermann und andere Anhänger einer Geographie als quantitativer Wissenschaft beziehen sich auf ihn als Vorgänger, und die allgemeine Fachgeschichte von Preston E. James und Geoffrey J. Martin bis David N. Livingstone erinnert an seine Vision der Geographie. In diesem Aufsatz wird die These vertreten, dass Varenius' kanonischer Status in der Geschichte der Grundgedanken über die Geographie sich nicht nur aus der philosophischen Kompetenz seiner Arbeit ableitet, nicht einmal aus der Verbindung seiner Arbeit mit Isaac Newton; vielmehr wird hier vorgeschlagen, dass zwei verwandte Besonderheiten der Druckgeschichte seiner englischen Ausgaben im 18. Jahrhundert diesen Status erklären: Die *Geographia Generalis* wurde veröffentlicht als *A Compleat System of General Geography* (1733; erweitert und verbessert 1743, 1736 und 1765). Zum einen wurde Varenius hier als ‚klassischer' Text ediert, und zwar genau in der Zeit, als man mit Shakespeares Dramen genauso verfuhr. Varenius wurde nicht mehr plagiiert und stillschweigend emendiert (wie es in der vergessenen Übersetzung von Varenius 1682 durch Richard Blome geschehen war), sondern er wurde als Persönlichkeit gewürdigt, dessen Worte zu bewahren und der Nachwelt intakt und unverändert zu überliefern waren. Zum zweiten, und als Folge davon, war das Druckformat von *A Compleat System* anders als in den Geographiebüchern vor dieser Zeit, denn es wurde mit ausführlichen Fußnoten und Scholien versehen. Die stoffliche Form von *A Compleat System* hebt das Werk hervor als einen geographischen Text von noch nie dagewesener Bedeutung, das mit besonderer Ehrfurcht zu behandeln war. Diese Hochachtung wird ihm in der Tat seit mehr als 200 Jahren bezeugt. Daher werden hier die Ausstattung des Drucks und die Herausgabe des Textes als Schlüssel für die Konstruktion der Identität und Geschichte der geographischen Disziplin gesehen.

The Canonization of Varenius

Bernhard Varenius' *Geographia Generalis* has been accorded an extraordinary status in the history of geographical thought, especially since his work was singled out for praise by Humboldt and Vidal de la Blache.[1] The critical reception of Varenius from Humboldt to the present has emphasised his modernity; that his mathematically grounded form of geographical enquiry leads us into a version of the discipline recognisable to geographers practicing today. The claim is put in its starkest form in an excellent paper by Fred Lukermann; "the intellectual divide separating ancient and medieval geography from modern geography is generally taken to be the publication of the *Geographia Generalis* by Bernhard Varenius."[2] Likewise, the much-respected 1911 edition of the *Encyclopaedia Britannica* (sub "History of Geographical Theory") enshrined the judgement that "the treatise of Varenius is a model of logical arrangement and terse expression; it is a work of science and of genius; one of the few of that age which can still be studied with profit". This argument has been strengthened by Varenius' linkage with Isaac Newton, Newton having edited the *Geographia Generalis* in 1672. The upshot of this is that the name of Varenius has become a signifier for the signified of 'scientific geography', a status clearly evidenced in the decision of the US's National Center for Geographic Information and Analysis to name its new project to advance GIS 'Project Varenius'.

In sum, a huge rhetorical weight has been placed upon the shoulders of Varenius. He is seen as ushering in or prefiguring or standing as a symbol for geography as a science, that is both mathematical and numerical, while the *Geographia Generalis* is further depicted as the most popular textbook in seventeenth and eighteenth century European and North American geographical education.[3] This paper suggests that this canonization of Varenius begins in the eighteenth century and that it

---

[1] Alexander von Humboldt, *Cosmos: A Sketch of the Physical Description of the Universe*, vol. 1, transl. E.C. Otte, introd. Nicholaas Rupke, Baltimore 1997, 66–67; Paul Vidal de la Blache, "Le Principe de la Géographie Générale," *Annales de Géographie* 5 (1895–6): 129–42.

[2] Fred Lukermann, "The *Praecognita* of Varenius: Seven Ways of Knowing," *Text and Image: Social Construction of Regional Knowledges*, eds. Anne Buttimer, Stanley D. Brunn and Ute Wardenga, Leipzig 1999, 7.

[3] Preston E. James, "On the Origin and Persistence of Error in Geography," *Annals of the Association of American Geographers* 57 (1967): 21; William Warntz, "*Geographia Generalis* and the Early Development of American Academic Geography," *Origins of American Academic Geography*, ed. Brian Blouet, Hamden CT 1981, 245–263.

can be discerned by looking at the ways in which the text was edited in England.

The English Print History of Varenius: Quantitative Issues

Varenius' *Geographia Generalis* had a long and involved print history in England, spanning from Newton's Latin edition for Cambridge University Press in 1672 to the last impression of the Dugdale-Shaw English translation in 1765. This is clearly a print life far longer than that accorded to most English geography books, although it is by no means unique; in the realm of digests of special geography, for example, both Patrick Gordon's *Geography Anatomiz'd* and William Guthrie's *New Geographical, Historical and Commercial Grammar* went through new editions and impressions for a similar span of time.[4]

The two 'editions' produced by Cambridge University Press in 1672 and 1681 as edited by Newton have identical pagination and only seem to have changed in their typeface and the correction of a few typographical errors. There is no evidence to indicate that they had exceptional print runs: David McKitterick[5] suggests a run of 1000–1200 copies as a maximum, something which fits well with the recorded fact that the third edition to emerge from Cambridge, that annotated by James Jurin in 1712, had a print run of 1000 copies, an entirely standard print run for a mathematical-cum-scientific text with a potential student audience.[6]

The two major English translations of the *Geographia Generalis*, that by Richard Blome (1682 and 1693) and that by Dugdale and Shaw (1733, 1734, 1736 and 1765) reinforce the impression which the Latin editions yield of a work of some but by no means exceptional popularity. Blome entered a copyright claim on a translation of Varenius in the Stationers' Company register in 1663, renewing this in 1668.[7] Yet it was only in 1682 that this translation appeared in print as part 1, "The General and Absolute Part," of *Cosmography and Geography*, conjoined to Nicolas

[4] O.F.G. Sitwell, *Four Centuries of Special Geography*, Vancouver 1991.

[5] Private communication 3 Dec. 2004.

[6] Donald McKenzie, *The Cambridge University Press, 1696–1712: A Bibliographical Survey*, 2 vols., Cambridge 1966, 1: 340f, 100.

[7] Henry Plomer, *A Dictionary of Printers and Booksellers at Work in England, Scotland and Ireland from 1668 to 1725*, Oxford 1922, 39.

Sanson's *Geographical Description of the World*, an English translation which Blome had first issued in 1670. Blome's 1693 'edition' of Varenius again appears with his Sanson translation and is clearly a reissue of the 1682 project, to which country maps of England derived from John Speed have also been appended. Whereas the dedication was changed in 1693—perhaps unsurprisingly given that in 1682 the whole had been dedicated to the Duke of Abermarle, who had staunchly supported James II and was now dead—the print ornaments which still bear the Royal Crest and the initials 'CR' (Carolus Rex) point to a product designed to flatter a monarch long since dead and from a line deposed five years previously. This, of course, was not treason but a result of the exigencies of Blome's need to clear his stocks and recoup his costs on a project which had manifestly failed to sell its print run. In short, we can infer that Blome finally published his translation of Varenius to help clear stocks of his Sanson translation, but that this project itself failed to meet his expectations, leading to its reissue in 1693.

A similar print history appears to hold good for the Dugdale-Shaw translation, which first appeared in 1733 and is—unlike Blome's project—an edition in its own right, not part of a broader bookseller's strategy. A second edition appeared in 1734, this clearly being a 'real' edition, in the sense that it contains substantial amounts of new material, including new footnotes and substantial additions to the text such as a collection of geographical paradoxes taken from Patrick Gordon's *Geography Anatomiz'd* and a long discussion of tidal flows.[8] Subsequently two further 'editions' of the Dugdale-Shaw translation appeared, in 1736 and 1765, but in both cases they are reissues of the second, 1734, text, presumably designed to shift remaining copies. In both cases the text is identical, and whilst the 1765 'fourth' edition has a new publishing group—L. Hawes, W. Clarke and R. Collins—for the simple reason that Stephen Austen, publisher of the first three editions, died in 1750,[9] it still refers in the text to 1734 as the year of composition[10] and retains a flyleaf on the final page advertising Austen's other titles. In short, it would appear that a successful first edition in 1733 led to

[8] Varenius 1765: 779ff and 811ff.

[9] Plomer et al., *A Dictionary of Printers and Booksellers who were at Work in England, Scotland and Ireland from 1726 to 1775*, Oxford 1932, 10; *British Book Trade Catalogue*, 2005: sub "Austen" (http://www.bbti.bham.ac.uk/Details.htm?TraderID=112714. Accessed 12 July 2005).

[10] Varenius 1765: 839 [mispaginated as 339].

the rapid development of a second, expanded edition of 1734, which then proved hard to shift and was still being cleared when Hawes *et al.* purchased the property rights on the death of Stephen Austen.

Simply put, there were three main versions of Varenius in English print culture, the three Latin editions from Cambridge University Press, the two issues of Blome's translation, and the four issues of two editions of the Dugdale-Shaw translation. It would appear that the Latin editions from Cambridge sold best of all, in good part perhaps because their audience was not restricted to the Anglophone world[11] and because it was required reading in Cambridge, yet in each case it would be implausible to detect a huge demand amongst students or the learned. General claims made by historians of geography about Varenius being a standard textbook clearly do not hold good in the English speaking world, if that means it was unusually widely available or exceptionally profitable as a piece of intellectual property.

The English Editing of Varenius: Qualitative Issues

Yet another element of the print history of English editions of Varenius suggests that the *Geographia Generalis* was indeed a pivotal text in the history of geographical thought. In short, the ways in which Varenius came to be edited in the century over which his text was in circulation in the Anglophone world shows that he came to be accorded a special—if not quite a unique—place in the geographical culture of the time. Varenius started as an author whose text was edited and reissued as any other author might be in the contemporary Grub Street culture of English geographical writing and publishing,[12] but in the course of time his work came to be accorded a special status wherein his authorial voice was carefully preserved, even when it was being corrected or questioned in the light of Newtonian natural philosophy. The result of this editing was that the print space of the *Geographia Generalis* came to resemble that of 'classic' texts—ancient and vernacular—printed in England during this period, an honour accorded only to Varenius amongst English geographical textbooks, and perhaps only to Varenius

[11] David McKitterick, *A History of Cambridge University Press*, 2 vols., Cambridge 1992 and 1998, vol. 1, 381ff; vol. 2: 41, 82, 86.

[12] Robert Mayhew, "The Character of English Geography, c. 1660–1800: A Textual Approach," *Journal of Historical Geography* 24 (1998): 385–412.

and Camden amongst the broader gamut of works of geographical knowledge at this time.

Varenius as Hackwork: Newton and Blome, 1672–1693

At the start of the *Geographia Generalis*'s print history in England, it was not accorded any special status, something we can see clearly if we look at the two Cambridge editions attributed to Newton and in particular at Blome's translation of that edition. Whereas, as stated at the outset, much commentary has been devoted to Varenius—and thereby geography—having a direct link with Isaac Newton, the verifiable facts about Newton's connection with the text are rather less impressive. While Warntz claims Newton's editorial contributions "were many and substantial", he only in fact points out that Newton added a series of 31 figures or schemes, promised in but lacking from the Elsevier edition of 1671, and in addition maintains that Newton corrected grammatical and typographical errors and perhaps reworked a few paragraphs.[13] In fact, from Newton himself we have only the assertion that "the Book here in Presse is *Varenius* his Geography, for which I have described Schemes",[14] something which he also imparted to others: According to John Conduitt the book's title is misleading, for "Sir I.N. told Mr Jones all he did in the edition of Varenius before wch is put Curante Isaaco Newtono was to draw the schemes".[15] Newton may well, as Warntz asserts, have cleaned up the text to a greater extent than merely by adding figures, but it is also possible that a vigilant corrector at the press did much of this work.[16]

Although the details of Newton's input to the 1672 edition cannot be ascertained with certainty, it is clear that even on Warntz's reading he only tinkered with small elements in the text. Above all, Newton chose not to correct Varenius, nor did he take on what he would have seen as manifest errors in Varenius' trigonometry,[17] nor in his Cartesian

---

[13] William Warntz, "Newton, the Newtonians, and the *Geographia Generalis Varenii*," *Annals of the Association of American Geographers* 79:2 (1989): 178–81.

[14] Isaac Newton, *Correspondence*, vol. 1: *1661–1675*, ed. H.W. Turnbull., Cambridge 1959, 161.

[15] Newton, *The Mathematical Papers of Isaac Newton*, vol. 2: *1667–1670*, ed. D.T. Whiteside, Cambridge 1968, 288f, n. 43.

[16] See William St Clair, *The Reading Nation in the Romantic Age*, Cambridge 2004.

[17] Newton 1968: 10, n. 28.

natural philosophy (although we are still, of course, over a decade before Newton's *Principia* so that we cannot know the full depth of his disagreement with Descartes at this time). As such, Newton's editorial labours were not great and the Varenius edition, as it was Newton's first published book,[18] did not involve any great input of mental or editorial labour on his part. Thus, while Newton intended the text for students at the Cambridge of the 1670s and 1680s, the text retained the original dedication to eight Dutch dignitaries, dated 1650, which could hardly have been relevant. Furthermore, if Newton hoped this text would appeal to students, he did not choose to include even basic finding devices to help them: there are no indices and no marginal glosses describing the progress of the argument. In an era of burgeoning methods to aid students in searching and organising data to keep 'information overload' at bay,[19] Newton's *Geographia Generalis* was strangely spartan, a text with no editorial signposts to aid the reader. In this, Newton was following his original, but hardly pitching the text at its purported audience.

To the extent that Newton or his corrector did update the text, they produced as much confusion in Varenius as they rectified. Thus, a number of the branching diagrams which adorn the book do not actually replicate the claims made in the text itself. Looking at Book 1 chapter 8, for example, the text tells us that the world has ten islands which fit into the category 'large', whilst the supporting tree diagram tells us of 'magnas novem' and duly lists nine.[20] Clearly, whatever editing Newton did to the text, it was 'silent' in that we are not alerted to the fact that he is departing from the original text and furthermore it was partial, in that the text and diagrams do not always tally, alterations not having been worked through the entirety of the text for consistency. In short, as a project aimed at updating Varenius for a student audience, Newton's edition of 1672 and 1681 was really rather deficient: it lacks finding aids for the student, its updating of basic mathematical propositions is partial and its consistency where points are adjusted is uneven. As such,

[18] A.R. Hall, "Newton's first book," *Archives Internationales d'Histoire des Sciences* 13 (1960): 55–61.

[19] See Ann Blair's three articles: "Annotating and Indexing Natural Philosophy," *Books and the Sciences in History*, eds. Marina Frasca-Spada and Nick Jardine, Cambridge 2000, 69–89; "Reading Strategies for Coping with Information Overload, ca. 1550–1700," *Journal of the History of Ideas* 64 (2003): 11–28; and "Note Taking as an Art of Transmission," *Critical Inquiry* 31 (2004): 85–107.

[20] Varenius 1681: 52f, 48.

Newton's edition neither cleaves to the authorial original, nor does it thoroughly update it; above all, it does not make clear which policy is being adopted at any time by way of textual or paratextual signposts. While one would hesitate to call the result 'hackwork', one would have to admit that the occasional nature of the project—to produce a workable student edition—and Newton's own low key comments about his input to the Cambridge Varenius seem borne out by the text itself.

Still more orientated in the direction of hackwork was publisher Richard Blome's English edition of Varenius, published as the first part of *Cosmography and Geography* in 1682. This project looks much grander than Newton's Cambridge editions, an 18 cm octavo being replaced by a 35 cm folio. Furthermore, the text also uses more print ornaments such as fancywork and initial drop lettering and contains marginal glosses of the progress of the argument as a finding aid. Yet in most respects Blome's approach to Varenius was that of a hack publisher, struggling to sell off copies of a floundering geographical project, his Sanson, which was first published in 1670 and was here repackaged as the second or special part of *Cosmography and Geography*. By 1693 and the 'third' impression, Varenius and Sanson were complemented by a set of county maps of England derived from John Speed. This creates an unavoidable mishmash, wherein Blome appears to bring together several of his geographical and cartographical publishing ventures in the hope of selling them all off by yoking them together.

With respect to Varenius specifically, Blome's translation appears fairly faithful to the Latin text as received from Newton's 1672 edition. The unfortunate result of this is that Blome also uncritically takes over and perpetuates the sorts of editorial inconsistency which we have already highlighted as present in Newton's edition, including for instance the nine large islands in his diagrams, while retaining reference to ten (indeed eleven, with "California, if that be an Island").[21] Clearly, the translator (unnamed) was as casual about consistency as the Cambridge edition he used as his base text, retaining the single voice and silent alterations which created inconsistency.

In fact, Blome's edition was more confusing and contradictory than was Newton's, because of the way in which it spliced together two separate projects, Sanson's and Varenius'. Varenius, of course, was

[21] Richard Blome, *Cosmography and Geography*, third impression, London 1693, 37, 34.

distinguished by his lack of interest in special or descriptive geography in the *Geographia Generalis*, yet Blome then conjoined this with a long special geography of the world by Sanson. Thus we have Varenius stating he only deals with political or special geography because "we must yield somewhat to Custom and the Profit of Learners",[22] while the book itself—in the form of Sanson's *A Geographical Description of the World*—in fact devotes 493 folio pages to this topic! More generally, the splicing together of the two texts creates a confusion, when reference is made in the Varenian sections to 'special geography', as to whether this refers the reader to special geography inasmuch as it is treated in the *Geographia Generalis* or to Sanson's *Geographical Description* as appended.

Finally, and in a move characteristic of English hack geographies, Blome also expanded the British sections of Sanson's description so as to make the text more directly appealing to its English audience. Thus, where the other nations of Europe receive 10 to 20 folio pages of description, Britain covers some 108 pages. The British section is clearly taken from Camden's *Britannia*, following his division of Britain according to the kingdoms of the Saxon heptarchy and then into modern counties, hundreds and parishes. This was no doubt itself excerpted from Blome's earlier publication, *Britannia* (1673), which William Nicolson described in his *English Historical Library* as "a most entire piece of theft out of Camden and Speed".[23]

By 1682, then, Blome's *Cosmography and Geography* conjoined Varenius with Sanson, despite the contradictory nature of their geographical projects, further failing to correct inconsistencies in Newton's Varenius and inserting material he had already lifted from Camden for other purposes into Sanson's special geographical description of the world. Taken as a whole, Varenius was clearly positioned as just another geographical author ripe for plagiarism, silent emendation and sloppy adjustment in the pursuit of profit, or more probably in this case in the avoidance of loss. *Cosmography and Geography*, as a polyvocal text whose inconsistencies had not been ironed out due to indifference and haste on the part of its anonymous editor(s), places Varenius in a hackwork project *par excellence*. Here we see little of that reverence for Varenius as a canonical author which we have seen enshrined in modern histories

[22] Blome 3.

[23] *Dictionary of National Biography*, London 1885–1900, sub "Blome, Richard".

of geographical thought. The transition in Varenius' status comes in the annotated editions produced by Jurin and by Dugdale and Shaw.

Varenius as a Latinate Classic: Jurin's 1712 Edition

In many—indeed perhaps most—respects, James Jurin's 1712 Latin edition of the *Geographia Generalis* for Cambridge University Press continued the editorial indifference of the Newtonian editions which it was designed to update. Jurin's text remains in 511 octavo pages and retains the inconsistencies we have noted previously in Newton's edition. In short, the bulk of the book is simply a reissue of the earlier work, presumably to meet continuing student demand in Cambridge and beyond.

And yet Jurin's edition can also be seen as marking a new stage in the English reception of Varenius. More precisely, Jurin's edition begins for the first time to treat Varenius' work as a modern classic of geographical writing, which needs to be respected even as it is emended. Where both Newton and Blome had been happy to adjust the text silently and had not been too concerned about the textual inconsistency resulting from such interventions, Jurin is more careful in these respects.

Before we look at Jurin's editorial labours themselves, it is worth noting that the 1712 edition, while in most respects identical with that of 1682 in pagination and the like, is at least reset in a far clearer typeface, part of Cambridge University Press's policy of using Dutch printers and typefaces.[24] Furthermore, the edition was adorned with ornamental printwork by the Dutch engraver Simon Gribelin, whose skills were also deployed on other prestigious projects such as Richard Bentley's edition of Horace.[25] Finally, Newton's figures or 'schemes' were redistributed onto 8 leaves as opposed to the original five, the result being that these figures were far more easily consulted by the reader. As part of the campaign to make Cambridge University Press's editions respected in the *respublica litterarum*, there had been a drive to emulate Dutch typography, so that the 'look' of the print space appeared to be of the highest standard, and Jurin's edition clearly benefited from a

[24] McKitterick 1998: 62.
[25] McKenzie 1: 363.

cleaner and more professional typography than had its Cantabridgian predecessors.

The editorial element in Jurin's work which matches the new ambitions symbolised by the typography of the 1712 edition is contained in his "Appendix". This "Appendix" is a newly paginated 54 page supplement to Newton's edition, the main aim of which is to emend, correct and update statements made in the main text in the light of over half a century of investigations in natural philosophy.[26] The "Appendix" takes the form of a series of notes keyed to chapters or pages in the main text. The appended endnotes are further adorned with footnotes, these being references for claims made in the text of the Appendix, and with some marginal notes directing the reader to three new foldout pages with a total of seven new figures which are keyed to the text of the appended commentary. The content of the notes is as Jurin's "Praefatio" leads us to expect: he takes the findings of the post-Cartesian generation of natural philosophers, especially English and French ones, and uses them to update Varenius on a series of basic issues in mathematical geography such as the shape of the earth, tides, the hydrological cycle and global wind patterns. In the course of the Appendix such luminaries of modern science as Boyle, Halley and Huygens are cited and other papers, notably from the Royal Society's *Philosophical Transactions*, are quoted with regularity; but pride of place is given to "Felicissime *Newtonus*",[27] whose *Principia* forms the backbone of the adjustments made by Jurin.

Whereas the content of Jurin's Appendix is well known and has been studied for the way in which it brought the *Geographia Generalis* within the ambit of Newtonian science,[28] the style of editing Jurin deployed has received less attention. Jurin's edition included a new dedication to Richard Bentley, Master of Trinity College, Cambridge where Jurin (as his title page announced) was a Fellow. This was not just (although it was no doubt in part) a call for patronage, as Bentley had suggested the project to Jurin, nor was it simply a sign of his gratitude to Bentley for checking the Latin style of his Appendix.[29] Bentley was famed for his skills as a classical editor, "the greatest scholar that England, or

---

[26] Varenius 1712: Appendix 1–2.

[27] Varenius 1712: Appendix 13.

[28] Warntz 1989: 182–88; Andrea Rusnock, ed., *The Correspondence of James Jurin (1684–1750)*, Amsterdam 1996, 10f.

[29] Rusnock 62f.

indeed Europe, had ever bred",[30] and in particular for his belief in emendatory editing, or the adjustment of established classical texts not in the light of new manuscript evidence but of logic and editorial intuition. Importantly, Jurin's approach to editing Varenius clearly bears some parallels to Bentley's then celebrated efforts. Like Bentley, Jurin does not seek merely to reproduce a text which Varenius as the author penned, instead he is interested in using subsequent knowledge to update Varenius. Furthermore, just as Bentley was to suggest numerous alterations to Milton in his 1732 edition but only actually inscribed a very small number of them in the text itself,[31] so Jurin makes a clear separation between the text and his additions or corrections, the result being a set of annotations which subvert the text they aim to support. Moreover, Jurin does not actually annotate Varenius' original Latin text at all, but instead the already altered text which Newton's edition had established, with all its faults and inconsistencies. In this, Jurin followed the standard editorial practice of the age, wherein new editions were based not on the original text but on the *textus receptus* as handed down by a sequence of editors and printers, "a procedure so well calculated to foster the survival of old and persistent error"[32] of the sort we have already seen bedevilled Newton's Cambridge editions.

Jurin's editorial method not only perpetuates the textual inconsistencies and problems of Newton's received text, but also creates some editorial problems of its own due to the decision to merely add an Appendix. This decision was probably taken on cost grounds: it allowed the Newtonian text simply to be reset from an earlier printing with no alterations. The decision not to tamper with the received text means that while the Appendix refers the reader to the specific pages of the text it glosses, the main text has no diacritical marks to direct the reader to the Appendix. As such, intertextual reference only works from apparatus to text, not in both directions as would normally and preferably be the case. Furthermore, the separation of text and commentary in Jurin's Varenius makes cross-referencing cumbersome, even if the reader is aware (as he cannot be from the text itself) that a passage has an appended comment by Jurin. The Appendix adds further confusion by

30 E.J. Kenney, *The Classical Text: Aspects of Editing in the Age of the Printed Book*, Berkeley 1974, 114.

31 Marcus Walsh, *Shakespeare, Milton and Eighteenth Century Literary Editing: The Beginnings of Interpretative Scholarship*, Cambridge 1997, 67.

32 Kenney 68.

its previously-mentioned addition of 3 new fold-out leaves with 7 new figures, because they were numbered anew, rather than sequentially after Newton's original 33 schemas. As a result, there is more than one figure 1 in Jurin's edition, making referencing somewhat confusing.

Simply put, Jurin's 1712 edition of the *Geographia Generalis* was part of the broad thrust at Cambridge University Press initiated by Richard Bentley to produce scholarly editions aimed at a market which would extend beyond England to the European republic of letters, an aim that was inflected in the case of this work by the need to keep costs down for a student audience which led to some corners being cut.[33] Its typography reflected this desire, and the way in which it was edited suggested that Varenius was to be taken seriously as an author; that here was a modern classic of geographical writing. For Jurin, as for Bentley, this sense of reverence meant not a desire to produce an edition as close as possible to Varenius' *ipsissima verba*, but on the contrary, a willingness to carry forward the *textus receptus*, however altered it had been by the labours of subsequent editors and publishers, coupled to a desire to update that received text in the light of new findings in mathematical geography and natural philosophy. The result is a text which continues some confusions from Newton's edition and generates a few of its own by the lack of integration of the newly appended material with the Newtonian received text. Despite this, however, the overall effect of Jurin's edition is to give an unprecedented sense of Varenius as someone worthy of serious editorial labours, as a classic that needs to be transmitted in updated form to a new generation. We also get, for the first time, a clear distinction between the received text and new editorial additions. With Jurin, the canonization of the *Geographia Generalis* which has become such a feature of modern writing in the history of geography begins.

Varenius as a Vernacular Classic: The Dugdale-Shaw Editions

The final great English edition of the *Geographia Generalis*, the Dugdale-Shaw translation, first appeared in 1733. It is by some margin the most thoroughly edited and meticulously typeset edition of Varenius in the English print history of the work and can claim furthermore to be

[33] McKitterick 1998: 55f, 78.

one of the most professionally edited texts in the English geographical tradition more generally in this era. This is achieved by two main sets of work: first, the typography and notation are rendered clearly, consistently and accurately for the first time in the print history of the work; and secondly, annotation is practiced in a new way for this edition. The cumulative effect of these editorial labours is to create what is to my knowledge a unique print space for a geography book, and it aligns the *Geographia Generalis* with the format of canonical texts, both classical and vernacular.

Looking at these two sets of editorial work in more detail reveals the achievement of the Dugdale-Shaw Varenius. Primarily, then, this edition was the first to achieve a consistent mode of annotation and cross-referencing, allowing the reader to access information with ease and reliability by means of a series of linked changes. First, the inconsistencies between texts and tables which had ramified down from Newton's editions were corrected. Thus, for example, the previously mentioned inconsistency as to how many 'large' islands existed was rectified here for the first time in any edition.[34] Second, the figures or schema which Newton had added to the text and Jurin had expanded were integrated more thoroughly both by marginal references in the text to the figures and by references to the relevant pages in the text inserted next to each figure. A third major change was the addition of an "Index" of topics and place names, this being the first real 'finding aid' the text had been graced with in any edition, the running marginal glosses of the Blome translation excepted. This move was paralleled by the final great change in the typesetting, which was to move the notes from a collection of unkeyed endnotes in Jurin's Latin edition to footnotes with clear keying to the text in the Dugdale-Shaw translation. Most notes are keyed by means of a bracketed letter in the text mirrored at the foot of the text, although some notes were merely keyed by an asterisk mark.[35] The reason for this inconsistency is not clear: while some notes were added to the second edition of 1734, this does not explain the asterisked notes, some of which are present in this form in both editions. It can only be conjectured that this is a result of the interaction of the two editors, Dugdale and Shaw, not being fully ironed out in the typesetting, or of late notes being added at the printers. Yet

[34] Varenius 1765: 101, 109ff.

[35] For example, Varenius 1765: 37f.

the vast majority of the notes are alphabetically keyed to the text, and this is an enormous advance on previous editions in terms of ease of reference and also one, of course, which completely changes the print format of Varenius from one bloc of continuous text to that standard for editions of canonical authors at this time, a large typeface being used for the main text, coupled to a two-column commentary in a smaller typeface at the foot of the page.

The change in annotation for this edition is not simply about format—the switch to footnotes from endnotes—but also relates to their nature. Dugdale-Shaw presents notes which are in different 'voices', rather than a univocal or anonymous editorial 'voice'. More specifically, there are in many footnotes both Jurin's appended comments of 1712 and further updated portions of commentary taking into account new findings in the twenty years intervening between Jurin and this translation or (less frequently) offering differing interpretations. The result is a page which replicates the appearance of the 'variorum' edition wherein the opinions of different commentators stood shoulder to shoulder.[36] Unlike the variorum, which only fully emerges for English vernacular texts with Samuel Johnson's edition of Shakespeare in 1765,[37] what Dugdale-Shaw provides is not a set of competing interpretations of a contested point, but a sequence of additions to Varenius' by-now-obsolete geographical and natural-philosophical positions. Textually, however, the print format which results is identical and bespeaks the degree of reverence in which the author, Varenius, is held by the extent of the editorial labours which have been lavished upon his text.

One final typographic change needs to be noted, this time in the text itself rather than its many new and reformatted 'paratexts'.[38] As "The Translator's Preface" notes:

> There is often not the least *Consonance* or *Similitude* between the *Latin* and *modern English* [geographical] *Names* [...] And very often their Names are changed by *later Discoverers*, and their *Figures* and *Situations* better discovered since our Author's Time. Wherefore, in such Cases, we have taken

[36] Margreta De Grazia, *Shakespeare Verbatim: The Reproduction of Authenticity and the 1790 Apparatus*, Oxford 1991; Joanna Gondris, "'All this Farrago': The Eighteenth-Century Shakespeare Variorum Page as a Critical Structure," in: Gondris, Joanna ed. *Reading Readings: Essays on Shakespeare Editing in the Eighteenth Century*, London Press 1998, 123–39; Walsh.

[37] Simon Jarvis, *Scholars and Gentlemen: Shakespearian Textual Criticism and Representations of Scholarly Labour, 1725–1765*, Oxford 1995, 60; Walsh 170.

[38] Girard Genette, *Paratexts: Thresholds of Interpretation*, Cambridge 1997.

> the Liberty to *alter their Names, Situations, and Descriptions*, in order to make them conformable to our latest and best *English* Maps; *deviating* as little as possible from our Author's Sense; and making use of the *same Words* as 'tis likely he would have done, had he writ at the *same Time*, and in the *same Language*. We have done this to avoid, in some Measure, *Marginal Notes*, which must necessarily have been inserted to have explained a *strict Translation* [...] These Alterations are included in *Brackets*, and for the most part distinguished by a different Character.[39]

Two main transformations are announced in this excerpt, both of which are remarkable in the light of the previous editing of Varenius and in the context of geographical print culture more generally at this time, and both of them bespeak the editors according to Varenius' classic authorial status. First, there is a clear recognition that Varenius wrote at a different and earlier point in the history of geography so that errors he makes in the light of subsequent knowledge do not impair his authority as a geographer. In short, the historicity of geographical knowledge is recognised for the first time in this edition. This attitude manifests itself in the way in which this edition is updated for the first time in important respects. Varenius' questions for the student about tides, longitudes, and so forth are updated and made to centre on London in the 1730s rather than the Amsterdam of 80 years earlier. Some of this work was done for the first, 1733, edition, but much of this anglicisation of Varenius' text takes place in the second, 1734 edition.[40] Furthermore, and again as a concession to the audience for a vernacular edition of Varenius, the 1734 edition added a set of theorems and 'paradoxes' in mathematical geography, derived from Patrick Gordon's *Geography Anatomiz'd* on the grounds that

> to borrow from an Author, and to improve upon that Author, is not only justifiable, but has been the chief if not the only means by which the present Age has brought their Acquirements in Philosophy, and Mathematics, to the pitch to which they are now arrived; therefore to do justice to Mr *Gordon* in his Grammar, and yet to improve *Varenius* so as to make it the most excellent of it's Kind, the Publisher hath thought proper to add not only some *Theorems* [...] but also some of the most useful of *Gordon's Paradoxes*.[41]

---

[39] Varenius 1765: vii–viii.
[40] Varenius 1765: 266f, 692ff, 839.
[41] Varenius 1765: 779.

Thus Varenius is updated in recognition that both his mathematical geography and his pedagogic style may not suit an audience in mid eighteenth-century England, all this being done not silently, but with a clear respect for the *ipsissima verba* of the author. In other words, even as the text is adjusted, it is flagged for the reader typographically and the original is respected, something which marks a clear break with the editorial attitude of both the vast majority of geography books at this time and with the previous editions of Varenius. Yet the overall aim of the changes made in the light of the passage of time is to recast what had always been—both in the Dutch Elsevier and the English Cantabridgian editions—a Latinate student text for an erudite audience, as a vernacular work aimed at the burgeoning market for books of popular instruction. This shift helps to explain why the Dugdale-Shaw translation is the first English edition of Varenius to include no dedication, merely containing a "Translator's Preface" to the audience: instead of working in a system of patronage, whether aristocratic (as in Blome's editions) or academic (as Jurin's work), Dugdale and Shaw were targeting the more anonymous reader in the public sphere.

Secondly, and above all, there is a demand to respect the authorial intentions of Varenius even as they are superseded, in that all changes made in the text itself, as opposed to corrections in footnote commentaries, are marked typographically by the use of square brackets and italic typeface. Occasionally, the text even gives both Varenius' original geographical name and then a corrected name in brackets.[42] As such, any deviation from the copy text is immediately apparent to the reader, unlike all previous editions that had allowed for the silent alterations which had caused so many confusions. In part this simply reflects changing editorial fashions as the silent editor was replaced by the editor as commentator,[43] but it is also inevitably part of a process in which the author as creator is accorded a new reverence in editorial practice, while simultaneously being seen within the culture of his age. It should be noted, however, that as in the case of literary editing at this time, the authorial intention which is respected is not that of the original author in his own time and place, in this case Varenius in 1650

---

[42] As at Varenius 1765: 139.

[43] Lawrence Lipking, "The Marginal Gloss," *Critical Inquiry* 3 (1977): 625ff; De Grazia 210; Jarvis 45.

Amsterdam, but the composite of intentions as represented in the *textus receptus*, in this case Jurin's 1712 revision of Newton's edition.

Considering all these aspects, we can say that the Dugdale-Shaw translation of Varenius places the *Geographia Generalis* in two publishing contexts simultaneously. On the one hand, the respect for the intentions of the copy text even as it is altered bespeaks a new reverence for the idea of an original text produced in a certain time and place. In short, Varenius (as updated by Newton and Jurin) is seen as a classic text and is edited as such with an increasingly sophisticated awareness of the passage of time and an emergent historical consciousness. The result of this attitude is a wholly new print space for the geography book in the English publishing tradition, as it mirrors that for classical texts and becomes contemporaneous with the adoption of closely parallel typographical and editorial conventions for the first great editions of vernacular classics such as Theobald's Shakespeare of 1733. On the other hand, in making Varenius speak to an English audience in the mid-eighteenth century, the Dugdale-Shaw project also recasts the text as a work of useful knowledge in line with other successful publishing ventures of this ilk. A Latin text aimed at an erudite audience mutates into a work seeking broader appeal by being made relevant to the times, and by the attempt to make learning mathematical geography more entertaining. To modern eyes, the result is perhaps a bifurcated text which both respects and overrides authorial intention, a work which is at once an edition and a new book, a mishmash of historical respect, *receptus* adjustments and market-driven repackaging, but in this the Dugdale-Shaw Varenius is in fact representative of editorial attitudes at this time, which moved easily between these different approaches in treating both ancient[44] and vernacular[45] editing.

Conclusion

What emerges from this essay is the suggestion that the fusion of intellectual history and the history of science with an awareness of print culture as an historical phenomenon can be furthered by taking edit-

---

[44] L.D. Reynolds and N.G. Wilson, *Scribes and Scholars: A Guide to the Transmission of Greek and Latin Literature*, second edition, Oxford 1974; Rudolf Pfeiffer, *A History of Classical Scholarship: From 1300 to 1850*, Oxford 1976.

[45] De Grazia; Walsh.

ing seriously as a source of insights. In the case of Varenius, we have seen that how he was edited varied enormously in the course of time and resulted in very different positionings of his work on a spectrum between hackwork, as in the case of Blome's *Cosmography and Geography*, and the classics, as in the case of Dugdale and Shaw's *Compleat System*. Furthermore, editing could lead to widely varied appearances for the same text from a blank text in Newton's edition to a complex page with marginal directions, footnotes, keyed labels and multiple typefaces and sizes, as in Dugdale and Shaw's translation. As such, the way Varenius was edited and the contexts in which this occurred had important consequences for the costs of producing the book, for the sense of the authorial status of the Varenian copy text, and for the implied audience at which the *Geographia Generalis* was pitched.

In the case of Varenius, English print history initially appears to dissolve many of the platitudes about the importance of the text: clearly in quantitative terms, Varenius was a successful but not exceptional title in comparison with other geographical works of the time. Yet the present analysis of the editing of Varenius at least partially reinstates his centrality in the history of geography at the qualitative level: in both Jurin's work and especially in the labours of Dugdale and Shaw, we find Varenius being accorded an exceptional status. His text is deemed of sufficient worth to warrant editing as a modern classic. Equally, this process works the other way around as well: by editing Varenius with such lavish commentary and complex typography, the print format of the *Geographia Generalis* helps to define its status and in the end to accord it canonical standing. The printed space of the annotated editions announces a claim for the central importance of Varenius' work, a position which seems to have been taken up by modern historians of geography from Alexander von Humboldt to the present day.

THE INFLUENCE OF BERNHARD VARENIUS IN THE GEOGRAPHICAL WORKS OF THOMAS JEFFERSON AND ALEXANDER VON HUMBOLDT[1]

Sandra Rebok

Abstract

Der Einfluss von Bernhard Varenius auf die geographischen Werke von Thomas Jefferson und Alexander von Humboldt

Varenius' *Geographia Generalis* war das erste Textbuch in der allgemeinen Geographie, das in einem amerikanischen College nachweislich benutzt wurde. Daher wird in diesem Beitrag der Frage nachgegangen, in welchem Ausmaß die in diesem Buch vermittelte Geographiekonzeption zwei seiner berühmten Nachfolger auf diesem Gebiet beeinflusst haben könnte: den Staatsmann und Wissenschaftler Thomas Jefferson (1743–1826) und den preußischen Forschungsreisenden und Naturwissenschaftler Alexander von Humboldt (1769–1859). Jeffersons *Notes on the State of Virginia* sind ein denkwürdiges Zeugnis der amerikanischen Aufklärung, und sein geographisches Interesse bewog ihn, Humboldt nach seiner Expedition in Spanisch-Amerika in die Vereinigten Staaten einzuladen. Ihre Begegnung 1804 und die wachsende Freundschaft beförderte gemeinsame Forschungsinteressen. In den einschlägigen Indizes, die sowohl Jeffersons wie auch Humboldts Schriften (leider nicht umfassend) erschließen, findet sich allerdings kein direkter Beleg für ihren Bezug auf Varenius, und Humboldts ausführliche Würdigung von Varenius in seinem großen Alterswerk *Kosmos* liegt sehr viel später als der Amerikaaufenthalt. Er sieht sich in der Nachfolge von Varenius sowie der des spanischen Jesuiten José de Acosta; auf Acosta (*Historia Natural y Moral de las Indias*) beziehen sich Humboldt und auch Jefferson mehrfach. Obwohl bislang kein unmittelbarer Einfluss von Varenius auf Jefferson und Humboldt nachzuweisen ist, zeigt sich jedoch deutlich die Ähnlichkeit des Geographieverständnisses in dem, was Varenius ‚geographia specialis' nennt, wenn Varenius wie Jefferson oder Humboldt sich etwa mit einer bloßen Aufzählung von Beschreibungen nicht zufrieden geben mögen, sondern immer nach Auswirkungen und Erklärungen fragen und humangeographische Belange einbeziehen.

[1] This article is the result of my fellowship undertaken at the Robert H. Smith International Center for Jefferson Studies at Monticello, Virginia, awarded by the Thomas Jefferson Foundation.—John Maciuika from the Baruch College of CUNY very kindly read the first draft of this paper.

Introduction

Bernhard Varenius' *Geographia Generalis* (1650) is considered one of the first works to introduce a more disciplined approach to the subject of geography. It was also the first textbook on record to be used in general geography in an American college.[2] Therefore, it is interesting to note the extent to which the ideas, as well as the conception of geography expressed in this book, might have influenced two of his famous successors in this field: the Virginian politician and scientist Thomas Jefferson (1743–1826) and the Prussian explorer and scientist Alexander von Humboldt (1769–1859).

Both men are considered the founders of modern geography. But whereas Humboldt—who, along with Carl Ritter (1779–1859), established the *Berlin Geographical Society* in 1828 and turned geography into a formal academic discipline—is recognized as the originator of modern physical and biological geography, Jefferson's case is different. Due to Jefferson's numerous achievements as a politician, especially as the third president of the United States (1801–1809), relatively few scholars have dealt explicitly with him as a geographer.[3] Nevertheless, especially with his geographical study *Notes on the State of Virginia*,[4] the Virginian contributed greatly to science in general and to American geography in particular. This book, upon which much of his fame as a scientist-philosopher was based,[5] is regarded as the most important scientific and political work written by an American before 1785. These notes constitute a memorable expression of the ideas of the Enlightenment in a wider sense, but are also, and more particularly, an expression of the American Enlightenment. Providing much data about Virginia specifically, the *Notes* also develop Jefferson's ideas concerning religious freedom, the separation of church and state, an analysis of representative government versus dictatorship, theories on art and education, his attitude toward slavery, as well as his interest in science generally.

---

[2] Georg Kish, ed., *A Source Book in Geography*, Cambridge, Mass. 1978, 364.

[3] Albert Ellerly Bergh, ed., *The Writings of Thomas Jefferson*, 20 vols., Washington 1907, vol. 8, i–vii; John Logan Allen, "Imagining the West: The View from Monticello," *Thomas Jefferson and the Changing West: From Conquest to Conservation*, ed. James P. Ronda, St. Louis 1997.

[4] Thomas Jefferson, *Notes on the State of Virginia*, ed. William Peden, Chapel Hill 1982.

[5] Peden in Jefferson 1982, xi.

For many years, Jefferson would be especially interested in the scientific exploration of the unknown western regions of the North-American continent, until in 1803 he finally dispatched the famous Lewis and Clark expedition to the Pacific (1803–1806).[6] In this context it must also be mentioned that geographical interest was the primary reason Jefferson invited Humboldt to Washington DC in spring 1804. After the Louisiana Purchase in 1803, Jefferson was particularly interested in Humboldt's materials and maps from the Spanish colonial archives, which contained hitherto unknown data on the disputed establishment of borders between the U.S. and New Spain. The Prussian generously complied with Jefferson's request and gave the American president the latest geographic and statistical information on Mexico, which was of great value to the U.S. government.

Humboldt, for his part, had just finished his world-famous expedition through Spanish America (1799–1804) and was on his way back to Europe. He was very interested in getting to know the enlightened president of the first American republic and therefore was delighted to visit the United States from 20 May to 30 June 1804.[7] From this brief meeting, a lifelong friendship and a lively exchange of ideas ensued.[8] This is expressed clearly in the correspondence between the two men (extant letters from 1804–1825), as well as in their mutual exchange of books from their areas of common interest (among others, Humboldt sent him his *Mexico* work as well as the *Tablaux de la nature*, and asked Jefferson for a signed copy of *Notes on the State of Virginia*, which he already had in his possession before the expedition). Geographical questions were the motivation for the initial as well as continuing mutual interest of the two correspondents, as is evident throughout their communications.[9]

---

[6] Hartmut Wasser, *Die große Vision: Thomas Jefferson und der amerikanische Westen*, Wiesbaden 2004.

[7] Hermann R. Friis, "Baron Alexander von Humboldt's Visit to Washington," *Records of the Columbia Historical Society* 44 (1963): 1–35; Ingo Schwarz, "Alexander von Humboldt's Visit to Washington and Philadelphia: His Friendship with Jefferson, and His Fascination with the United States," *Alexander von Humboldt's Natural History Legacy and Its Relevance for Today*, in: *Northeastern Naturalist* 8, special issue no. 1 (2001): 43–56.

[8] Sandra Rebok, "Two Exponents of the Enlightenment: Transatlantic Communication by Thomas Jefferson and Alexander von Humboldt," *The Southern Quarterly, Imagining the Atlantic World*, 43:4 (2006): 126–152.

[9] Schwarz, "From Alexander von Humboldt's Correspondence with Thomas Jefferson and Albert Gallatin," *Berliner Manuskripte zur Alexander-von-Humboldt-Forschung* 2 (1991): 1–20; Schwarz, *Alexander von Humboldt und die Vereinigten Staaten von Amerika*, Berlin 2004;

On account of their respective importance for the further development of geography, and possibly also as a result of their personal contact, there are several references that, in one way or another, link both Jefferson and Humboldt to Bernhard Varenius. These references should be explored, because both men appear to be explicitly or implicitly inspired by Varenius, as the following quotations make clear:

> However Jefferson, 1781, may be said to stand in geographical tendencies between Bernhard Varenius, who in "Geographia generalis", 1650, essayed the interpretation of the climatic conditions and the physical changes of the earth's surface, and Humboldt's "Kosmos", 1845. The latter supplemented Varenius by pointing out the connection of climate and soil formations with the distribution of plant and animal life, and even more important the relation of geographic environment to the development of mankind, especially as to colonization, commerce and industry. Jefferson's "Notes on Virginia", fifty years in advance of Humboldt, is along lines definitely formulated by the latter in scientific geography.[10]

And a second example:

> Jefferson was himself an amateur scientist who must stand with the foremost men of his time. *Notes on the State of Virginia*, his only book, was one of the first examples of Scientific Geography, preceding the work of Humboldt by fifty years, and drawing upon Bernhard Varenius' *Geographia generalis* (1650). By 1803, Jefferson was probably the most informed American on the totality of the geography of the American West, and had the largest library anywhere on this subject.[11]

Although connections among these men are evoked in quotations such as these, no broader study of this question or an analysis of these links has to my knowledge been conducted. Filling this gap has been the motivation for the present contribution.

Explicit Reference to Varenius

The first step in an analysis of possible influence, or at least in the establishment of conceptual similarities, consists in searching for direct

---

Helmut de Terra, "Alexander von Humboldt's Correspondence with Thomas Jefferson," *Proceedings of the American Philosophical Society* 103 (1959): 783–806.

[10] Schwarz 1991; Schwarz 2004; Terra.

[11] http://www.nps.gov/jeff/LewisClark2/TheJourney/ScienceofExpedition.htm.

references to Varenius in Humboldt's and Jefferson's works. With regard to the famous representative of the New World, this research produced no positive evidence. Specifically, consulting the different editions of Jefferson's papers[12] and their indexes revealed no reference to the German geographer.[13] Similarly, an analysis of Jefferson documents at the Library of Congress, of different data bases and transcripts at the Robert H. Smith International Center for Jefferson Studies at Monticello,[14] and of other sources such as the Jeffersonian Cyclopedia,[15] yielded the same result. It must be mentioned, however, that since the Jefferson papers are still not all published,[16] there might yet be a chance of finding comments about Varenius' work.

These results correspond with an analysis of all the references to previous works that could be found in Jefferson's only book-length study, his *Notes on the State of Virginia*. In this work, Jefferson demonstrates that he had conducted major research on geographical questions, and that he knew numerous works from all over the world in different languages, which he quoted, mentioned, or commented upon. Among authorities cited in the text and footnotes the most varied personalities[17] emerged, but again, there is not a single mention of Varenius.

Finally, the catalogue of books purchased from Jefferson by the Library of Congress was consulted to see if he owned *Geographia*

[12] Bergh; Julian P. Boyd, ed., *The Papers of Thomas Jefferson*, 31 vols., Princeton 1952–2004; Paul Leicester Ford, ed. *The Writings of Thomas Jefferson*, 10 vols., New York and London 1892–1999.

[13] Since these indexes usually only list Jefferson's correspondents, not the themes discussed or the persons mentioned in the letters, references to Varenius may actually exist.

[14] At this point I want to express my gratitude towards the staff of the ICJS for all their help, particularly to Lisa Francavilla.

[15] http://etext.lib.virginia.edu/jefferson/quotations/foley/.

[16] Two edition projects are in progress: The "Presidential Papers Project" (1801–1809) at Princeton and the "Retirement Paper Project" (1809–1826) at the Robert H. Smith International Center for Jefferson Studies.

[17] Jefferson refers to the Italian scholar and traveller Giambattista Ramusio (c. 1485–1557), the Latin historian Caius Plinius Segundus, Pliny the Elder (23–79 AD), the classical historian Herodotus (c. 484–425 BC), the Spanish naval officer and scientist Antonio Ulloa (1716–1795), the Mexican historian Francisco Javier Clavijero (Clavigero) (1731–1787), the Italian navigator Amerigo Vespucci (1451–1512), the French traveller and mathematical geographer Charles Marie de La Condamine (1701–1774), the Spanish Jesuit missionary and author of *Historia Natural y Moral de las Indias* (1590) José de Acosta (1539–1600), and he makes it clear whose works he definitely refuted—the French authors and scientists Guillaume Thomas Francois Raynal (1713–1796) and particularly George-Louis Leclerc, the famous Comte de Buffon (1707–1788).

Generalis. Once again, in a very large list of different books on geography,[18] surprisingly, this work did not appear.

The situation was not much different in regard to Humboldt: in all his works only one reference to Varenius could be found;[19] this appears in his last synthesis of his life-long studies, the five-volume work, *Cosmos.*[20] Since this reference to Varenius is very extensive and interesting, besides being the only one, it is instructive to take a closer look at it. In the first volume of this famous work, in the chapter on the "physical description of the world" ("Begrenzung und wissenschaftliche Behandlung einer physischen Weltbeschreibung"), Humboldt writes about the great geographer Bernhard Varenius:

> Er unterscheidet sehr scharfsinnig *allgemeine* und *specielle* Erdbeschreibung, und theilt die erstere wieder in die absolut *tellurische* und die *planetarische* ein, je nachdem man betrachtet die Verhältnisse der Erdoberfläche in den verschiedenen Zonen, oder das solarisch-lunare Leben der Erde, die Beziehung unseres Planeten zu Sonne und Mond. Ein bleibender Ruhm für Varenius ist es, daß die Ausführung eines solchen Entwurfes der *allgemeinen* und *vergleichenden* Erdkunde *Newton's* Aufmerksamkeit in einem hohen Grade auf sich gezogen hatte; aber bei dem mangelnden Zustande der Hülfswissenschaften, aus denen Varenius schöpfte, konnte die Bearbeitung nicht der Größe des Unternehmens entsprechen. Es war unserer Zeit vorbehalten, die *vergleichende Erdkunde* in ihrem weitesten Umfange, ja in ihrem Reflex auf die Geschichte der Menschheit, auf die Beziehung der Erdgestaltung zu der Richtung der Völkerzüge und der Fortschritte der Gesittung, meisterhaft bearbeitet,[21] zu sehen.[22]

Thus, after a short explanation of Varenius' basic concept and a reference to the importance of Newton in the diffusion of these ideas, Humboldt writes that his predecessor laid the foundation for a new focus on geography, though it could not be performed comprehensively during his time. In this way Humboldt sees his own work as a continuation of that of Varenius.

---

[18] E. Millicent Sowerby, ed., *Catalogue of the Library of Thomas Jefferson*, Charlottesville 1983, vol. 4: 85–356.

[19] Since the indexes are selective, references may well exist.

[20] First edition 1845–1862. The version used here is: Alexander von Humboldt, *Kosmos: Entwurf einer physischen Weltbeschreibung*, eds. Ottmar Ette and Oliver Lubrich, Frankfurt a. M. 2004.

[21] Here Humboldt adds a note to refer to Carl Ritter and his work on geography.

[22] *Cosmos* 1, 32.

In a long footnote Humboldt adds to this comment a more detailed view on the importance as well as the limits he observes in Varenius' work:

> [...] Das überaus wichtige Werk des Varenius ist im eigentlichen Sinne des Wortes eine *physische Erdbeschreibung*. Seit der vortrefflichen Naturbeschreibung des Neuen Continentes, die der Jesuit Joseph de Acosta (*Historia natural de las Indias* 1590) entwarf, waren die tellurischen Phänomene nie in solcher Allgemeinheit aufgefaßt worden. Acosta ist reicher an eigenen Beobachtungen; Varenius umfaßt einen größeren Ideenkreis, da ihn sein Aufenthalt in Holland, als dem Mittelpunkt eines großen Welthandels, in Berührung mit vielen wohlunterrichteten Reisenden gesetzt hatte. *"Generalis sive universalis Geographia dicitur, quae tellurem in genere considerat atque affectiones explicat, non habita particularium regionum ratione"*. Die *allgemeine Erdbeschreibung* des Varenius (pars absoluta cap. 1–22.) ist in ihrem ganzen Umfange eine *vergleichende*, wenn gleich der Verfasser das Wort Geographia comparativa (cap. 33–40.) in einer viel eingeschränkteren Bedeutung gebraucht. Merkwürdig sind die Aufzählung der Gebirgssysteme und die Betrachtung der Verhältnisse ihrer Richtungen zu der Gestalt der ganzen Continente (p. 66–76. ed. Cantabr. 1681); die Liste der brennenden und ausgebrannten Vulkane; die Zusammenstellung der Resultate über die Vertheilung der Inseln und Inselgruppen (p. 220), über die Tiefe des Oceans in Vergleich mit der Höhe naher Küsten (p. 103), über den gleich hohen Stand der Oberfläche aller offenen Meere (p. 97), über die Strömungen in ihrer Abhängigkeit von den herrschenden Winden, die ungleiche Salzigkeit des Meeres und die Configuration der Küsten (p. 139), die Windrichtungen als Folge der Temperatur-Verschiedenheit u.s.f. Auch die Betrachtungen über die allgemeine Aequinoctial-Strömung von Osten nach Westen als Ursache des, schon am Cap San Augustin anfangenden und zwischen Cuba und Florida ausbrechenden Golf-Stromes (p. 140) sind vortrefflich. Die Richtungen der Strömung längs der west-afrikanischen Küste zwischen dem Grünen Vorgebirge und der Insel Fernando Po im Golf von Guinea werden äußerst genau beschrieben. Die sporadischen Inseln hält Varenius für "gehobenen Meeresgrund": *magna spirituum inclusorum vi, sicut aliquando montes e terra protrusos esse quidam scribunt* (p. 215)". [...][23]

Here, Humboldt discusses Varenius alongside José de Acosta, whom he considers to be a clear predecessor of his own ideas on the physical description of the earth, which he calls "Physique du monde"—as expressed several times in different works, and which can be deduced

[23] *Cosmos* 1, 32f.

from Humboldt's specific approach to geographical questions.[24] Humboldt clearly states his conviction that Varenius' work, like that of Acosta, represented a physical description of the earth, according to his understanding of the term, and furthermore, that since Acosta no such good and general description of the telluric phenomena of the earth had yet been written.

At this point the following question arises: Why, in the cases of both Jefferson and Humboldt, were the Jesuit José de Acosta and his *Historia Natural y Moral de las Indias* taken into account far more than Bernhard Varenius and his *Geographia Generalis*?[25] In addition to Humboldt, Jefferson also used Acosta as a point of orientation and referred to the Spanish Jesuit in his *Notes on the State of Virginia* several times.[26] We also know that Jefferson possessed his own copy of de Acosta's famous work.[27] Although these references may not be very rich in content, they demonstrate Jefferson's occupation with Acosta and indicate that he considered his work important enough to cite.

Nevertheless, the citation above also reveals some general statements Humboldt makes about *Geographia Generalis*, along with particular evaluations and critical considerations of its content. The exact references to specific chapters show that Humboldt really had dealt with this text and proves his abiding interest in Varenius. Further in the footnote, he considers Varenius to be an astute geographer and contrasts biographical details found in different sources with specific indications he localized in Varenius' writings.

Similarities in the Scientific Concepts

Besides Humboldt's long and explicit reference to Varenius, implicit parallels between their respective descriptive models can be detected (in spite of the lack of direct reference, and to a lesser degree, this is also the case with Jefferson). At this point the primary question is not

[24] A more detailed analysis of this question can be found in: Sandra Rebok, "Alejandro de Humboldt y el modelo de la *Historia Natural y Moral*". *Humboldt im Netz* II, 3 (2001): http://www.uni-potsdam.de/u/romanistik/humboldt/hin/rebok-HIN3.htm.

[25] Unfortunately, this question cannot be answered in the frame of this article, but should be an inspiration for future studies.

[26] Jefferson 1982, ed. Peden: 267, 272.

[27] Apparently he possessed it in two different editions, because Sowerby lists them as nos. 4096 [1590] and 4097 [1608], vol. 4: 254f.

whether this might prove a particular influence of Varenius on Jefferson, but if it demonstrates a similarity in their ideas and concepts.

From the beginning of his scientific activities, the Prussian scientist clearly expressed how he defined the aim of his research: to understand and describe the New World, as well as to develop his idea of the *Physical Geography*. His holistic concept of science envisaged the Earth as an inseparable organic whole, all parts of which were mutually interdependent. He regarded this synthesis as a harmonious unity and derived satisfaction through the scientific analysis of the ways in which things and phenomena on the Earth depend upon each other, an idea which he had clearly in mind as early as 1797 and developed further in his final work, of *Cosmos*, in the first volume:

> Was mir den Hauptantrieb gewährte, war das Bestreben, die Erscheinungen der körperlichen Dinge in ihrem allgemeinen Zusammenhange, die Natur als ein durch innere Kräfte bewegtes und belebtes Ganzes aufzufassen. Ich war durch den Umgang mit hochbegabten Männern früh zu der Einsicht gelangt, daß ohne den ernsten Hang nach der Kenntniß des Einzelnen alle große und allgemeine Weltanschauung nur ein Luftgebilde sein könnte. Es sind aber die Einzelheiten im Naturwissen ihrem inneren Wesen nach fähig, wie durch eine aneignende Kraft sich gegenseitig zu befruchten. [...] So führt den wißbegierigen Beobachter jede Klasse von Erscheinungen zu einer anderen, durch welche sie begründet wird oder die von ihr abhängt.[28]

Here we can detect an obvious similarity with Varenius, who also saw a basic unity in a single earth system with close linkages among all terrestrial, celestial, physical and human forces.[29] As Varenius lamented, in his time geography was criticized as being either too narrowly descriptive or too widely extensive, since readers were rather bored with a bare enumeration and description of regions without an explanation of the customs of the people. In his work *Geography of Plants*, Humboldt also condemned the numerical listing of the flora by continents; what he was interested in was the distribution of vegetation and its relationship to climatic zones as well as other factors that affected the way it spread.

As regards Jefferson, we see a scientific curiosity on the part of an enlightened scientist, a researcher with a pragmatic orientation also directed toward the whole cosmos. Jefferson did not confine himself to a mere enumeration of towns, rivers, boundaries, inhabitants, industries,

---

[28] "Vorrede", 3–4.

[29] Amando Melon y Ruiz de Gordejuela, "Esquema sobre los modeladores de la moderna ciencia geográfica", *Estudios geográficos* n. 20–21 (Aug./Nov. 1945): 396.

production, and the form of government in Virginia. He described not only its rivers, but their relations to commerce and especially to their possible utility in trade, and he classified the plants and trees according to their value for ornamental, medicinal and esculent purposes. He included comparative views of America's native birds and animals with those of Europe.[30] As a result, his *Notes* touched all phenomena of the natural region, with its people and its government. Jefferson's approach to geography, in the final analysis, reflects what Varenius calls 'special geography', which is the description of particular places, versus 'general geography', which denotes the study of general and universal laws or principles that apply to all places. Furthermore, Varenius insisted on the practical importance of the type of knowledge included in 'special geography'—which is exactly Jefferson's focus in his function as a representative of the American Enlightenment. Jefferson did not elaborate on abstract concepts or universal laws characteristic of the general geography. He studied them, but he had a specific geographical interest deriving from his work as a politician, especially relevant during his term as president of the United States.

Humboldt, by contrast, wrote two regional studies, essays about Cuba and New Spain respectively. These were not limited to physical descriptions of these regions. In fact, they provide explanations of general geography terms as well. Finally, in his last work, *Cosmos*, he elaborated more profoundly on these theoretical aspects of geography. Therefore, and like Varenius, he was not regarded as an outstanding contributor to any specific field of knowledge, but to the organisation of knowledge.

It is in connection with this focus that the attitude of the three scientists regarding the position of man towards nature becomes clear. Humboldt included human beings in his concept of 'physical geography', regarding them as integral parts of nature; he did not, however, consider men a primary determinant, or assign them a special place in his research. According to his holistic view and his understanding of human society and nature as a harmonious system, he looked for the interconnection of human and physical nature, which later on inspired the term 'ecology'. In Varenius' concept of special geography, the "human properties of a place" form a third category, in addition to the celestial and terrestrial properties of a locale. He was convinced that

[30] Bergh iii.

the description of the cultures, the language, government, religion etc. of the inhabitants were to be examined alongside factors like the climate, the surface features, the minerals, animals and plants. Like Humboldt, Jefferson did not assign a special position to men in his geographical description of Virginia, for in the topics he treated in the 23 queries, only some are dedicated to human aspects.

Another common criterion shared by these three geographers is their comparative approach to science. As the founder of plant geography, Humboldt recognized the interdependence of areal phenomena and the need for explaining any one set of spatially distributed phenomena in relation to its spatial context. Therefore, he repeatedly compared areas with similar landscapes in different parts of the world. This is an obvious parallel with Varenius, who divides his *General Geography* into an absolute part, a respective part, as well as a comparative part, which contains an explanation of those properties that in turn arise from the comparison of diverse places on the earth. In Jefferson's case, too, the comparative method is manifested in his refutation of Buffon's ideas on the supposed inferiority of America, where he contrasts the properties of the Old and the New World.[31]

Finally, another similarity in the work of the three authors can be observed in their attitudes towards the role of theology in science. Although geography in Varenius' time was closely linked to religious questions, he dissented from that view, and his work can be considered the first scientific approach to this discipline.[32] Humboldt and Jefferson also continued in this line in spite of all the problems they had to face as a consequence of this conviction: the former in trying to disconnect theology from science in the first case, and the latter in establishing a new secular form of education that culminated in the creation of the University of Virginia, inaugurated in 1824.

Conclusion

Humboldt's direct reference to Varenius is very limited, though the one large comment that could be located is quite expressive and reveals Humboldt's opinion about Varenius. In Jefferson's writing there were

[31] Especially in query VI.
[32] Kish 370.

no comments about Varenius; nonetheless, similarities between the scientific concepts of the two can be detected. In any case, it is necessary to emphasize that these obvious parallels in the contributions of the three men to systematizing the knowledge available about our planet at their time may not necessarily be the result of a direct transmission of ideas. They integrated previous conceptions and ideas, which in some cases even derived from the Greek and Roman epochs, and enriched contemporary geographical understanding, due to European expeditions and discoveries, with their own observations and conclusions.

Nevertheless, what can be shown is that general inspiration through other works, through the implicit elaboration of similar intellectual models as well as explicit references; in brief, the evolution and continuation of concepts, called the *chain of ideas*, is a very fascinating topic. As is well known, Varenius, too, included and processed the ideas of previous as well as contemporary thinkers, particularly those of Bartholomäus Keckermann (1571–1608),[33] just as in several of Humboldt's texts references and comments about Thomas Jefferson and his work can be found.

Finally, it must be noted that the themes dealt with in this article could not be treated exhaustively. Therefore, this account should be regarded more as a first approach, especially with respect to the ideological and theoretical relationships among these three founders of modern geographical science.

[33] About the predecessors and sources of Varenius see: Horacio Capel Sáez, "La Personalidad geografica de Varenio", Introduction to Varenius': *Geografia General, en la que se explican las propiedades generales de la* tierra, ed. H. Capel Sáez, transl. José María Requejo Prieto, Barcelona 1974, 38–42; Hanno Beck, *Geographie: Europäische Entwicklung in Texten und Erläuterungen*, Freiburg, Br. 1973, 115–116.

BERNHARD VARENIUS' *GEOGRAPHIA GENERALIS* AND THE RISE OF MODERN GEOGRAPHICAL STUDIES IN RUSSIA

Denis J.B. Shaw

Abstract

Bernhard Varenius' Geographia Generalis *und der Aufstieg moderner geographischer Studien in Russland*

In diesem Beitrag wird der allgemeine geographische Kontext der Übersetzung und Veröffentlichung der *Geographia Generalis* in Russland untersucht. Unterschiede zwischen Russland und Westeuropa in der kulturellen Entwicklung werden aufgezeigt, und die weit verbreitete Vorstellung, dass die russische Kultur vor Peter dem Großen (reg. 1682–1725) im Grunde noch mittelalterlich gewesen sei, wird kurz erörtert. Es ist jedoch festzuhalten, dass Russland nicht unberührt von den Strömungen der Modernisierung geblieben ist, die durch Europa seit dem Ausgang des Mittelalters fegten, und dass diese Strömungen in Russland einen Einfluss auf das geographische Denken und seine Praxis selbst vor Peter hatten. Viel wichtiger im Blick auf die Modernisierung im allgemeinen und den Aufstieg der modernen Geographie im besonderen waren Peters Reformen, in denen die Übersetzung und Veröffentlichung von Varenius' Buch einen wichtigen Abschnitt markiert. Es zeigt sich, dass Peter und seinen Mitarbeitern Varenius deshalb zusagte, weil er sowohl eine offenbar ‚moderne' Weltsicht bot wie auch eine praktische Verfahrensweise zur Behandlung geographischer Belange.

The publication in Russia of Bernhard Varenius' *Geographia Generalis* in 1718 has been rightly seen as a significant step in the introduction of geography as a scientific enterprise in that country and as an important contribution to Peter the Great's campaign to modernize his realm. The aim of the present paper is to place the book's publication into its historical context, to suggest a few of the possible reasons for its perceived importance to the ruler of Russia, and to consider some of the ramifications of its appearance.

It is important to place Varenius' book into what may appear to be the rather unfamiliar context of geography and of science more generally in Russia before Peter the Great's reign (1682–1725). It is a truism that Russia before Peter had been to a considerable degree sheltered from many of the great intellectual movements which had influenced western Europe since the close of the medieval period, notably the Renaissance and the Scientific Revolution. In the words of one scholar, prior to Peter Russian culture was still essentially medieval.[1] The reasons for Russia's partial isolation were partly political and geographical: from the thirteenth to the fifteenth centuries the country had been ruled as part of the Khanate of the Golden Horde, an off-shoot of the former Mongol Empire, which served to cut Russia off to some degree from European influences. Situated far from western Europe and the Mediterranean, Russia was hardly known to medieval Europeans and had only limited contacts even with central and eastern Europe prior to the establishment of maritime trading links with the West in the latter part of the sixteenth century. Another reason for Russia's isolation was cultural. The country's Byzantine and Orthodox heritage made Russians less than welcoming to intellectual innovations stemming from the Catholic West and caused them to treat such innovations with suspicion. After the fall of Constantinople to the Ottoman Turks in 1453, Moscow's donning of the mantle of the 'Third Rome', citadel of Orthodox truth, probably only exacerbated such tendencies.[2]

However, to suggest that Russia was entirely unaffected by the changes which affected Europe at the close of the medieval period would be a considerable exaggeration. In the fifteenth and sixteenth centuries Russia experienced two processes which were also operative in the West in the same period. These were: the unification of the state under the rule of a single dynasty, in this case centred on Moscow; and the country's enormous territorial expansion (in Russia's case, overland rather than overseas as in the case of western European states) to become a multinational empire. Inevitably such changes had important implications for government and for the ways in which the empire was ruled and defended. Central to the perceived need for reform in such matters was a growing consciousness of threats to the country which might emanate from the aggressive and expanding European states situated to the country's west. Also significant were the longer-standing threats stemming from the steppe peoples and their states situated to the south, now backed by the formidable rising power of the Ottoman Turks. From the fifteenth century Russian rulers began to hire foreign

[1] W.F. Ryan, *Magic in Russia: The Bathhouse at Midnight*, Stroud 1999, 10.
[2] Geoffrey Hosking, *Russia: People and Empire, 1552–1917*. London 1997, 3–8.

specialists with the aim of modernizing their activities in such areas as military technique, fortification and building, metallurgy, medicine and administration. New technologies like gunpowder and printing were adopted. And there were inevitable ramifications in such areas as architecture, art and literature. Russia, in other words, was in the process of change, albeit failing to keep pace with the dynamic societies of western Europe in many important respects.

A number of these developments had important implications for geographical endeavour, as we shall see. But before discussing these, it is important to say something of the social context in which science and related pursuits were conducted, since the context was very different from that in western Europe.[3] For example, pre-Petrine Russia had no universities and, apart from an academy which taught classical learning of very recent origins, no centres of higher education or research. Scholars have traced this fact to the very different role which the church and the monasteries played in Russia by comparison with those in western Europe.[4] Largely cut off by its use of Old Church Slavonic from the Greek and Latin learning of the classical world, the Russian church did not assume the role of protector of classical learning that had been the case in the West. By and large Russia remained ignorant of the classical tradition, or knew of it only in garbled form through the barrier of inadequate translations. There was, in other words, no strongly developed tradition of ecclesiastical or secular learning in Russia, and the church regarded the maintenance of doctrinal truth and the eradication of heresy as its primary tasks, by comparison with which the search for truth in any broader sense was regarded as insignificant and even dangerous. Printing was introduced in the sixteenth century but remained virtually a church monopoly and the few books and materials published were mainly devotional or governmental in character.

Given the significance of printing to the spread of the New Learning in the West, it is important to note the different conditions for science and learning which existed in Russia before Peter. By and large, Russia's was not a literate culture (even by comparison with the very restricted degree of literacy which characterized the West). The new discoveries

---

[3] D.J.B. Shaw, "Mapmaking, Science and State Building in Russia Before Peter the Great." *Journal of Historical Geography* 31 (2005): 409–429.

[4] Ryan 9–24.

associated with the Renaissance in the West remained largely unknown in Russia until a later period. In geography, for example, the rediscovery in the West of Ptolemy's *Geography* early in the fifteenth century, which was to have such important repercussions for Western cartography, was probably unknown in Russia for at least two centuries. In any case Russia largely lacked the mathematical skills necessary for the adoption of Ptolemy's techniques at this period. Down until the eighteenth century popular conceptions of geography were largely based on a series of traditional texts which had been translated into Russian at an earlier period. One very celebrated one was the *Christian Topography* of Cosmas Indicopleustes, a sixth century Byzantine text which may have been introduced to Russia as early as the tenth century. Noted for its presentation of the concept of a flat earth situated at the centre of the universe, a copy of this text is said to have been in the possession of most monasteries in Peter's day. Another was the so-called *Elucidarius* which may have been translated in the early sixteenth century from German but based on a medieval Latin source. It is a detailed geography but lacks knowledge of the existence of the New World and contains numerous fantastic tales and legends. Similarly, the *Cosmography* of 1670 is a mixture of the factual and the fantastic and appears to be based on a variety of sources.[5]

Given the lack of a tradition of secular learning in Russia, and the underdevelopment of a commercial market whereby secular books and materials might have been acquired by some of those eager to learn, it was probably inevitable that the leading agency in the development of learning and science in Russia should have been the state. One aspect of the changes which began to affect Russia from the fifteenth century, and which were alluded to above, was the growing complexity of the state. As their realm expanded and as its defence and administration became more complicated, Russian rulers found it increasingly necessary to surround themselves with the administrative machinery and educated individuals which governance now required. No longer could government be conducted as in the past on the basis of largely informal and interpersonal relationships. What was required was a nascent bureaucracy. Before Peter this bureaucracy developed and expanded in

---

[5] T. Rainov, *Nauka v Rossii, XI–XVII veko*, Moscow-Leningrad 1940, 42ff, 224ff, 420ff; B.E. Raikov, *Ocherk po istorii geliotsentricheskogo mirovozreniya v Rossii*, Moscow-Leningrad 1947, 14ff.

an *ad hoc* fashion, but it proved to be an important instrument in the country's intellectual modernization.

As noted above, a number of the changes through which Russia began to go from the fifteenth century onwards had important implications for geographical endeavour. One was the beginnings of land inventory and survey. The unification of the Russian state was accompanied by a policy to systematize and centralize its military forces. As new lands were acquired, Russian rulers began to allocate them to their nobles and servitors on condition of service to the sovereign, usually meaning military service. By the middle of the sixteenth century this policy had been applied to all of Russia's landholders, including those who held their land by hereditary right. Such a policy naturally required a process of land inventory and recording in order to ensure that all land was being held in accordance with the state's requirements. The extent and character of landholdings, and the identity of the landholders, were recorded in cadastral documents which often described the detailed topography of the landholdings in addition. Needless to say, such documents were of key importance to the state in such matters as land allocation, the maintenance of property rights, the specification of service duties, and taxation. Cadastral surveys also permitted the drawing of detailed topographical maps in certain cases.[6]

Mapping itself soon came to be regarded as an important part of the state's functions. Not only was this seen as a significant aid to land inventory, but also as an aid to other aspects of government including defence and general administration. In Russia mapmaking appears to have been a late development, the first reference to a map (*chertezh*) dating from the late fifteenth century, though it is often forgotten that the art of mapmaking was a relatively uncommon pursuit even in medieval Europe.[7] Russia appears to have lacked the tradition of *mappaemundi* which was found in some parts of western Europe as well as of portolan or sea charts, but route maps (*dorozhniki*) were recorded from the later fifteenth century. Some scholars believe that the same period witnessed the first attempts to map the realm as a whole, based perhaps on Italian models (Italian specialists were hired to rebuild the Moscow kremlin at this period). Whilst almost none of the maps drawn

[6] A.V. Postnikov, *Razvitie krupnomasshtabnoi kartografii v Rossii*, Moscow 1989, 13ff.

[7] P.D.A. Harvey, *The History of Topographical Maps: Symbols, Pictures and Surveys*, London 1980, 155.

before the late seventeenth century have survived, inventories of the maps which were formerly in the possession of the various departments of state have done so, and it is clear that many regional maps, as well as large-scale maps often composed as part of legal processes involving land, were drawn up at this period.[8] Particularly significant was the so-called Great Map of the Russian realm as a whole, including parts of its surrounding territories, which appears to date from the 1590s and to have been redrawn around 1627. Unfortunately both these maps subsequently perished, but their accompanying textual description, known as the Book of the Great Map, has survived.[9] From these and other maps and descriptions it is apparent that the Russian state attempted to map much of its territory to some extent or other but that certain strategic parts, notably borders and vulnerable frontier regions, and especially the newly acquired territory of Siberia, were of particular interest.

As noted already, pre-Petrine Russia lacked developed mathematical skills and was probably ignorant of Ptolemy (though some elements of mathematical geography did begin to percolate in the seventeenth century in connection with trade and the acquisition of some Western publications like those of Mercator, Ortelius and Blaeu). The regional and local maps drawn before Peter's reign were, in fact, what Harvey calls 'picture maps', like most European maps of the late medieval or even early modern periods. In other words, they were not drawn to a fixed scale or projection (to say nothing of astronomically determined points—something which came in quite late even in Europe) but generally around a network of routes—usually rivers in the Russian case. The accent was on general direction (there is limited evidence of the use of the compass in Russia before the mid-seventeenth century), relative location of important features, and distance. The celebrated Siberian cartographer, Semen Remezov, who began to work on maps of the region in the 1680s, did begin to apply elements of the new mathematics to the maps and atlases he produced in the early part of Peter's reign.[10] But by this stage Peter's policies were already beginning to open the country up to the far reaching effects of modernization.

---

[8] Cf. Shaw 2005; V. Kivelson, "Cartography, Autocracy and State Powerlessness: The Uses of Maps in Early Modern Russia," *Imago Mundi* 51 (1999): 83–105.

[9] Shaw 2005.

[10] Leo Bagrow, "Semen Remezov: A Siberian Cartographer," *Imago Mundi* 11 (1954): 111–125.

Before turning to consider the immediate context for the publication of Varenius' book in Russia, which is provided by Peter the Great's reforms, two other aspects of pre-Petrine geographical endeavour deserve a brief mention. One is the translation into Russian of a number of foreign works of scientific significance. Much of this was done by specialists employed by departments of state, particularly the office dealing with foreign affairs and ambassadors in Moscow, known as the *Posol'skii prikaz*. The translation work can be seen as part of the limited programme of modernization which characterized pre-Petrine Russia, alongside the employment of foreign technical and military specialists and the selective adoption of technological innovations. Among the European works having geographical significance which were translated in whole or in part at this period were the atlases or cosmographies of Mercator, Ortelius and Blaeu, and Hevelius' *Selenography*. Such works were important not only for the factual information they contained but also because of the way they pointed in the direction of a new mathematical understanding of the world and universe and towards the new science of Copernicus, Tycho Brahe, Galileo and others. Unfortunately, most foreign translations were available only to government officials and perhaps to certain high ranking aristocrats and churchmen, since few were published. There was, however, a lively tradition in copying manuscripts.

The other geographical endeavour which should be mentioned is expeditionary work. This had a long pedigree in Russia, particularly in journeys to Siberia and other eastern territories undertaken by Cossacks, merchants and others from at least the fifteenth century. Most journeys had a severely practical purpose, such as enhancing trade, seeking out new fur bearing territories, or pursuing some military or diplomatic goal. Before the end of the seventeenth century, however, the idea of scientific discovery was also beginning to prove significant. Particularly important in this regard is N.G. Spafarius' expedition to China in 1675–6.[11] This was an official embassy from the tsar charged with the task of easing political tensions with China along the Amur River, facilitating trade and gathering information about the geography of the routes across Siberia to that country. Spafarius himself was a Moldavian of Greek descent and his expedition included several other foreigners. It was notable for the fact that it took along a number of scientific instruments, including

[11] Rainov 413ff.

astronomical instruments and compasses which aided the drawing of maps, and for its gathering of a wide variety of geographical and other information about eastern Siberia, China and other regions. In many ways it was the forerunner of Peter the Great's later expeditions, notably those of Messerschmidt and Bering.

It is apparent, therefore, that Peter the Great's reforms, as far as they concerned geography, did not build on a *tabula rasa* but rather on a variety of geographical endeavours which reflected the beginnings of Russia's rise as a major European state. Those endeavours became particularly noticeable during the half century or so which preceded the beginning of Peter the Great's reign, reflecting the changes through which Russia was now starting to pass in the context of an even more dynamic Europe. Yet these changes were, of course, as yet insufficient to transform Russia or to make it into a truly European power.

It is probably a mistake to think that Peter the Great's reforms were motivated by any clearly thought out ideology. Equally, however, there can be little doubt about the connection in his mind between power and defence on the one hand and the instrumental value of the New Learning on the other. Peter's interests in the new European science were primarily practical, particularly at first. We know of his keen interest in geographical matters, derived not only from the general European intellectual milieu of his day but also, no doubt, from the more immediate influence of both foreign and Russian acquaintances, including Jacob Bruce (Yakov Brius), a Russian of Scottish descent. Peter was clearly well aware of the significance of the new geography for his projects of reform, including military and frontier defence, territorial administration and surveillance, resource inventory, foreign relations, trade, education and much else. Eventually he also became aware of the significance of geographical achievement for enhancing his own international standing and that of the realm over which he ruled. Equally, however, he proved himself most dissatisfied with Russian geographical progress prior to his reign. A geography which was not solidly based on the latest mathematics would do little to meet the state's needs for accurate mapping and comprehensive survey. A geography which was not comprehensive would do little to enhance accurate knowledge either of Russia's own sprawling territories, peoples and resources, or of those of its neighbours and enemies. A geography which mixed fact and fable in indiscriminate fashion would hardly foster a rational understanding of the world on the part of his subjects. A geography which paid scant regard to the world's oceans could not possibly serve Russia's growing naval ambitions. And a geography which remained unpublished and was known to only a handful of government and church officials and the tiny minority of intellectuals would certainly fail to lift the veil of mental darkness and superstition which was the source of so much opposition to his reforms. The reform of both the theory and practice of geography therefore became a vital part of Peter's policies.

Space forbids a detailed exposition of Peter's many geographical enterprises, and only a brief synopsis can be made here.[12] One important endeavour, for example, was in the area of surveying and mapping, beginning with the tsar's determination to map in as accurate manner as possible such objects as defensive works, overland and river routes, and the general theatre of war in his military campaign against the Turks in the late 1690s. In this effort Peter employed both talented Russian associates like Jacob Bruce and foreign specialists. The resulting maps were published in Amsterdam under a special contract and were among the first Russian maps ever to be published (later, secular presses capable of publishing maps were established in Russia itself). Other cartographic enterprises encompassed coasts and theatres of war along the Black, Baltic and Caspian Seas and state frontiers with Turkey, Sweden and China. Eventually Peter embarked on the extremely demanding goal of mapping the empire as a whole, an enterprise so ambitious that the tsar did not live to see its completion.

Needless to say, such extensive cartographic endeavours would have been impossible without the necessary numbers of trained personnel who were sufficiently well acquainted with European mathematical and cartographic methods to meet the tsar's stringent demands. There were insufficient foreigners for this purpose. Peter therefore established special training institutions, beginning with the Moscow School of Mathematics and Navigation, founded in 1701, and continuing with the St Petersburg Naval Academy, established in 1715. For the Moscow School Peter hired the Scottish mathematician Henry Farquharson whom he had met on his famous 'Great Embassy' to the German lands, Holland and England in 1697–8, a journey which had done much to open the tsar's eyes to western European scientific and technical accomplishments. Farquharson and his two English assistants taught

[12] D.J.B. Shaw, "Geographical Practice and Its Significance in Peter the Great's Russia," *Journal of Historical Geography* 22 (1996): 160–176; Shaw, "A Strong and Prosperous Condition: The Geography of State Building and Social Reform in Peter the Great's Russia." *Political Geography* 18 (1999): 991–1015.

a mathematical syllabus including geography, allegedly based on that of the Royal Mathematical School at Christ's Hospital in London. The graduating 'geodesists' were employed on a variety of mapping and surveying projects. Peter also founded a series of other specialist as well as general schools, in many of which geography seems to have found a place in the mathematical curriculum. Finally, at the end of his life, the tsar decreed the establishment of the Academy of Sciences in which geography also found a significant place. Varenius' book, with its emphasis on geography as a mathematical and cartographical enterprise, fully accorded with this practical approach.

Mapping was clearly linked to the needs of defense but also more broadly to the whole process of modernization and development in which Peter was engaged and which also engaged the energies of other European states of the day. For Russia to survive and prosper in the competitive and rapidly changing world in which it now found itself, the country needed to discover itself, in other words to know its own territories and their potentials for contributing to national wealth and strength. Mapping was but one part of this project. Like other European rulers, Peter therefore engaged in a whole series of endeavours to collect information of all kinds about the different parts of the empire, including information of a statistical and systematic kind. In various ways this process had preceded Peter but it was now pursued in a much more systematic and comprehensive way. Similarly expeditions of various kinds were sent forth, especially to the little known eastern parts of the empire, to discover and describe new lands. Again, as we have seen, this was not a new initiative but, as illustrated by the expeditions of Messerschmidt and Bering towards the end of the reign, they were now conducted on the planned and more disciplined basis which had begun to characterize similar European endeavours. The one thing which Peter lacked, but from which other European rulers greatly benefited, was the presence of a lively commercial class and a developed intellectual elite who did so much to stimulate and disseminate geographical knowledge in other countries, often without the state's direct involvement. In Russia most of the initiative at this period had to come from the state. It was only later in the eighteenth century that other stimuli to geographical understanding became significant but even then in the context of formidable barriers to rapid change.

How, then, are we to understand the significance of the translation into Russian and the publication in 1718 in Moscow of Varenius'

Geographia Generalis?[13] Some historians have subscribed to what has been termed a 'moving goalposts' theory of Russian change and development. This is to say the contention that, throughout much of its history, no sooner has Russia set its sights on the adoption of some innovation from the West, then the West moves on to its next novelty, leaving the Russians to adopt something which is now outdated. Thus Russia is fated to trail permanently behind.

Was this true of Varenius' book, published in Russia nearly seventy years after it first appeared in Amsterdam? This seems hardly the case. By the early eighteenth century Varenius was something of a standard text in western Europe, with several editions appearing in the second half of the seventeenth century and the first part of the eighteenth. Two Cambridge Latin editions were edited by Isaac Newton in 1672 and 1681. Newton was highly esteemed by Peter's associate Jacob Bruce who had received mathematics training in London and who collected Newton's books. Bruce's library also contained a Russian edition of Varenius (Bruce may well have been the instigator behind the book's Russian publication, but there is also the possibility that the initiative came from Peter himself.).[14] Peter's adoption of Varenius' book is therefore by no means a confirmation of the 'moving goalposts' theory. On the contrary, its contemporary importance to western Europeans would in and of itself have confirmed its significance to Peter. Its translation and publication were undoubtedly a sign of his desire to see its influence disseminated among his (elite) subjects. Few Russians, it should be noted, could read Latin in Peter's day.

The Russian edition of *Geographia Generalis*, or *Geografiya general'naya* as it was called, was published in Moscow in 1718.[15] It appears to have been based on the Amsterdam edition of 1664. The translator and editor, Fedor Polikarpov, was an official of the Moscow printing office. Cracraft has discussed Polikarpov's introduction and some of the many difficulties he faced in his translation. Lamenting his ignorance of the "geographical and mathematical sciences", for example, Polikarpov also

[13] This issue, which I briefly discussed in 2006 ("The Translation and Publication of Bernhard Varenius' *Geographia Generalis* in Russia, 1718," *Study Group on Eighteenth Century Russia Newsletter* 34 (2006): 9–13) will be further explored in a forthcoming paper.

[14] James Cracraft, *The Petrine Revolution in Russian Culture*, Cambridge, Massachusetts 2004, 206.

[15] Cracraft 205–210.

Fig. 12. *Geografiya general'naya*, 1718. Frontispiece of the Russian translation

Fig. 13. *Geografiya general'naya*, 1718.
Engraved title-page

apologises for his inability to reproduce Varenius' elegant Latin in the Slavonic tongue of his day. He evidently felt that Varenius' erudite prose demanded employment of only the most scholarly Slavonicisms, whilst admitting that such practices would hinder the book's ready comprehension by young Russian students of science. Indeed, Cracraft's archive evidence shows that Polikarpov's initial manuscript draft of 1716 was rejected by the tsar on the grounds of poor and inadequate translation. He was ordered to use "the simple Russian language [...] of the Ambassadorial Office"—in other words the new, simplified, bureaucratic language of government. Perhaps nothing speaks so eloquently of the pragmatism of Peter's goals. The tsar was determined to modernize at all costs and no considerations of scholarly taste and refinement were to be permitted to interfere with his plans.

Peter the Great's intellectual interests, it is sometimes argued, were narrow and practical in focus. There is some truth in this assertion. The tsar himself was a man of action, taken up with military and other practical matters, consumed by his desire to reform his realm along European lines, and with little time and probably not much patience for intellectual pursuits. There is much in Varenius which would have appealed to the tsar's practical side. Varenius' description of geography as a 'mathematical science' must have appealed strongly to a ruler who was all too conscious of Russia's mathematical backwardness and the hindrance this posed to the implementation of many of Peter's cherished reforms. In contrast to the largely descriptive cosmographies of the past, with their mixture of fact and fable, here was a text steeped in the modern European science, whose explanations and descriptions were based on logic and seemingly solid empirical evidence. No doubt the treatment of 'general geography' was regarded as providing much of importance for cartography, navigation, trade and much else. In the dedication to *Geographia Generalis*, Varenius had written of the usefulness of what he called "a knowledge of the earth", among many other things, to "the arts of trading and navigation". "Mariners", he asserted, "acknowledge the enormous [...] aid which geography can give when they undertake to cross distant seas and a raging ocean, relying on the accuracy of the maps and principles which geography has bestowed on them for their use".[16] Peter cannot have been insensible

[16] Quoted in J.N.L. Baker, "The Geography of Bernhard Varenius," *The History of Geography*, Oxford 1963, 111.

of such sentiments. He was not, as is sometimes asserted, interested only in military power. Power was certainly close to his heart, as it was to that of all European rulers, but the link between power, trade and commerce was also evident to all.

Yet to argue that Peter's interests were primarily practical is not to assert that they were exhausted by the practical. Peter, after all, was a tsar who had been obsessed by the European spirit ever since he had consorted with the denizens of Moscow's foreign suburb in the days of his youth, a free association with foreigners which lasted for the whole of his life.[17] He was a tsar to whom modern or European fashions and appearances mattered deeply, who insisted his aristocratic associates dressed in European clothes and discarded their Muscovite beards, who wanted them to copy European manners, who adopted European timekeeping and the European calendar, who encouraged European architecture, and who hired European specialists so that his new capital of St Petersburg should take on the appearance (and if possible assume the spirit) of a European city. Whatever his practical intellectual interests may have been at first, Peter gradually began to see the connection between science and fame, knowledge and earthly glory. Here was a tsar who evidently took great delight in his election to the Paris Academy of Sciences in 1717 and pride in the sensation caused by the presentation to the Academy in 1721 of the latest Russian map of the Caspian Sea. At the launching of a warship in 1719 he is said to have spoken of the sciences which, having migrated from their original home in Greece, possibly quitting their present abode in England, France and Germany and coming "to settle for some centuries among us". Pray and work, he exhorted his hearers, and "in that case be persuaded, you may happen even in our lifetime to put other civilized nations to the blush, and to carry the glory of the Russian name to the highest pitch".[18] In an age in which geography, together with all its many associated activities such as survey, mapping, astronomy and navigation, was seen as central to the modern way of viewing, controlling and changing the world, of course a book like Varenius' *Geographia Generalis* would have appealed to Peter and his associates. They will have regarded it as essential to their whole modernizing project.

[17] Lindsey Hughes, *Russia in the Age of Peter the Great*, New Haven 1998; and Hughes, *Peter the Great: A Biography*, New Haven 2002.

[18] L.R. Lewitter, "Peter the Great and the Modern World," *Russia and Europe*, ed. Paul Dukes, London 1991, 95.

It is, of course, impossible to trace the significance of one book only in a situation in which so many influences were at work. Russia, or at least its higher echelons, was modernizing; geography and hence Varenius' celebrated book were seen as essential to that modernization. Other books, like Philip Cluverius' *Introductio in Universam Geographiam*, Christian Huygens' *Kosmotheoros*, and Johann Hübner's *Kurtze Fragen*, were also translated and published in this period. Together they no doubt helped open Russian eyes to the significance of geographical space, as viewed and analysed in a European perspective.

Much has been made of the importance of the translation of Varenius' text in introducing to the Russian language many entirely new scientific terms (though not all scholars have regarded the translation of Varenius in particular as entirely successful in this respect). Cracraft has listed several hundred scientific and technical terms[19] which were permanently added to the Russian language in Peter's reign, not a few of which came from the Russian edition of *Geographia Generalis*. As Cracraft notes,[20] Polikarpov successfully struggled to find suitable Slavonic terms for "the technical vocabulary of the science [of geography] as it had developed in Europe and particularly Holland by the later seventeenth century"—"translation, not simply transliteration, was taking place".

The translation and publication of Varenius' work was in fact part of the whole difficult process whereby modern science was assimilated into a traditional culture to which it was, in its essentials, entirely alien. We know that Varenius' book found its way into the libraries of some of the principal elite of Petrine Russia. No native Russian geography was published under Peter, and the first geographical textbook suitable for use in schools became available only in the 1740s. As noted already, geography appeared in the curriculum of many schools as a practical discipline, but after Peter's death in 1725 secular education in Russia suffered a reverse. The influence of Varenius on education in Russia must have been an indirect one. According to Aleksandrovskaya,[21] however, his importance for some intellectual proponents of geographical knowledge and practice in eighteenth century Russia was more apparent. Such persons included Vasilii Tatishchev and Mikhail Lomonosov, often regarded as fathers of Russian geography. As Cracraft notes: "*Geografiia*

[19] Cracraft 430–471.
[20] Cracraft 209f.
[21] O.A. Aleksandrovskaya, *Stanovlenie geograficheskoi nauki v Rossii v XVIIIv*, Moscow 1989.

had become a Russian science, one fully practicable in Russian, within a few decades of the publication of the Russian Varenius".[22]

In conclusion, then, Russia differed from most European states to its west because of its very different intellectual and cultural heritage. At the same time, as we have seen, the country's unification and expansion from the fifteenth century did confront it with significant challenges which could only be met by a limited degree of modernization. Peter the Great, however, set out to close the gap which had by now developed between his realm and the principal European states by attempting to transform Russia into a modern European power. Part of that transformation included a reformation of the Russian geographical understanding of the world, and a key role in that reformation was played by Varenius. Varenius' view of the world was that of modern European science, grounded in mathematical logic and empirical evidence, a far cry, it seemed, from the traditional cosmographies and unreliable historical geographies of Russia's past. It is this fact which, no doubt, appealed to Peter and his associates. Varenius not only underpinned the practical emphasis which lay at the heart of Petrine geography, but helped open Russian eyes to the modern way of viewing and explaining the world. It helped move Russia, in other words, closer to Europe. In Peter's eyes, that could only be to the good.

[22] Cracraft 210.

APPENDIX

Extracts from *Geographia Generalis (1650)*

Epistula Dedicatoria (*2v–*6v)	Widmungsschreiben
Pars absoluta: Sectio prima, caput 1 (1–9)	...Erster Abschnitt, Kapitel 1
2 folded tables inserted after p. 8, in facsimile	
["Pag. 8"] Generalis Geographia in tres partes...	Die Allgemeine Geographie ist...
["Pag. 9"] Specialis Geographia considerat...	Die Spezielle Geographie betrachtet...

The Latin text of the beginning sections of *Geographia Generalis* in the first edition of 1650 is juxtaposed with the German translation by Ernst-Christian Volkmann. The original pagination of the Latin text is indicated in the transcription. Ligatures or the long "s" are not reproduced. Obvious misprints and superfluous commas are corrected, usually in accordance with subsequent editions, and the original version is given in brackets [◊...]. The original Latin text is usually not broken into paragraphs.

German readers may also want to consult a) Ulrich Staffhorst's translation of the letter of dedication (in Kaminske/Staffhorst 1986/87: 95–99), and b) Hanno Beck's translation and résumé of significant parts of chapter 1 (1973: 116–122).

English readers are referred to Dugdale's translation in Varenius, *A Compleat System* (1712), and in particular to Appendix I in Bowen (276–283). (Baker comments on translated and paraphrased passages, but does not give a continuous translation.)

[*2v]

EPISTOLA DEDICATORIA

NOBILISSIMI, AMPLISSIMI, SPECTATISSIMI, ET PRUDENTISSIMI CONSVLES

Tota haec rerum Universitas, quam Mundum appellare consuevimus, à longo jam tempore communi quasi hominum consensu divisa fuit in Tellurem & Coelum. Quae partitio etsi minus regulis Logicis & ipsi naturae conveniens videatur, partim quod essentialem Telluris à coelo differentiam supponit, partim quod exiguam Tellurem toti coelo, hoc est centrum orbi comparat, tamen ab omnibus ferè jam recepta, & tanquam legitima approbata fuit, idque varias ob causas. Plerique ideo Mundum ita dividendum censent, quod tellus sit diversae admodum à coelestibus corporibus naturae: illam enim variis mutationibus & corruptionibus obnoxiam esse: haec verò incorruptibilia sine omni jactura persistere autumant. Deinde ajunt, Tellurem adeo parvam esse si comparetur ad Coelum, ut sit instar puncti vel centri: centrum autem diversam valde naturam habere ab ipso orbe, / [*3r] ac proinde rectè Tellurem & Coelum opponi, tanquam diversa & contraria. Caeterum hae rationes non sunt veritati consentaneae. Etenim neque excepta sunt à mutationibus & corruptionibus caelestia corpora, ut demonstrarunt hujus & praecedentis seculi Mathematicorum observationes: neque Tellurem centrum coeli occupare firmis ostensum est hactenus argumentis. Quod si ob parvitatem opponenda esset coelo, multò rectius Luna & quaedam aliae stellae, quae minores sunt Tellure, eidem opponerentur. Cum itaque hae rationes falsitatibus laborent, aliae afferendae sunt, propter quas totius hujus Vniversi unam partem Coelum, alteram esse Tellurem dixerint Majores nostri. Illae videntur esse hae. Prima, quoniam Tellus

WIDMUNGSSCHREIBEN

HOCHVORNEHME, HOCHANGESEHENE, HOCHBEWÄHRTE UND ERFAHRENE KONSULN!

Dieses ganze Weltall, das wir Welt zu nennen pflegen, ist seit schon langer Zeit wie durch einhelligen Beschluss der Menschen in Erde und Himmel eingeteilt. Auch wenn diese Einteilung den Regeln der Logik, ja sogar der Natur selbst, recht wenig angemessen zu sein scheint, teils weil sie einen Wesensunterschied zwischen Himmel und Erde unterstellt, teils weil sie die winzige Erde mit dem gesamten Himmel, d.h. das Zentrum mit dem Himmelsgewölbe vergleicht, ist sie doch von fast allen Menschen nunmehr gutgeheißen und gewissermaßen gesetzlich anerkannt, und zwar aus mannigfaltigen Gründen. Die meisten meinen, die Welt deshalb so einteilen zu müssen, weil die Erde von völlig verschiedener Natur im Vergleich zu den Himmelskörpern sei: Jene nämlich sei mannigfachen Veränderungen und Zerrüttungen ausgesetzt, diese dagegen, so meint man, verharrten unvergänglich ohne jegliche Einbuße [in ihrem unwandelbaren Zustand]. Ferner sagt man, die Erde sei bei einem Vergleich mit dem Himmel so klein, dass sie so viel wie ein Pünktchen oder ein Kreismittelpunkt sei: das Zentrum aber habe im Vergleich zum eigentlichen Kreise eine sehr verschiedenartige Natur, und darum würden Erde und Himmel zu Recht einander gegenübergestellt, wie etwas Verschiedenes und Gegensätzliches. Nun aber befinden sich diese Begründungen gar nicht im Einklang mit der Wirklichkeit. Denn weder sind die Himmelskörper von den Veränderungen und Zerrüttungen ausgenommen, wie die Beobachtungen der Mathematiker dieses und des vorigen Jahrhunderts gezeigt haben, noch ist bisher durch zuverlässige Beweise dargelegt, dass die Erde den Mittelpunkt des Himmels einnehme. Wenn sie aber wegen ihrer Kleinheit dem Himmel gegenübergestellt werden müsste, dann würden mit viel mehr Recht der Mond und gewisse andere Gestirne, die kleiner als die Erde sind, ihr auch gegenübergestellt werden. Da also diese Begründungen an Unrichtigkeiten leiden, müssen andere beigebracht werden, derentwegen unsere Vorfahren gesagt haben, dass der eine Teil dieses gesamten Universums der Himmel, der andere die Erde sei. Jene Begründungen scheinen folgende zu sein: Erster Grund: Da ja die Erde die Heimstätte

est generis humani domicilium, ideoque à nobis non minoris aestimanda est, quam totum coelum. Altera, quia non tantum Tellus est domicilium nostrum, sed etiam ab illa prima nostra origo est, illa nobis praebet & conservationis & propagationis adminicula. Tertia, quoniam è Tellure nostra coelestium corporum motus contemplamur. Quarta, quia in Tellurem coelestia corpora vires suas exerere videntur.

Quinta, / [*3v] quoniam Telluris constitutio magis nobis cognita est, quam coelestium corporum. Verùm, utcunque se istae rationes habeant, id certè ab omnibus prudentibus conceditur, quod cognitio Telluris sit non tantùm homine dignissima, sed etiam tam in Republica literaria, quam ad vitae usum pernecessaria. Non jam dicam de usu quem praestat Historiarum studio, cujus lampades non immerito statuuntur Chronologia & Geographia: non de usu, quem praestat Theologiae, Physicae, Politicae, & aliis disciplinis: illas modo utilitates & commoditates consideremus, quibus mercaturam & rem nauticam, quae duo Vestrae Reipubl. aerarium, hoc est nervum rerum gerendarum constituunt, adjuvari à Geographia docet & testatur quotidiana in hac urbe experientia. Annon ad mercaturam feliciter & cum fructu, addo etiam jucundè, exercendam plurimum facit, scire regionum, ad quas merces mittendae, vel è quibus ad hunc vel alia loca deferendae sunt, situm, distantiam, interjecta maria, iter, vicina loca infensos vel amicos incolas habentia, & alia, quae quidem è Mappis Geographicis non sine honestâ animi voluptate hauriuntur? Nautae verò amplissimum imò divi-/[*4r]num Geographiae usum agnoscunt, quando remota maria & furentem Oceanum sulcare suscipiunt freti Mapparum Geographicarum & aliarum regularum, quas ad directionem navis Geographia illis largita est, certitudine. Quae etsi hodie in hac vestra civitate non tam à Mathematicis, quam ab artificibus lucri causa &

der Menschheit ist, darf sie von uns nicht geringer geachtet werden als der gesamte Himmel. Zweiter Grund: Da ja die Erde nicht nur unsere Heimstätte, sondern auch unsere uranfängliche Ursprungsstätte ist, gewährt sie uns Hilfsmittel für die Entfaltung und Fortpflanzung. Dritter Grund: Von unserer Erde aus beobachten wir ja die Bewegungen der Himmelskörper. Vierter Grund: Die Himmelskörper scheinen ja ihre Kraft auf die Erde auszuüben.

Fünfter Grund: Die Beschaffenheit der Erde ist uns ja bekannter als die der Himmelskörper.—Doch wie auch immer es um diese Begründungen steht, das eine wird sicherlich von allen einsichtsvollen Menschen eingeräumt, dass die Erkenntnis der Erde nicht nur des Menschen würdigstes Anliegen, sondern auch—wie in einem um die Wissenschaften bemühten Staat, so auch für die Praxis des Lebens—eine durch und durch notwendige Aufgabe sei. Ich will nicht mehr über den Nutzen sprechen, den [die Kenntnis der Erde] für das Studium der historischen Forschungen gewährt, als dessen Glanzlichter nicht ohne Verdienst die Chronologie und Geographie gelten; auch nicht über den Nutzen, den sie für Theologie, Physik, Politik und andere Disziplinen gewährt: Nur jene Vorzüge und Annehmlichkeiten wollen wir betrachten, durch die, wie die tägliche Erfahrung in dieser Stadt lehrt und bezeugt, Handel und Seewesen, zwei Dinge, die das Staatsvermögen, d.h. den Nerv aller Tätigkeiten und Vorhaben Eurer Republik begründen, von der Geographie her gefördert werden. Kenntnis von der Lage der Regionen, in die man Waren schicken will oder aus denen sie hierher oder in andere Orte gebracht werden sollen; Kenntnis von der Entfernung; Kenntnis, welche Meere dazwischenliegen; Kenntnis der Route; Gewissheit, ob die Bewohner benachbarter Orte freundlich oder feindselig eingestellt sind und dergleichen mehr, was jedenfalls sich aus den geographischen Karten nicht ohne ehrenvollen geistigen Genuss entnehmen lässt: Macht das alles etwa nicht am meisten aus, um glücklich und mit Gewinn und, so möchte ich hinzufügen, auch zur Freude Handel zu treiben? Die Seeleute vollends erkennen den ganz bedeutenden, ja geradezu göttlichen Nutzen der Geographie an, da sie es unternehmen, entlegene Meere und den tosenden Ozean zu durchpflügen im Vertrauen auf die Zuverlässigkeit geographischer Karten und anderer Regeln, die ihnen zu Steuerung ihres Schiffes die Geographie geschenkt hat. Auch wenn diese [Karten und Regeln] heute in eurem Staat nicht so sehr von Mathematikern als vielmehr von

nautarum informatoribus, imò ab ipsis nautis fiant & doceantur, nemo tamen ideo existimare debet, ab illis, non à Mathematicis profecta haec esse. Etenim inventio illarum rerum & demonstratio, unde certitudo dependet, multò difficilior fuit & majoris est operae, quam usurpatio: hanc Mathematici ita facilem reddidere, ut etiam illi, qui Mathematici non essent, capere possent: illam vero, demonstrationem dico, quoniam ex ipsis Mathesis penetralibus, nempe Theoretica Geometria, maxima ex parte petitur, soli Mathematici asservant, quandoquidem majorem postulat ingenii fatigationem, quàm homines, qui lucro tantùm intenti solam usurpationis & utilitatis cognitionem quaerunt, sustinere velint vel possint. Neque hoc tantum in Geographia locum habet, sed etiam in plerisque aliis Scientiis Mathe-/[*4v]maticis idem perspicuè licet videre, inprimis in Arith[◊e]metica.

Quid ejus usu & exercitio hodie vulgarius, praesertim in hac vestra civitate? quid facilius? quis è mercatoribus, imò mercatorum pueris, non novit promtissime calculum juxta regulas Arithmeticae instituere, imò etiam alios artem hanc docere? At verò, antequam regulae, juxta quas praxis instituitur, à Mathematicis essent è penetralibus Arithmeticae erutae & demonstratae, & methodus expedita exercitii monstrata, planè tunc nemo erat qui ista facere posset. Neque si ab illis, qui promtissimi sunt in usu, quaeras, cur hoc vel isto modo ad supputationem agendum sit, quicquam rationis vel fundamenti nancisceris, quoniam demonstrationes à Mathematicis tantum & fiunt & asservantur: neque si nova quaedam regula pro novo usu invenienda sit, ad alios quam Mathematicos recurritur. Et eodem modo cum Perspectiva, statica Mechanica, Musica, Arte muniendi, Geodaesia & aliis scientiis comparatum esse perquam facile est ostendere: ut inde manifestum sit, quanta beneficia à Mathesi in vitae humanae usum, inprimis vestram civitatem, profecta sint. Sed ut ad Geographiam redeam, etsi haec / [*5r] tantùm de Tellure agat, quae instar puncti est si cum Coelo conferatur, tamen non minus peculiarem partem Mathesis constituit quam Astronomia, idque propter causas supra allatas. Duplex autem est Geographia: una Generalis, altera Specialis. Illa considerat Tellurem in genere, varias

Handwerkern des Gewinns wegen und von Informanten der Seeleute, ja sogar von den Seeleuten selbst hergestellt und gelehrt werden, so darf darum doch niemand glauben, dass sie von jenen, und nicht von den Mathematikern, herrührten. Denn die Findung jener Angaben und ihr Nachweis, wovon die Zuverlässigkeit abhängt, war viel schwieriger und erfordert mehr Mühe als ihre Benutzung. Diese Benutzung haben die Mathematiker so leicht gemacht, dass auch diejenigen, die keine Mathematiker sind, damit umgehen können, denn die Findung erfordert ja mehr ermüdende geistige Anstrengung als Menschen, die nur unter dem Blickwinkel des Gewinns einzig die Kenntnis der Nutzanwendung erstreben, aushalten wollen oder können. Das ist freilich nicht nur in der Geographie so, sondern auch in den meisten anderen mathematischen Wissenschaften kann man dasselbe deutlich sehen, vor allem in der Arithmetik.

Was ist heute allgemein üblicher als ihre Anwendung und die Beschäftigung mit ihr, zumal in diesem eurem Staate? Was leichter? Wer von den Kaufleuten, ja sogar deren Kindern, verstünde es nicht, aufs rascheste entsprechend den Regeln der Arithmetik den Rechner einzusetzen, ja sogar diese Fertigkeit andere zu lehren? Dagegen gab es, bevor die Mathematiker die Regeln, nach denen die Praxis gelehrt wird, aus dem Heiligtum der Arithmetik ans Tageslicht gebracht und dargelegt und eine bequeme Übungsmethode gezeigt hatten, wirklich niemanden, der jene Dinge hätte tun können. Weder wird man, wenn man jene fragt, die mit der Anwendung sehr rasch bei der Hand sind, warum auf diese oder jene Weise bei einer Berechnung zu verfahren ist, eine vernünftige Begründung bekommen, da ja Beweisführungen nur von Mathematikern erbracht und überwacht werden, noch greift man, wenn irgendeine neue Regel für eine neue Anwendung zu finden ist, auf andere als auf Mathematiker zurück. Und dass es sich mit der Kunst der Perspektive, Statik, Mechanik, der Musik, der Befestigungskunst und den anderen Wissenschaften ebenso verhält, ist sehr leicht zu zeigen, so dass von daher ganz deutlich ist, wie umfangreiche Wohltaten von der Mathematik für die Bedürfnisse des menschlichen Lebens, besonders für euren Staat, herrühren. Um aber zur Geographie zurückzukehren: Zwar handelt sie nur von der Erde, die im Vergleich zum Himmelsgewölbe so groß wie ein Pünktchen ist, doch bildet die Mathematik einen ebenso bedeutenden Teil wie die Astronomie, und das wegen der oben angeführten Gründe. Doppelt aber ist die Geographie: die eine ist die allgemeine, die andere die spezielle. Jene betrachtet die

ejus partes & generales affectiones explicans: haec autem, specialis scilicet, observans regulas generales recenset singularum regionum situm, divisionem, terminos & alia cognitu digna. Caeterum qui hactenus Geographiam scripserunt, illi de sola fere speciali prolixè egerunt, pauca verò ad Generalem pertinentia explicarunt, plurimis iisque necessariis neglectis & omissis, ita ut juvenes specialem addiscentes fundamenta hujus disciplinae maxima ex parte ignorarent, & ipsa Geographia vix Scientiae titulum tueretur. Quod cum animadverterem, ut malo huic remedium facerem, caepi cogitationes meas ad supplendum hunc defectum hoc est, ad scribendam Generalem Geographiam convertere, neque à suscepto labore cessavi, donec opus absolutum esset, ut nimirum pro mearum virium modulo, quem ex diligenti Mathematicarum disciplinarum per multos an-/[*5v]nos cultura Deus mihi concessit, Reipubl. literariae atque juventutis studiis inservirem, vel saltem voluntatem inserviendi probarem. Hunc meum laborem cur vobis, Viri Magnifici, inscripserim & consecraverim, plurimis inductus sum rationibus. Prima, quoniam toto terrarum orbe nulla est civitas, quae magis quàm haec Vestra Geographiae cognitione & indigeat & utatur propter admirandas in omnes terrae angulos navigationes. Secunda, quia Vestrae Reipubl. navigationibus non parum crevit Geographiae studium, multa dubia & incertitudines sublatae sunt. Tertia, quoniam Vos cognovi esse omnis Eruditionis Fautores & Promotores, ideo nullus dubito, quin Geographicum quoque studium maximi aestimetis. Quarta, eaque praecipua causa est, quod hoc meum opus in Vestra urbe & confectum & in lucem editum fuit. Non itaque sub alterius Patroni favore, quam Vestro, in lucem prodire debuit. Etenim cum Patria mea variis prius belli injuriis afflicta tandem in favillas & cineres tota redacta esset, bono quodam fato vel potius Dei Opt. Max. providentia in Vestram civitatem me contuli, ubi commodam nactus sum scri-/[*6r]bendi occasionem de iis rebus, quas in alio loco agens vix potuissem absolvere, quandoqui-

Erde im Allgemeinen, indem sie ihre mannigfaltigen Teile und allgemeinen Merkmale darlegt; diese aber, d.h. die spezielle, mustert unter Beachtung der allgemeinen Maßstäbe die Lage der einzelnen Gegenden, ihre Gliederung, Grenzen und anderes Wissenswertes. Übrigens haben die bisherigen geographischen Schriftsteller fast ausschließlich über die spezielle [Geographie] Abhandlungen geschrieben und nur weniges mit Bezug auf die allgemeine Geographie dargelegt unter Vernachlässigung und Auslassung sehr vieler und zwar notwendiger Dinge, so dass junge Leute beim Hinzulernen der speziellen Geographie die Grundlagen dieser Disziplin größtenteils nicht kannten und die Geographie selbst nur mit Mühe den Titel „Wissenschaft" behauptete. Als ich das bemerkte, begann ich, um diesem Übel abzuhelfen, meine Gedanken auf das Schließen dieser Lücke, d.h. auf das Schreiben einer Allgemeinen Geographie, zu richten, und ich hörte mit der begonnenen Mühe nicht auf, bis das Werk vollendet wäre, um nach dem Maß meiner allerdings bescheidenen Kräfte, die mir Gott aufgrund der sorgfältigen Pflege der mathematischen Wissenschaften viele Jahre hindurch gewährt hat, der um die Wissenschaft bemühten Republik und den Studien der Jugend zu Diensten zu stehen oder doch wenigstens meinen guten Willen zum Dienen unter Beweis zu stellen. Dass ich diese meine Arbeit an Euch, Hochherzige Männer, adressiert und Euch gewidmet habe, dazu haben mich sehr viele Überlegungen bewogen. Erster Grund: Es gibt keine Bürgerschaft auf dem ganzen Erdenrund, die mehr als Eure nach Kenntnis der Geographie verlangt und sich ihrer bedient wegen der bewunderungswürdigen Seefahrten in alle entlegenen Winkel der Erde. Zweiter Grund: Dank den Seefahrten Eurer Republik nahm die Bemühung um die Geographie nicht unwesentlich zu, und viele Zweifel und Unsicherheiten wurden beseitigt. Dritter Grund: Ihr seid, wie ich weiß, die Förderer und Gönner jeglicher Bildung, und darum zweifle ich überhaupt nicht, dass Ihr auch den Wert des geographischen Studiums ganz hoch einzuschätzen wisst. Vierter Grund, und er ist der hauptsächlichste: Dieses mein Werk ist in Eurer Stadt sowohl vollendet als auch veröffentlicht worden. Es durfte daher unter der Gunst keines anderen Patrons als unter Eurer Schirmherrschaft erscheinen. Denn als mein Vaterland durch die Gewalttätigkeiten des Krieges zuerst nur mitgenommen, dann aber schließlich ganz in Schutt und Asche gelegt war, habe ich mich unter einem guten Geschick oder vielmehr dank des besten und höchsten Gottes Vorsehung in Euer Gemeinwesen begeben, wo ich angemessene Gelegenheit fand, über diejenigen Dinge zu schreiben, die ich trotz meiner Tätigkeit an anderem Ort wohl kaum

dem non pauca ab aliis Scriptoribus vel omissa vel perperam tradita de Terrae & Oceani partium conditione, situ, ventis, & populorum moribus exquirenda fuerunt à nautis vel aliis, qui in regionibus istis commorati fuerant. Multa in hoc libro loca id testantur: idem manifestum est ex commentariis, quos praecedenti anno de Iaponiae regno, religione, ut & diversa gentium religione edidi: & magis id perspicuum evadet, si favore Vestro & liberalitate studia mea juvabitis, ut perficere liceat, quae inter alias meas meditationes de naturalibus observationibus in variis Telluris partibus, quae de cibo & potu diversorum populorum, quae de cibi & potus essentia & materia, item de diversa diversarum gentium medicina, & quae de medicamentis paratu facilibus atque aliis rebus commentari incepi.

Laus vestra est, PROCERES, quod industrios & labore suo Reipublicae inservire cupientes atque hoc ipso opere demonstrantes libenter foveatis, alatis & promovere studeatis: date & mihi cujus conatus adversa fortuna Marte & Vulcano impedire satagit, ut vestram liberali-/[*6v]tatem laudare, Vobisque eorum, quae publico bono edere studeo, promotionem asscribere possim, quo tota Respubl. literaria intelligat, quantum Vestrae Prudentiae debeatur. Quod si feceritis, operam dabo ut intelligatis, non in hominem ignavum & Vestra liberalitate atque favore abutentem beneficium Vos contulisse. Deus Vos, vestramque Rempubl. diu incolumem servet. Valete. Ex meo musaeo Calendis Augusti, Anno M D C L.

V^{ae} . *Magf*^{ae} . *Ampl*^{is}. *& Prud*^{ae}.
Observantissimus
BERNHARDUS VARENIUS,
Med. D.

hätte vollenden können, da nun einmal andere Schriftsteller recht vieles ausgelassen oder die Erdteile und Teile des Ozeans, ihre Lage, die Winde und die Sitten der Völker betreffende Dinge falsch überliefert haben, Angaben, nach denen man Seeleute und andere Personen fragen musste, die sich in jenen Gegenden aufgehalten hatten. Viele Stellen in diesem Buch bezeugen das; dasselbe wird auch deutlich aus meinen Aufzeichnungen, die ich im vorangegangenen Jahr über das Japanische Reich, die Religion wie auch über die unterschiedliche Religion der Volksstämme herausgegeben habe. Und das wird noch deutlicher hervortreten, wenn Ihr mit Eurer Gunst und Freigebigkeit meine [wissenschaftlichen] Bemühungen unterstützt, damit ich vollenden kann, was ich unter anderem bei meiner gedanklichen Arbeit über die Naturbeobachtungen in den verschiedenen Erdteilen, über Speise und Trank und deren Konsistenz bei den verschiedenen Völkern, dergleichen was ich über ihre unterschiedliche Heilkunst und was ich über die leicht herstellbaren Heilmittel und andere Dinge zu kommentieren begonnen habe.

Euch zum Ruhme, vornehme Herren, gereicht es, dass Ihr Menschen, die fleißig sind und durch ihre eigene Arbeit dem Staate zu Diensten sein wollen, unterstützt, ihnen Unterhalt gewährt und sie zu fördern Euch bemüht: Gewährt auch mir, dessen Anstrengungen mit Feuer und Schwert zu hemmen ein widriges Geschick sich eifrig bemüht, dass ich Eure Hochherzigkeit loben und die Förderung dessen, was herauszugeben ich mich bemühe, Euch zuschreiben kann, damit der gesamte um die Wissenschaften verdiente Staat einsieht, wieviel man Eurer Einsicht schuldet. Wenn Ihr das tut, werde ich mich um Eure Einsicht bemühen, dass Ihr keinem untüchtigen Menschen, der Eure Großzügigkeit und Gunst missbraucht, eine Wohltat erwiesen habt. Gott möge Euch und Eure Republik lange wohlbehalten bewahren. Lebt wohl. Aus meinem Studierzimmer am 1. August des Jahres 1650.

Euch, hochherzige, angesehene und einsichtvolle Persönlichkeiten
sehr verehrender
Bernhardus Varenius, Dr. med.

[p.1]

GEOGRAPHIAE GENERALIS PARS ABSOLUTA.
SECTIO PRIMA.
DE PRAECOGNITIS GEOGRAPHIAE.

Caput I
De Geographiae Definitione, Divisione, Methodo & aliis prolegomenis

Dudum invaluit consuetudo, ut illi qui scientiam aliquam aut disciplinam integrè pertractant, de hujus doctrinae conditionibus atque methodo & constitutione aliisque proprietatibus quaedam praemittant. Neque hoc sine ratione, modò absque obreptione Sophistica fiat, existimo fieri, cum per talem informationem Lectorum intellectus Ideam quandam totius disciplinae vel saltem argumentum illius concipiat, & quomodo versari debeat in hac disciplina, cognoscat. Nos itaque primo hoc capite de Geographiae Constitutione, & natura, pauca praefabimur.

Definitio

Geographia dicitur scientia Mathematica mixta, quae Telluris, partiumque illius affectiones à quantitate dependentes, nempe figuram, locum, magnitudinem, motum, caelestes apparentias, atque alias [◊aliis] proprietates affines docet.

[p. 2] A quibusdam nimis strictè sumitur pro sola regionum Telluris descriptione & distributione. Ab aliis contra nimis late extenditur, qui descriptionem politicam singularum regionum adjungunt. Sed hi facile excusantur, cum illud faciant ad retinendos & excitandos lectorum animos, qui nudâ regionum enumeratione & descriptione sine morum gentis explicatione somnolenti plerunque redduntur.

DER ALLGEMEINEN GEOGRAPHIE FÜR SICH BESTEHENDER TEIL. ERSTER ABSCHNITT. ÜBER DIE VORKENNTNISSE ZUR GEOGRAPHIE.

Kapitel I
Über die Definition von Geographie, Einteilung, Methode und andere Vorbemerkungen

Längst hat sich die Gewohnheit eingebürgert, dass diejenigen, die eine Wissenschaft oder ein wissenschaftliches Fach vollständig behandeln, einiges über die Verhältnisse, Methode und Beschaffenheit und andere Eigentümlichkeiten dieser Lehre vorausschicken. Und wie ich glaube, geschieht dies nicht ohne Grund, wenn sich nur keine sophistischen Spitzfindigkeiten einschleichen, da durch eine solche Information das Begriffsvermögen der Leser eine gewisse Vorstellung von der gesamten Wissenschaft erhält oder der Leser wenigstens den stofflichen Gegenstand in sich aufnimmt und erkennt, wie er mit diesem wissenschaftlichen Fach umgehen muss. Deshalb wollen wir in diesem ersten Kapitel einige Vorbemerkungen über die Beschaffenheit und das Wesen der Geographie machen.

Begriffsbestimmung

Die Geographie wird eine gemischte mathematische Wissenschaft genannt, welche die von der Quantität abhängenden Merkmale der Erde und ihrer Teile, also Gestalt, Gelände, Größe, Bewegung, Himmelserscheinungen und andere verwandte Eigentümlichkeiten lehrt.

Von einigen Leuten wird sie zu eng gefasst, nämlich nur für die Beschreibung der Erdregionen und ihre Verteilung. Von anderen dagegen wird sie zu weit ausgedehnt, die eine politische Beschreibung der einzelnen Regionen hinzufügen. Diese aber werden leicht entschuldigt, da sie jenes tun, um die Leser zu fesseln und anzuregen, die von der bloßen Aufzählung der Regionen und der Beschreibung ohne Darlegung der Bräuche eines Volkes meist in Schläfrigkeit verfallen.

Divisio

Dividimus Geographiam in Generalem & Specialem, sive Universalem & Particularem. (Golnitzius explicationem Geographiae ait duplicem esse, Exteriorem & Interiorem, sed impropria & Catachristica est haec appellatio, & sine ratione assumta, cum termini [◊,] Generalis & Specialis sint aptiores.) Generalis sive Universalis Geographia dicitur, quae Tellurem in genere considerat atque affectiones explicat non habitâ particularium regionum ratione. Specialis sive Particularis dicitur, quae singularum regionum Telluris constitutionem docet, estque duplex, Chorographia & Topographia. Chorographia regionis alicujus, magnitudinem saltem mediocrem habentis, descriptionem proponit. Topographia parvum aliquem Telluris tractum seu locum describit.

Nos hoc libro Generalem Geographiam exhibebimus, quam in tres partes distribuendam esse existimavimus, quae sint *Pars Absoluta, Pars Respectiva, Pars Comparativa.* In Parte absoluta contemplabimur corpus ipsum Telluris & partes ejus atque affectiones ejus proprias, ut figuram, magnitudinem, motum, terras, maria, fluvios, &c. In Parte Respectiva illas affectiones atque acci-/[p.3]dentia considerabimus, quae Telluri à caelestibus causis accidunt. Pars denique Comparativa continebit illarum proprietatum explicationem, quae ex diversorum locorum Telluris comparatione existunt.

Objectum

Objectum Geographiae, sive subjectum circa quod est Tellus, inprimis superficies ejus & partes.

Einteilung

Wir teilen die Geographie in eine allgemeine und eine besondere oder [mit anderen Worten] eine universale und partikulare. (Golnitzius sagt, die Auslegung der Geographie sei eine doppelte, eine nach außen und eine nach innen gerichtete, aber unpassend und missbräuchlich ist diese Benennung und ohne Überlegung gewählt, denn die Ausdrücke „allgemein" und „speziell" sind geeigneter.) Allgemeine oder universale Geographie wird diejenige genannt, die die Erde im Allgemeinen betrachtet und ihre Merkmale darlegt, ohne Rücksicht auf die einzelnen Regionen zu nehmen. Spezielle oder partikulare [auf einen Teil gerichtete] heißt diejenige Geographie, die die Beschaffenheit der einzelnen Regionen der Erde lehrt, und zwar ist sie doppelter Art, nämlich die Chorographie [Länderkunde] und Topographie [Ortskunde]. Die Chorographie gibt die Darstellung irgendeiner Gegend von wenigstens mittlerer Größe. Die Topographie beschreibt irgendeinen nur kleinen Landstrich der Erde oder einen Ort.

Wir werden mit diesem Buch die Allgemeine Geographie darbieten, die, wie wir geglaubt haben, in drei Teile eingeteilt werden muss, den „absoluten" Teil, den „berücksichtigenden" Teil und den „vergleichenden" Teil. Im „absoluten" Teil werden wir den Erdkörper selbst betrachten sowie seine Teile und charakteristischen Beschaffenheiten, d.h. Gestalt, Größe, Bewegung, Länder, Meere, Flüsse usw. Im „berücksichtigenden" Teil werden wir jene Eigenschaften und äußeren Umstände [Symptome] betrachten, die der Erde durch kosmische Ursachen widerfahren. Der „vergleichende" Teil schließlich wird eine Darlegung jener Charakteristika enthalten, die sich aus dem Vergleich verschiedener Gegenden der Erde ergeben.

Vorliegender Gegenstand

Vorliegender Gegenstand der Geographie oder das Zugrundeliegende hinsichtlich dessen, was die Erde ist, besonders ihre Oberfläche und Teile.

Affectiones

Triplicis generis esse videntur illa, quae in singulis regionibus considerari merentur, ac proinde in Speciali Geographia ad singulas regiones explicari possunt cum fructu discentium & Lectorum, nempe *Terrestria, Caelestia, & Humana.* Caelestes affectiones voco illas, quae ex apparenti solis & stellarum motu dependent, & octo esse videntur.

Primo, *Elevatio poli, distantia loci ab Aequatore & à polo.* Secundo, *Obliquitas motus diurni stellarum supra Horizontem illius loci.* Tertio, *Quantitas diei longissimi ac brevissimi.* Quarto, *Clima & Zona.* Quinto, *Calor, frigus atque anni tempestates: item pluvia, nix, venti & alia meteora.* Etsi enim haec ad terrestres proprietates possint referri, tamen quoniam cum quatuor anni temporibus & Solis motu magnam habent connexionem ideo ad caelestium classem retulimus ea. Sexto, *stellarum ortus, apparentia, & mora supra Horizontem.* Septimo, *stellae per verticem loci transeuntes.* Octavo, *quantitas vel celeritas motus,* quo juxta Copernicanam hypothesin singulis horis locus quilibet rotatur. Juxta Astrologos *nona affectio* addi posset, quoniam illi singulis regionibus praeficiunt unum ex duodecim Zodiaci signis at-/[p.4]que hujus signi planetam. Sed mihi vana haec doctrina semper visa fuit, neque ullum ejus doctrinae fundamentum perspicio: in fine tamen Specialis Geographiae recensebimus eorum hanc distributionem.

Haec de caelestium affectionum classe. *Terrestres* voco, quae in ipsius regionis loco considerantur, quarum numero decem. Primo, *Limites & circumscriptionem.* Secundo, *Figuram.* Tertio, *Magnitudinem.* Quarto, *Montes.* Quinto, *Aquas, nempe fluvios, fontes, sinus maris.* Sexto, *Sylvas & deserta.* Septimo, *Fertilitatem & sterilitatem, item fructuum genera.* Octavo, *Mineralia, sive fossilia.* Nono, *Animalia.* Decimo, Longitudinem loci, quae primae terrestri proprietati nempe circumscriptioni adjungi potest.

Merkmale

Von dreifacher Art scheinen jene Dinge zu sein, die in den einzelnen Regionen betrachtet zu werden verdienen und daher in einer speziellen Geographie zu den einzelnen Regionen dargelegt werden können mit Gewinn für die Lernenden und die Leser, nämlich die terrestrischen, kosmischen und vom Menschen ausgehenden Dinge. Kosmische Merkmale nenne ich jene, die von der sichtbaren Bewegung der Sonne und der Gestirne abhängen, und (deren) scheinen acht zu sein.

Erstens: Polhöhe, Entfernung eines Ortes vom Äquator und Pol. Zweitens: Neigung[swinkel] der täglichen Bewegung der Gestirne oberhalb des Horizonts jenes Ortes. Drittens: Dauer des längsten und des kürzesten Tages. Viertens: Klima und Zone. Fünftens: Wärme, Kälte und Jahreszeiten, ebenso wie Regen, Schnee, Winde und andere Witterungserscheinungen. Zwar lassen sich diese Dinge auch auf die terrestrischen Besonderheiten beziehen; dennoch haben wir sie, da sie mit den vier Jahreszeiten und der Sonnenbewegung in engem Zusammenhang stehen, ebendeshalb auf die kosmische Abteilung bezogen. Sechstens: Aufgang der Gestirne, ihr Sichtbarwerden und ihre Verweildauer über dem Horizont. Siebtens: Gestirne, die durch den Scheitelpunkt des Ortes hindurchgehen. Achtens: Umfang oder Schnelligkeit der Bewegung, mit der sich gemäß der Annahme des Kopernikus zu jeder einzelnen Stunde jeder beliebige Ort im Kreise bewegt.—Gemäß den Astrologen könnte man ein neuntes Merkmal hinzufügen, da ja jene [Astrologen] einem der zwölf Tierkreissternbilder und einem Planeten dieses Sternbildes die Aufsicht über die einzelnen Regionen übertragen. Mir aber ist diese Lehre immer unbegründet erschienen, und ich erkenne keinerlei Grundlage für diese Lehre. Dennoch werden wir am Ende der Speziellen Geographie diese Einteilung von ihnen begutachten.

Soviel über die Abteilung der kosmischen Merkmale. Terrestrische Merkmale nenne ich diejenigen, die man direkt im Gelände einer Region betrachtet, Merkmale, von denen ich zehn aufzähle. Erstens: Begrenzungen und Breite. Zweitens: Gestalt. Drittens: Größe. Viertens: Berge. Fünftens: Gewässer, nämlich Flüsse, Quellen und Meeresbuchten. Sechstens: Wälder und Wüsten. Siebtens: Fruchtbarkeit und Unfruchtbarkeit, ebenso die Fruchtarten. Achtens: Mineralien oder Fossilien. Neuntens: Lebewesen [Tierwelt]. Zehntens: Geographische Länge, die dem ersten terrestrischen Charakteristikum, nämlich der Breite, hinzugefügt werden kann.

Tertium affectionum, quae in singulis regionibus spectari merentur, genus facio *Humanas*, quae ab hominibus sive incolis regionum dependent, quarum etiam decem circiter constitui possunt. Primo, *Incolarum statura, conformatio, color, vitae diuturnitas, origo, cibus, potus.* Secundo, *Quaestus & Artes, quibus occupantur incolae, mercatura & merces, quas mittit illa regio in alias.* Tertio, *Virtutes, vitia* [◊vitiae], *eruditio, ingenium, scholae*, &c. Quarto, *Consuetudines circa puerperia, nuptias, funera.* Quinto, *Sermo seu lingua, qua utuntur incolae.* Sexto, *Regimen politicum.* Septimo, *Religio & status Ecclesiasticae rei.* Octavo, *Vrbes & loca celebriora.* Nono, *Historiae memorabiles.* Decimo, *Viri illustres, artifices & inventa à singularum regionum incolis.*

Haec sunt tria affectionum genera in Speciali Geographia explicanda, etsi illae, quae tertiam classem faciunt, minus rectè ad Geographiam refe-/[p.5]rantur: Sed dandum est aliquid consuetudini & utilitati discentium. Nos praeter illa adjungemus Speciali Geographiae multa capita de Usu Geographiae.

In Generali vero Geographia, quam hoc libro explicabimus, considerantur primo affectiones absolutae Telluris, & partium constitutio. Deinde caelestes proprietates in genere contemplabimur, quae deinde in Speciali Geographia ad singulas regiones applicandae [◊applicanda] sunt. Denique in Parte comparativa ea proponentur, quae in unius loci cum alio comparatione offeruntur.

Principia

Triplicia sunt principia, quibus Geographia utitur ad confirmandam Propositionum veritatem. Primo, Geometricae, Arithmeticae atque Trigonometricae propositiones. Secundo, Astronomica praecepta & theoremata, etsi miraculo simile videatur, quod ad Telluris, in qua

Zur dritten Art von Merkmalen, die in den einzelnen Regionen eine Betrachtung verdienen, mache ich die auf den Menschen bezogenen, die von den Menschen oder Einwohnern der Regionen abhängen, Merkmale, von denen man auch ungefähr zehn aufstellen kann. Erstens: Statur der Einwohner, ihre körperliche Bildung, Farbe, Lebensdauer, Ursprung, Speise und Trank. Zweitens: Erwerbstätigkeiten und Fertigkeiten, mit denen sich die Einwohner befassen, Handel und Waren, welche jene Region in andere schickt. Drittens: Tugenden und Laster [Vorzüge und Schwächen], Erziehung, Veranlagung, Schulen usw. Viertens: Gewohnheiten im Zusammenhang mit Geburt, Hochzeit und Beerdigung. Fünftens: Sprache oder Mundart, die die Einwohner gebrauchen. Sechstens: Politische Herrschaft. Siebtens: Religion und Organisation des Kirchenwesens. Achtens: Städte und bekanntere Orte. Neuntens: Denkwürdige historische Berichte. Zehntens: Hervorragende Männer, Künstler und Erfindungen von den Bewohnern der einzelnen Regionen.

Diese drei Arten von Merkmalsbereichen müssen in einer Speziellen Geographie dargelegt werden, auch wenn dasjenige, was die dritte Abteilung ausmacht, sich nicht so gut auf die Geographie beziehen lässt. Man muss aber der Gewohnheit und dem Vorteil für die Lernenden irgendein Zugeständnis machen. Wir werden außer jenen Dingen an die Spezielle Geographie (noch) viele Kapitel über die Anwendung der Geographie anfügen.

In der Allgemeinen Geographie aber, die wir mit diesem Buch darlegen wollen, werden zuerst die für sich bestehenden Merkmale der Erde und die Beschaffenheit ihrer Teile betrachtet. Danach werden wir die kosmischen Eigentümlichkeiten im Allgemeinen betrachten, die dann in der speziellen Geographie auf die einzelnen Regionen angewendet werden müssen. Danach wird im Vergleichenden Teil dasjenige vorgestellt werden, was sich im Vergleich eines Ortes mit einem anderen darbietet.

Grundlagen

Dreifach sind die Prinzipien, derer sich die Geographie zur Bestätigung der Wahrheit ihrer Darlegungen bedient. Erstens: Die geometrischen, arithmetischen und trigonometrischen Sätze. Zweitens: Astronomische Lehren und Lehrsätze, auch wenn es einem Wunder zu gleichen scheint, dass wir uns zum Kennenlernen des Wesens der Erde, auf der wir

habitamus, naturam cognoscendam uti debeamus caelestibus corporibus, quae tot myriadibus milliarium à nobis remota sunt. Tertio, Experientia. Etenim maxima Geographiae pars, inprimis specialis, solâ experientiâ & observatione hominum nititur, qui regiones singulas descripserunt.

Ordo

De Ordine, quem in disciplina hac Geographica observare commodum esse existimo, dictum est in Divisione & Affectionum explicatione. Occurrit tamen quaedam difficultas de ordine in affectionum harum explicatione observando, nimirum, utrum singulis regionibus suae [◊suas] affectiones, an verò affectionibus generaliter explicatis ipsae regiones attribuendae sint?

Aristoteles libr. 1. / [p.6] Histor. Animal. ut etiam libr. 1. de part. Anim. similem movet dubitationem, & multis disputat: utrum ad singulas animalium species enumerandae sint proprietates eorum, an verò proprietates hae generaliter explicandae & animalia, in quibus eae reperiantur, subjungenda. Similis difficultas in aliis Philosophiae partibus occurrit. Nos in Generali Geographia [◊,] quasdam affectiones generaliter explicavimus, quas in speciali Geographia ad singularum regionum explicationem accommodabimus.

Methodus

De Methodo, hoc est Modo probandi dogmatum Geographicorum veritatem, sciendum est, quod in Generali Geographia plurima propriè dictis Demonstrationibus confirmentur, inprimis caelestes affectiones: in Speciali autem Geographia omnia ferè sine Demonstratione explicantur (exceptis caelestibus affectionibus, quae demonstrari possunt) quo experientia & observatio, hoc est sensuum testimonium illa confirmat, neque possunt alio modo probari. Etenim scientia tripliciter sumitur.

wohnen, der Himmelskörper bedienen müssen, die so viele Myriaden von Meilen von uns entfernt sind. Drittens: Die Erfahrung. Der größte Teil der Geographie nämlich, besonders der speziellen, beruht einzig und allein auf der Erfahrung und Beobachtung der Menschen, die die einzelnen Regionen beschrieben haben.

Ordnung

Über die Ordnung, die wie ich meine, in diesem geographischen Fach zu beachten vorteilhaft ist, ist in der Einteilung und Darlegung der Merkmale gesprochen worden. Dennoch taucht eine gewisse Schwierigkeit auf hinsichtlich der Einhaltung der richtigen Reihenfolge bei der Darlegung dieser Merkmale, nämlich [bei der Frage], ob den einzelnen Regionen ihre eigenen Merkmale oder aber den allgemein erklärten Merkmalen die Gegenden selbst zuzuweisen sind.

Aristoteles (Tiergesch., B. I, wie auch Tierklassifikation, B. I) erwägt einen ähnlichen Zweifel und macht sich viele Gedanken: Ob zu den einzelnen Tierarten ihre Charakteristika gezählt werden müssen oder aber ob diese Charakteristika allgemein erklärt und die Tiere, bei denen sie sich finden, [den Charakteristika] untergeordnet werden müssen. Eine ähnliche Schwierigkeit begegnet auch in anderen Teilen der Philosophie. Wir haben in der Allgemeinen Geographie gewisse Merkmale allgemein dargelegt, die wir in der Speziellen Geographie zur Erklärung der einzelnen Regionen anwenden werden.

Methode

Zur Methode, d.h. zur Art und Weise, die Wahrheit geographischer Lehrsätze zu erweisen, muss man wissen, dass in der Allgemeinen Geographie das meiste vorzugsweise durch die vorgetragenen Beweisführungen erhärtet wird, besonders die kosmischen Merkmale: In der Speziellen Geographie dagegen wird fast alles ohne Beweisführung dargelegt (ausgenommen die kosmischen Merkmale, die bewiesen werden können), soweit die Erfahrung und Beobachtung, d.h. das Zeugnis der Sinne, jene Dinge bestätigt, und auf andere Art und Weise können sie nicht bewiesen werden. Wissenschaft nämlich wird dreifach gefasst: Erstens als jegliche Erkenntnis ohne Unterschied, auch wenn sie nur aus dem Bereich des Wahrscheinlichen kommt.

Primo, pro qualicunque cognitione etiamsi ex probabilibus tantum sit. Secundo, pro cognitione certa, sive haec certitudo à Demonstrationum robore, sive à sensuum testimonio dependeat. Tertio, pro sola cognitione per Demonstrationem: quae usurpatio strictissima est, convenitque Geometriae, Arithmeticae, reliquisque Scientiis Mathematicis, exceptâ Chronologia, Astrologia & Geographia, quibus Scientiae vocabulum in secunda significatione competit.

Probantur etiam vel ostenduntur potius plurimae propositiones per Globum terrestrem artifi-/[p.7]cialem, per mappas quoque Geographicas: atque harum propositionum, quae ita ostenduntur, quaedam possunt legitimis Demonstrationibus confirmari (quod tamen propter captum Lectorum omittitur) aliae nequaquam ita probari possunt, sed ideo recipiuntur, quia supponimus, omnia loca in globo & mappis ita esse disposita ut in ipsa Tellure jacent. In hisce tamen Descriptiones potius ab autoribus Geographis confectas sequimur: globus & mappae ad illustrationem & faciliorem comprehensionem inserviunt.

Origo Geographiae

Origo Geographiae nova non est, neque uno partu in lucem edita est, neque ab uno homine profecta: sed dudum ante multa secula ejus primordia jacta sunt, etsi antiqui Geographi tantùm in describendis regionibus, hoc est Chorographia & Topographia occupati fuerunt. Romani solebant, devictá & subjugatá aliquâ provinciâ, hujus Chorographiam in Tabula expressam & claris signis illuminatam monstrare in triumpho spectatoribus. Erant praeterea Romae in porticu Luculli multae Geographicae Tabulae omnium contemplationi expositae. Senatus Romanus anno ante natum Christum centesimo circiter [◊,] miserat Geodaetas & Geographos in diversas mundi plagas, ut totam Tellurem dimetirentur, sed vix vigesimam partem perlustrarunt. Neco Aegyptiorum Rex [◊,] multis ante Christum natum seculis per Phaenices universum Africae externum latus integro triennio explo-

Zweitens als sichere Erkenntnis, sei es, dass diese Sicherheit von der Kraft der Beweisführungen, sei es, dass sie von dem Zeugnis der Sinne abhängt. Drittens als reine Erkenntnis durch Beweisführung: Diese Anwendung ist die strengste, und sie ziemt sich [nur] für die Geometrie, Arithmetik und die übrigen mathematischen Wissenschaften mit Ausnahme der Chronologie, Astrologie und Geographie, auf die das Wort „Wissenschaft" [nur] in der zweiten Bedeutung zutrifft.

Auch bewiesen oder vielmehr gezeigt werden die meisten Angaben durch einen terrestrischen Kunstglobus, auch durch geographische Karten: und von diesen so gezeigten Angaben lassen sich einige durch treffende Beweisführungen bestätigen (was freilich wegen der Auffassungsgabe der Leser unterbleibt), andere lassen sich keineswegs so beweisen, aber sie werden darum übernommen, weil wir unterstellen, dass alle Orte auf dem Globus und den Karten so verteilt sind, wie sie auf der Erde selbst liegen. Gleichwohl folgen wir in diesen Dingen mehr den von geographischen Schriftstellern verfassten Beschreibungen: Globus und Karten stehen zur anschaulichen Darstellung und zum leichteren Verständnis zu Diensten.

Ursprung der Geographie

Der Ursprung der Geographie ist nicht jung, und weder ist sie in einem einzigen Geburtsvorgang ans Licht gebracht worden, noch rührt sie von nur einem Menschen her, sondern schon lange, viele Jahrhunderte, liegen ihre Anfänge zurück, auch wenn sich die antiken Geographen in der Beschreibung von Gegenden, d.h. in Chorographie und Topographie, erschöpften. Die Römer pflegten nach Besiegung und Unterwerfung einer Provinz deren Chorographie auf einer Tafel dargestellt und mit deutlichen Zeichen ausgeschmückt im Triumphzug ihren Zuschauern zu zeigen. Es gab außerdem in Rom in der Säulenhalle des Lucullus viele geographische Tafeln, die zur Betrachtung durch die Öffentlichkeit ausgestellt waren. Der römische Senat hatte ungefähr 100 Jahre vor Christi Geburt Landvermesser [Geodäten] und Geographen in die verschiedenen Gegenden der Welt ausgesandt, um die gesamte Erde zu vermessen, die aber kaum den 20. Teil durchwanderten. Der Ägypterkönig Necho ließ viele Jahrhunderte vor Christi Geburt durch Phönizier die gesamte Außenflanke Afrikas in einem vollen Zeitraum von drei Jahren erforschen. Darius gab den Auftrag, die Mündungen

rare jussit. Darius Indi ostia & mare Aethiopicum Orientali scrutari mandavit. Alexander Magnus in expeditione sua Asiatica duxit secum duos itinerum mensores & descri-/[p.8]ptores, Diognetum & Betonem, teste Plinio, ex quorum annotationibus & Itinerariis Geographi sequentium seculorum multa desumserunt. Etenim cum omnium fere aliarum artium studium à bellis afficiatur damno, sola ferè Geographia ab illis crescit una cum Fortificatoria, quam vocant.

Caeterum valde manca, imperfecta, & falsitatibus plurimis scatens Geographia veterum erat, quoniam ignorabant has [◊haec], quae non infimae sed praecipuae sunt cognitionis Telluris partes, (vel saltem non habebant certam de iis experientiam.) 1. Totam Americam. 2. Terras Septentrionales. 3. Terram Australem & Magellanicam. 4. Terram circum navigari posse & Oceanum continuo tractu ambire Tellurem, (quosdam tamen ex antiquis in hac opinione fuisse non nego, at illud certo scivisse, id nego.) 5. Zonam Torridam habitabilem esse & ab infinitis populis habitari. 6. Telluris dimensionem veram, etsi multa in hoc negotio scripserint. 7. Africam posse circumnavigari, & australem Africae partem. 8. Regionum remotarum veris Descriptionibus carebant tam Graeci quam Romani, atque innumera falsa & fabulosa de populis in Asiae termino atque Septentrionalibus locis scripta reliquerunt. 9. Motum maris sive fluxus diversitates & generalem illum fluxum ignorabant. 10. Imo Graeci atque ipse Aristoteles Fluxum & Refluxum maris non habuerunt cognitum. 11. Ventorum diversitates paucas habebant perspectas, generalis omnino illis incognitus. 12. Magnetis insignis proprietas, quà Septentrionem & Austrum monstrat, ipsis occulta fuit, etsi alteram illam, quâ ferrum trahit, cognoverint. Anaximander autem, qui vixit anno / [p.9] ante Christum cccc circiter, primus scribitur tentasse Telluris dimensionem.

des Indus und das im Osten gelegene Äthiopische Meer zu untersuchen. Alexander der Große führte nach dem Zeugnis des Plinius auf seinem Feldzug nach Asien die beiden Routenvermesser und Beschreiber Diognetus und Beto mit sich, aus deren Anmerkungen und Reisebeschreibungen die Geographen der folgenden Jahrhunderte vieles entnommen haben. Während nämlich die Beschäftigung mit fast allen anderen Wissenschaften unter kriegerischen Einwirkungen Schaden nimmt, wird die Geographie fast als einzige in der Folge jener Wirren gefördert gemeinsam mit der sogenannten Festungsbaukunst.

Übrigens war die Geographie der Alten recht mangelhaft, unvollständig und von ganz vielen falschen Vorstellungen wimmelnd, da sie diejenigen Erdteile, die nicht den untersten, sondern einen vorzüglichen Erkenntniswert haben, nicht kannten (oder wenigstens darüber keine sichere Erfahrung hatten): 1. ganz Amerika, 2. die nördlichen Länder, 3. das Südland [Terra Australis] und Magellanland, 4. wussten sie nicht, dass man die Erde umsegeln könne und dass der Ozean in durchgehender Ausdehnung um die Erde herumgehe. (Dass allerdings einige von den antiken Geographen diese Meinung hatten, verneine ich nicht, aber dass sie es mit Bestimmtheit gewusst hätten, das verneine ich). 5. Sie hatten keine Kenntnis davon, dass die heiße Zone bewohnbar ist und von unzähligen Völkern bewohnt wird. 6. Ihnen war die wirkliche Ausdehnung der Erde unbekannt, auch wenn sie vieles in dieser Angelegenheit geschrieben haben. 7. Sie wussten nicht, dass man Afrika und den südlichen Teil Afrikas umsegeln kann. 8. verfügten Griechen ebensowenig wie Römer über wirklichkeitsgetreue Beschreibungen entfernter Gegenden, und sie haben unzählige falsche und sagendurchwobene schriftliche Angaben über die Völker innerhalb Asiens und über die nördlichen Gegenden hinterlassen. 9. Sie kannten nicht die Meeresbewegung oder die gegensätzlichen Strömungsverhältnisse und [überhaupt] jene allgemeine Strömung. 10. Ja sogar die Griechen und selbst Aristoteles kannten Fluss und Rückfluss des Meeres [noch] nicht. 11. Von den unterschiedlichen Winden hatten sie nur wenige erkannt; der allgemeine [Wind] war ihnen völlig unbekannt. 12. Die berühmte Eigentümlichkeit des Magneten, mit der er die Nord- und Südrichtung zeigt, war ihnen selbst verborgen, auch wenn sie jene andere Eigentümlichkeit, mit der er Eisen anzieht, kannten. Anaximander aber, der ungefähr im Jahre 400 vor Christus lebte, soll als erster den Erdumfang untersucht haben.

Geographiae Praestantia

Commendat studium Geographicum. 1. Dignitas, quia hominem Telluris incolam & ratione prae aliis animalibus praeditum maximè decet. 2. Jucundum quoque est & quidem honesta recreatio contemplari Telluris regiones & proprietates. 3. Insignis utilitas & necessitas, cum neque Theologi, neque Medici, neque Juris-consulti, neque Historici neque Literatores alii carere possint Geographiae cognitione, si absque remora progredi voluerint in suis studiis. Haec ab aliis satis ostensa sunt & multis exemplis illustrari possunt.

Subjungo hic Tabulas duas, quarum prior Contenta hujus libri nempe Geographiae Generalis ante oculos ponit, altera ordinem in Speciali Geographia ad singularum regionum explicationem observandum.

Vortrefflichkeit der Geographie

Das Studium der Geographie empfiehlt sich 1. wegen der Würde, weil sie sich für den Menschen, der ja ein Bewohner der Erde und vor den anderen Lebewesen mit Vernunft begabt ist, am meisten schickt. 2. Auch angenehm ist das Studium, und gewiss ist es eine achtbare Erholung, die Regionen der Erde und ihre Eigentümlichkeiten zu betrachten. 3. Hervorragend ist der Nutzen und die Unvermeidlichkeit, da weder die Theologen noch die Mediziner noch die Rechtsgelehrten noch die Historiker noch andere gebildete Menschen auf die Bekanntschaft mit der Geographie verzichten können, wenn sie, und zwar ohne Verzögerung, Fortschritte in ihren eigenen Studien machen wollen. Das ist von anderen genügend dargelegt worden und kann durch viele Beispiele veranschaulicht werden.

Ich füge hier zwei Verzeichnisse an, deren erstes den Inhalt dieses Buches, nämlich der Allgemeinen Geographie, vor Augen stellt; das zweite bringt die Gliederung, die man in einer Speziellen Geographie bei der Darlegung der einzelnen Regionen beachten muss.

- GENERALIS GEOGRAPHIA in tres Partes in hoc libro divisa est, quæ sunt
  - PARS ABSOLUTA, divisa in Sectiones sex.
    - SECTIO PRIMA, de præcognitis capita duo.
      - Cap. I. De Prolegomenis.
      - Cap. II. De quibusdam necessariis ex Geometria.
    - SECTIO SECUNDA, affectiones totius Telluris explicat capitibus V.
      - Cap. III. De Telluris Figura.
      - Cap. IV. De ejus dimensione & magnitudine.
      - Cap. V. De ejus motu.
      - Cap. VI. De ejus loco in Systemate Mundi.
      - Cap. VII. De ejus substantia & materia.
    - SECTIO TERTIA, in qua Terræ Constitutio & partes explicantur capitibus quatuor.
      - Cap. VIII. De Divisione Terræ per Aquas.
      - Cap. IX. De Montibus in genere.
      - Cap. X. De Montium differentiis.
      - Cap. XI. De Sylvis, desertis, & fodinis.
    - SECTIO QUARTA, HYDROGRAPHIA in qua Aquarum constitutio & proprietates explicantur, capitibus VI.
      - Cap. XII. De Divisione Aquarum.
      - Cap. XIII. De Oceano & Mari.
      - Cap. XIV. De Motibus maris, inprimis de Fluxu & Refluxu.
      - Cap. XV. De Fluviis.
      - Cap. XVI. De Lacubus, stagnis & paludibus.
      - Cap. XVII. De aquis mineralibus.
    - SECTIO QUINTA.
      - Cap. XVIII. De mutatione locorum aridorum in Aquosa & contra.
    - SECTIO SEXTA, de Athmosphæra.
      - Cap. XIX. De Athmosphæra & Aëre.
      - Cap. XX. De Ventis in genere.
      - Cap. XXI. De Ventorum differentiis & in specie.
  - PARS RESPECTIVA explicans affectiones cæelestes capitibus.
    - Cap. XXII. De affectionibus cælestibus in genere.
    - Cap. XXIII. De latitudine loci & elevatione poli.
    - Cap. XXIV. De Divisione Telluris in Zonas.
    - Cap. XXV. De longitudine dierum & divisione Telluris in Climata.
    - Cap. XXVI. De luce, calore, & tempestatibus anni.
    - Cap. XXVII. De Umbris & divisione incolarum respectu umbrarum.
    - Cap. XXVIII. De comparatione Affectionum cælestium in diversis locis, ubi de Antœcis, Periœcis & Antipodibus.
    - Cap. XXIX. De Diversitate Temporis in diversis locis.
    - Cap. XXX. De diverso ortu Solis, Lunæ, &c. atque aliis apparentiis.
  - PARS COMPARATIVA, affectiones illas considerans, quæ ex comparatione unius loci ad alium oriuntur.
    - Cap. XXXI. De longitudine locorum.
    - Cap. XXXII. De situ locorum ad invicem.
    - Cap. XXXIII. De distantia locorum mutua.
    - Cap. XXXIV. De Horizonte visibili.
    - Cap. XXXV. De arte Nautica in genere & navium structura.
    - Cap. XXXVI. De onere navibus imponendo.
    - Cap. XXXVII. De Directoria artis Nauticæ parte prima, Cognitio intervalli.
    - Cap. XXXVIII. Secunda parte Cognitio plagarum.
    - Cap. XXXIX. Tertia pars de Histriodromia, sive via navis.
    - Cap. XL. Quarta pars de loco navis in itinere.

Fig. 14 [Tabula ad „pag. 8."]

Die Allgemeine Geographie ist in diesem Buch in drei Teile eingeteilt, und zwar: Absoluter [für sich bestehender] Teil, geteilt in sechs Abschnitte

Erster Abschnitt: Zwei Kapitel über die Vorkenntnisse

Kap. I	Einleitung
Kap. II	Über einige Unverzichtbarkeiten aus der Geometrie

Zweiter Abschnitt: Darlegung der Eigenschaften der gesamten Erde in fünf Kapiteln

Kap. III	Über die Gestalt der Erde
Kap. IV	Über ihre Abmessung und Größe
Kap. V	Über ihre Bewegung
Kap. VI	Über ihren Standort im System der Welt
Kap. VII	Über ihre Substanz und ihren Grundstoff

Dritter Abschnitt, in dem der Zustand der Erde und ihre Teile in vier Kapiteln erklärt werden

Kap. VIII	Über die Teilung der Erde durch Gewässer
Kap. IX	Über Berge im Allgemeinen
Kap. X	Über die Unterschiede von Bergen
Kap. XI	Über Wälder, Wüsten und Bergwerke

Vierter Abschnitt: Hydrographie, worin in sechs Kapiteln die Beschaffenheit der Gewässer und ihre Eigentümlichkeiten dargelegt werden

Kap. XII	Über die Einteilung der Gewässer
Kap. XIII	Über den Ozean und das Meer
Kap. XIV	Über die Bewegungen des Meeres, insbes. über Fluss und Rückfluss (Gezeiten)
Kap XV	Über die Flüsse
Kap. XVI	Über Binnenseen, stehende Gewässer und Sümpfe
Kap. XVII	Über Mineralwässer

Fünfter Abschnitt

Kap. XVIII	Über den Wandel von trockenen Orten in wasserreiche und umgekehrt

Sechster Abschnitt: Atmosphäre

Kap. XVIIII	Über die Atmosphäre und die Luft
Kap. XX	Über die Winde im Allgemeinen
Kap. XXI	Über die Unterschiede von Winden und (über Winde) im einzelnen

Pars Respectiva explicans affectiones cælestes capitibus.

- Cap. XXII. De affectionibus cælestibus in genere.
- Cap. XXIII. De latitudine loci & elevatione poli.
- Cap. XXIV. De Divisione Telluris in Zonas.
- Cap. XXV. De longitudine dierum & divisione Telluris in Climata.
- Cap. XXVI. De luce, calore, & tempestatibus anni.
- Cap. XXVII. De Umbris & divisione incolarum respectu umbrarum.
- Cap. XXVIII. De comparatione Affectionum cælestium in diversis locis, ubi de Antœcis, Periœcis & Antipodibus.
- Cap. XXIX. De Diversitate Temporis in diversis locis.
- Cap. XXX. De diverso ortu Solis, Lunæ, &c. atque aliis apparentiis.

Pars Comparativa, affectiones illas considerans, quæ ex comparatione unius loci ad alium oriuntur.

- Cap. XXXI. De longitudine locorum.
- Cap. XXXII. De situ locorum ad invicem.
- Cap. XXXIII. De distantia locorum mutua.
- Cap. XXXIV. De Horizonte visibili.
- Cap. XXXV. De arte Nautica in genere & navium structura.
- Cap. XXXVI. De onere navibus imponendo.
- Cap. XXXVII. De Directoria artis Nauticæ parte prima, Cognitio intervalli.
- Cap. XXXVIII. Secunda parte Cognitio plagarum.
- Cap. XXXIX. Tertia pars de Histriodromia, sive via navis.
- Cap. XL. Quarta pars de loco navis in itinere.

Fig. 14a. Detail [Tabula ad „pag. 8.“]

Berücksichtigender Teil, der die kosmischen Merkmale darlegt

Kap. XXII	Über die kosmischen Beziehungen im Allgemeinen
Kap. XXIII	Über die [geographische] Breite eines Ortes und die Polhöhe
Kap. XXIV	Über die Einteilung der Erde in Zonen
Kap. XXV	Über die Länge der Tage und die Einteilung der Erde in Klimazonen
Kap. XXVI	Über Licht, Wärme und die Jahreszeiten
Kap. XXVII	Über die Schatten und über die Einteilung der Erdbwohner hinsichtlich der Schatten
Kap. XXVIII	Über den Vergleich der kosmischen Merkmale an verschiedenen Standorten, desgleichen über Antöken, Periöken und Antipoden
Kap. XXIX	Über die Verschiedenheit der Zeit an unterschiedlichen Standorten
Kap. XXX	Über den unterschiedlichen Aufgang von Sonne, Mond usw. und andere Erscheinungen

Vergleichender Teil, der jene Merkmale betrachtet, die sich aus dem Vergleich eines Standortes zu einem anderen ergeben

Kap. XXXI	Über die [geographische] Länge von Orten
Kap. XXXII	Über die Lage von Orten zueinander
Kap. XXXIII	Über die wechselseitige Entfernung von Orten
Kap. XXXIV	Über den sichtbaren Horizont
Kap. XXXV	Über die Nautik im Allgemeinen und über die Bauart von Schiffen
Kap. XXXVI	Über das Beladen von Schiffen mit Lasten
Kap. XXXVII	Über den ersten—steuernden—Teil der Nautik: Erkennen des Abstands
Kap. XXXVIII	Über den zweiten Teil: Erkennen der Richtungen
Kap. XXXIX	Dritter Teil: Über die Histiodromie [Segelkunde] oder den Kurs des Schiffes
Kap. XL	Vierter Teil: Über den Standort des Schiffes unterwegs

Pag. 9.

- Specialis Geographia considerat in singulis regionibus triplicia.
  - Terrestria decem.
    1. Limites & Circumscriptionem.
    2. Longitudinem loci & situm.
    3. Figuram.
    4. Magnitudinem.
    5. Montes, { eorum appellatio & situs altitudo, proprietas & contenta.
    6. Fodinas.
    7. Sylvas & deserta.
    8. Aquas { Mare, lacus, paludes, Fluvii. Horum fontes, ostium, tractus, latitudo, aquæ copia, celeritas, aquæ qualitas, cataractæ, &c.
    9. Fertilitatem, vel sterilitatem, & fructus.
    10. Animalia.
  - Cœlestia octo.
    1. Distantiam loci ab Æquatore & Polo.
    2. Obliquitatem motus supra Horizontem.
    3. Quantitatem dierum.
    4. Clima & Zonam.
    5. Calorem, atque anni tempestates, ventos, pluvias & alia meteora.
    6. Stellarum ortum & moram supra Horizontem.
    7. Stellas per verticem loci transeuntes.
    8. Quantitatem vel celeritatem motus juxta Copernicanam hypothesin.
  - Humana decem.
    1. Incolarum statura, vita, cibus & potus, origo, &c.
    2. Quæstus & artes, mercatura, merces.
    3. Virtutes & vitia, ingenium, eruditio, &c.
    4. Consuetudines circa puerperia, nuptias, funera.
    5. Sermo & lingua.
    6. Regimen Politicum.
    7. Religio & status rei Ecclesiasticæ.
    8. Urbes.
    9. Historiæ memorabiles.
    10. Viri illustres vel feminæ, artifices, inventa.

Fig. 15 [Tabula ad „pag. 9."]

Die Spezielle Geographie
betrachtet in den einzelnen Regionen drei Beziehungsbereiche:

Zehn terrestrische Bezüge
1. Grenzen und Umfang
2. [Geographische] Länge eines Ortes und seine Lage
3. Gestalt
4. Größe
5. Berge (ihre Benennung, Lage, Höhe, Eigentümlichkeiten und Bestandteile)
6. Bergwerke
7. Wälder und Wüsten
8. Gewässer (Meer, Seen, Sümpfe, Flüsse, deren Quellen, Mündung, Verlauf, Breite, Wassermenge, Fließgeschwindigkeit, Beschaffenheit des Wassers, Wasserfälle usw.)
9. Fruchtbarkeit oder Unfruchtbarkeit und vorkommende Früchte
10. Tiere

Acht kosmische Bezüge
1. Entfernung des Standortes vom Äquator und Pol
2. Schrägstellung einer (Gestirn)bewegung oberhalb des Horizonts
3. Dauer der Tage
4. Klima und Zone
5. Wärme [Temperatur] und Jahreszeiten, Winde, Regenfälle und anderes Witterungskundliches
6. Sternenaufgang und ihre Verweildauer über dem Horizont
7. Sterne, die durch den Ortszenit hindurchgehen
8. Umfang oder Bewegungsgeschwindigkeit gemäß der kopernikanischen Hypothese

Zehn Humanbezüge
1. Statur der Einwohner, ihr Leben, Speise und Trank, Herkunft usw.
2. Gewerbe, künstlerische Fertigkeiten, Handel und Waren
3. Tugenden und Laster, Begabung, Erziehung usw.
4. Gewohnheiten bei Geburten, Heiraten und Beerdigungen
5. Mundart und Sprache
6. Politisches Regime
7. Religion und Zustand des Kirchenwesens
8. Städte
9. Geschichtliche Denkwürdigkeiten
10. Berühmte Männer oder Frauen, Künstler und Erfindungen

BIBLIOGRAPHY

Note

Not included in this bibliography is Johann Anselm Steiger's list of references, since he is not dealing with Bernhard Varenius, but with his father Heinrich Varenius. Steiger's bibliography may be found at the end of his essay.

Johann Anselm Steigers Bibliographie ist hier nicht enthalten, da sein Beitrag sich nicht auf Bernhard Varenius bezieht, sondern auf dessen Vater Heinrich Varenius. Steigers Angaben befinden sich im Anschluss an seinen Aufsatz.

Aa, Pieter van der, ed. *Naaukeurige versameling der gedenk-waardigste zee en land-reysen na Oost- an West-Indiën*... 28 Bde. [mit 127 jeweils separat paginierten Teilen]. Leiden 1706–07.

Acosta, José de, SJ. *Historia Natural y Moral de las Indias*. Sevilla 1590.

Alberts, G., E. Atzema and J.A. van Maanen. "Mathematics in the Netherlands: A Brief Survey With an Emphasis on the Relation to Physics." *A History of Science in the Netherlands. Survey, Themes and* Reference, eds. Klaas van Berkel, A. van Helden and L.C. Palm. Leiden 1999, 367–404.

Album Academiae Helmstadiensis, bearbeitet von Paul Zimmermann, Bd. 1: Studenten, Professoren etc. der Universität Helmstedt von 1574–1636. Hannover 1926.

Aleksandrovskaya, O.A. *Stanovlenie geograficheskoi nauki v Rossii v XVIIIv*. Moscow 1989.

Allen, John Logan. "Imagining the West: The View from Monticello." *Thomas Jefferson and the Changing West: From Conquest to Conservation*, ed. James P. Ronda. St. Louis 1997.

Allgemeine Historie der Reisen zu Wasser und zu Lande; oder Sammlung aller Reisebeschreibungen, welche bis itzo in verschiedenen Sprachen von allen Völkern herausgegeben worden, und einen vollständigen Begriff von der neuern Erdbeschreibung und Geschichte machen; worinnen der wirkliche Zustand aller Nationen vorgestellet. Übers., Hg. u. Red. Johann Joachim Schwabe, Abraham Gotthelf Kästner et al. 21 Bde. Leipzig 1747–74.

Alschner, Uwe. *Universitätsbesuch in Helmstedt 1576–1810*. Braunschweig 1998.

Avé-Lallemant, Robert C.B. *Des Dr. Joachim Jungius aus Lübeck Briefwechsel mit seinen Schülern und Freunden [...], aus den Briefen der Hamburger Stadtbibliothek zusammengestellt*. Lübeck 1863.

——. *Das Leben des Dr. med. Joachim Jungius*. Breslau 1882.

Bagrow, Leo. "Semen Remezov: A Siberian Cartographer." *Imago Mundi* 11 (1954): 111–125.

Baker, J.N.L. "The Geography of Bernhard Varenius." *The History of Geography*. Oxford 1963, chap. 7, 105–118.

——. "Nathanael Carpenter and English Geography in the Seventeenth Century." *The History of Geography*. Oxford 1963, 1–13. Reprint from *Geographical Journal* 71 (1928).

Bauer, Adolf. „Antike Ansichten über das jährliche Steigen des Nil." *Historische Untersuchungen*. Festschrift für Arnold [Dietrich] Schäfer. Bonn 1882, 70–97.

Baumgart, Peter. „Die Anfänge der Universität Helmstedt im Spiegel ihrer Matrikel, 1576–1600." *Braunschweigisches Jahrbuch* 50 (1969): 5–32.

——. *Universitäten im konfessionellen Zeitalter: Gesammelte Beiträge*. Münster 2006.

Bautz: → *Biographisch-Bibliographisches Kirchenlexikon (BBKL)*

Beck, Hanno. *Geographie: Europäische Entwicklung in Texten und Erläuterungen* (Orbis Academicus). Freiburg, Br. 1973, bes. 115–122.

——. „Drei Geographien: Drei Arten von Geographen." *Große Geographen: Pioniere—Außenseiter—Gelehrte.* Berlin 1982, 45–53.

Benzing, Josef. *Die Buchdrucker des 16. und 17. Jahrhunderts im deutschen Sprachgebiet.* Wiesbaden 1982.

Bergh, Albert Ellerly, ed. *The Writings of Thomas Jefferson.* 20 vols. Washington 1907.

Berns, Jörg Jochen. „Herzog August—Frömmigkeit und kirchliche Tradition." *Sammler, Fürst, Gelehrter: Herzog August zu Braunschweig Lüneburg 1579–1666* (Ausstellungskatalog Herzog August Bibliothek 27). Braunschweig 1979, 341–378.

Biographisch-Bibliographisches Kirchenlexikon (BBKL), Hg. Friedrich Wilhelm Bautz. Hamm, Herzberg und Nordhausen. Bd. 1– . 1990– . [http://www.bautz.de/bbkl/]

Bitterli, Urs. *Alte Welt—neue Welt: Formen des europäisch-überseeischen Kulturkontakts vom 15. bis zum 18. Jahrhundert.* München 1986.

Blair, Ann. "Annotating and Indexing Natural Philosophy." *Books and the Sciences in History,* eds. Marina Frasca-Spada and Nick Jardine. Cambridge 2000, 69–89.

——. "Reading Strategies for Coping with Information Overload, ca. 1550–1700." *Journal of the History of Ideas* 64 (2003): 11–28.

——. "Note Taking as an Art of Transmission." *Critical Inquiry* 31 (2004): 85–107.

Blanke, Horst Walter. „Marco Polo, Bernhard Varenius und Engelbert Kaempfer. Oder: Vom Hörensagen über die gelehrte Recherche zum Autopsiebericht: Drei Stationen der europäischen Japan-Kunde." *Bernhard Varenius (1622–1650): Der Beginn der modernen Geographie,* Hg. Wolfgang Griep u. Susanne Luber. Begleitband zur Ausstellung 2001 (Veröffentlichungen der Eutiner Landesbibliothek 5). Eutin 2001, 29–67.

——. *Politische Herrschaft und soziale Ungleichheit im Spiegel des Anderen: Untersuchungen zu den deutschsprachigen Reisebeschreibungen vornehmlich im Zeitalter der Aufklärung.* 2 Bde. (Wissen und Kritik 6). Waltrop 1997.

Blome, Richard. *Cosmography and Geography.* Third impression. London: Richard Blome 1693.

Blotius, Hugo. *Tabula Peregrinationis continens capita politica* [ca. 1570]. In: Paul Hentzer. *Itinerarium Germaniae, Galliae, Angliae, Italiae.* Nürnberg 1629.

Blussé, Leonard, W. Remmelink u. I. Smits, eds. *Bridging the Divide: 400 Years The Netherlands—Japan.* Leiden 2000.

Bödeker, Hans Erich. „Lessings Briefwechsel." *Über den Prozeß der Aufklärung in Deutschland im 18. Jahrhundert,* Hg. H.E. Bödeker u. Ulrich Herrmann (Veröffentlichungen des Max-Planck-Instituts für Geschichte 85). Göttingen 1987, 113–38.

Bowen, Margarita. *Empiricism and Geographical Thought: From Francis Bacon to Alexander von Humboldt.* Cambridge 1981.

Boxer, Charles R. *The Christian Century in Japan 1549–1650.* Berkeley 1951.

Boyd, Julian P., ed. *The Papers of Thomas Jefferson.* 31 vols. Princeton and Oxford 1952–2004.

Brendle, Franz et al., Hg. *Deutsche Landesgeschichtsschreibung im Zeichen des Humanismus* (Contubernium 56). Stuttgart 2001.

Breusing, A. „Lebensnachrichten von Bernhard Varenius." *Petermanns Geographische Mitteilungen* 26 (1880): 136–141.

British Book Trade Index. [Maintained online by University of Birmingham] http://www.bbti.bham.ac.uk/.

Buijnsters, Petrus J. "De gouverneur/gouvernante tussen adel en burgerij: Enkele beschouwingen naar aanleiding van een vertoog van Justus van Effen uit 1734." *Nederlandse literatuur van de achttiende eeuw: Veertien verkenningen.* Utrecht 1984, 86–98.

——. *Justus van Effen 1684–1735: Leven en werk.* Utrecht 1992.

Brown, Gary I. "The Evolution of the Term 'Mixed Mathematics'." *Journal of the History of Ideas* 52 (1991): 81–102.

Burke, Peter. *Papier und Marktgeschrei: Die Geburt der Wissensgesellschaft.* Berlin 2000.

Burkhardt, Hans. „Jungius, Leibniz und die Logica Nova." *Praktische Logik, Traditionen und Tendenzen*, Hg. Peter Klein (Veröffentlichung der Joachim Jungius Gesellschaft 61). Göttingen 1990, 57–83.

Büttner, Manfred. *Die Geographia Generalis vor Varenius: Geographisches Weltbild und Providentialehre* (Erdwissenschaftliche Forschung 7). Wiesbaden 1973.

——. „Begründung der wissenschaftlichen Geographie durch Bernhard Varenius (1622–1650)?: Gedanken zu den Defiziten der Varenius-Forschung aus Anlaß des Varenius-Jahres 2000." *Berichte zur Wissenschaftsgeschichte* 24 (2001), 237–254.

Callmer, Christian. *Königin Christina, ihre Bibliothek und ihre Handschriften*. Stockholm 1977.

Capel Sáez, Horacio, "La Personalidad geografica de Varenio." [Introduction to] Bernhard Varenius: *Geografia General, en la que se explican las propiedades generales de la tierra*, ed. H. Capel Sáez, transl. Jose Maria Requejo Prieto. Barcelona 1974, 11–84.

Caron, François. *Beschreibung des mächtigen Königreichs Japan*, eingeleitet u. erläutert v. Detlef Haberland (Fremde Kulturen in alten Berichten 10). Stuttgart 2000.

Caron. François u. Jod. Schouten. *Wahrhaftige Beschreibungen zweyer mächtigen Königreiche / Jappan und Siam*. Nürnberg 1663.

China illustrata: Das europäische China-Verständnis im Spiegel des 16. bis 18. Jahrhunderts, Hg. Hartmut Walravens (Ausstellungskataloge der Herzog August Bibliothek Nr. 55). VCH Acta humaniora. Weinheim 1987.

Chytraeus, David. *De Creichgoïa—Kraichgau*. Facs. von 1561, übers. u. kommentiert v. Reinhard Düchting u. Boris Körkel. Ubstadt-Weiher 1999.

Conring, Hermann. *Examen rerumpublicarum potiorum totius* orbis (1710). Nachdruck: Rassem u. Stagl, Hg. 1994, 227–269.

Cooper, Michael. *They Came to Japan: An Anthology of European Reports on Japan, 1543–1640*. Berkeley 1965.

Corsten, Severin et al., Hg. *Lexikon des gesamten Buchwesens*. Bd. 2. 2. Auflage Stuttgart 1989.

Cortazzi, Hugh. *Isles of Gold: Antique Maps of Japan*. New York u. Tokyo 1983.

Cortesão, Armando. "The First Account of the Far East in the Sixteenth Century: The Name 'Japan' in 1513." *Comptes rendus du Congrès International de Géographie Amsterdam 1938*. Vol. 2,7. Leiden 1938, 146–152.

Cracraft, James. *The Petrine Revolution in Russian Culture*. Cambridge, Massachusetts 2004.

Critchley, John. *Marco Polo's Book*. Aldershot 1992.

Davids, Carolus A. *Zeewezen en wetenschap: De wetenschap en de ontwikkeling van de navigatietechniek in Nederland tussen 1585 en 1815*. Amsterdam 1986.

——. "Ondernemers in kennis: Het zeevaartkundig onderwijs in de Republiek gedurende de zeventiende eeuw." *De zeventiende eeuw* 7 (1991): 37–48.

De Grazia, Margreta. *Shakespeare Verbatim: The Reproduction of Authenticity and the 1790 Apparatus*. Oxford 1991.

Deuser, Hermann. „Vorsehung I", *Theologische Realenzyklopädie* 35, Berlin 2003, 309 f.

Dictionary of National Biography. London 1885–1900.

Dongelmans, B.P.M., P.G. Hoftijzer, O.S. Lankhorst (red.), eds. *Boekverkopers van Europa: Het 17de-eeuwse Nederlandse uitgevershuis Elzevier*. Zutphen 2000.

Düchting, Reinhard u. Boris Körkel, Hg. *David Chytraeus, Kraichgau, De Creichgoia* (Heimatverein Kraichgau e.V., Sonderveröffentlichung Nr. 21). Ubstadt-Weiher 1999.

Elias, J.E. *De vroedschap van Amsterdam 1578–1795*. 2 vols. Haarlem 1903–1905. Reprint Amsterdam 1963.

Elsmann, Thomas, Hanno Lietz u. Sabine Pettke, Hg. *Nathan Chytraeus 1543–1598: Ein Humanist in Rostock und Bremen—Quellen und Studien*. Bremen 1991.

Encyclopaedia Britannica. Cambridge 1911.

Enoki, Kazuo. "Marco Polo and Japan." *Oriente Poliano: Studi e conferenze tenute all' Is. M.E.O. all' occasione del 7 centenario della nascita di Marco Polo (1254–1324)*, ed. Etienne Balasz. Rom 1957, 23–44.

Erler, G., Hg. *Die Matrikel der Universität Königsberg*, 3 Bde. Leipzig 1910–1917.

Felde, Dietrich zum. „Die Abhängigkeit der Kenntnisse und Auffassungen des Varen vom klassischen Altertum." *Geographische Zeitschrift* 38 (1932): 13–22 [based on his dissertation Heidelberg 1921].

Fisch, Stefan. „Forschungsreisen im 19. Jahrhundert." *Der Reisebericht: Die Entwicklung einer Gattung in der deutschen Literatur*. Hg. Peter J. Brenner. Frankfurt (M.) 1989, 383–405.

Ford, Paul Leicester, ed. *The Writings of Thomas Jefferson*. 10 vols. New York, London 1892–1999.

Forster, Georg. *A Voyage round the World*. London 1777. Reprint: *Werke*. Bd. 1. Berlin (DDR) 1968.

——. *Johann Reinhold Forster's... Reise um die Welt*. 2 Bde. Berlin 1778/80. Nachdruck: *Werke* Bd. 2 u. 3. Berlin (DDR) 1965.

Forster, Johann Reinhold. *Observations made during a Voyage Round the World*. London 1978.

——. *Bemerkungen über Gegenstände der physischen Erdbeschreibung, Naturgeschichte und sittlichen Philosophie auf seiner Reise um die Welt gesammlet*, Hg. u. Übers. Georg Forster. Berlin 1783. Nachdruck: *Beobachtungen während der Cookschen Weltumsegelung*. Stuttgart 1981.

Franke, Otto. *Geschichte des Chinesischen Reiches: Eine Darstellung seiner Entstehung, seines Wesens und seiner Entwicklung bis zur neuesten Zeit*. Bde. 4 u. 5. 2. Aufl. Berlin 2001.

Frick, Georg. *Die Elzevir'schen Republiken* (Hallesche Abhandlungen zur Neueren Geschichte 30). Halle 1892.

Friedrich, Markus. „Chorographica als Wissenskompilationen: Probleme und Charakteristika." *Sammeln, Ordnen, Veranschaulichen. Zur Wissenskompilatorik in der Frühen Neuzeit*, Hg. Frank Büttner, M. Friedrich u. H. Zedelmaier. (Pluralisierung und Autorität 2) Münster 2003, 83–110.

Friis, Hermann R. "Baron Alexander von Humboldt's Visit to Washington." *Records of the Columbia Historical Society* 44 (1963): 1–35.

Frijhoff, Willem. "Het Amsterdamse Athenaeum in het academische landschap van de zeventiende eeuw." *Athenaeum Illustre: Elf studies over de Amsterdamse doorluchtige school 1632*–1877, ed. Eco O.G. Haitsma Mulier. Amsterdam 1997, 37–65.

Garber, Daniel. "Descartes and the Scientific Revolution: Some Kuhnian Reflections." *Perspectives on Science* 9 (2001): 405–422.

Genette, Girard. *Paratexts: Thresholds of Interpretation*. Cambridge 1997.

Giesemann, Gerhard u. Xenja von Ertzdorff, Hg. *Erkundung und Beschreibung der Welt: Zur Poetik der Reise und Länderberichte*. Amsterdam 2003.

Gilreath, James and Douglas L. Wilson, eds., *Thomas Jefferson's Library: A Catalog with the Entries in His Own Order*. Washington DC 1989 [full-text version maintained by Library of Congress: http://www.loc.gov/catdir/toc/becites/main/jefferson/88607928.html].

Glaser Karl-Heinz et al., eds. *David und Nathan Chytraeus: Humanismus im konfessionellen Zeitalter*. Ubstadt-Weiher 1993.

Goel(l)nitz. Abraham. *Compendium Geographicum succincta methodo adornatum*. Amsterdam 1643.

Gondris, Joanna. "'All this Farrago': The Eighteenth-Century Shakespeare Variorum Page as a Critical Structure." *Reading Readings: Essays on Shakespeare Editing in the Eighteenth Century*, ed. Joanna Gondris. London 1998, 123–39.

Goodman, Grant Koon. *Japan and the Dutch 1600–1853*. Richmond 2000.

Gregor, Helmut. *Das Indienbild des Abendlandes (bis zum Ende des 13. Jahrhunderts)*. Wien 1964.

Griep, Wolfgang u. Susanne Luber, Hg. *Reiseliteratur und Geographica in der Eutiner Landesbibliothek* (Kataloge der Eutiner Landesbibliothek). 2 Bde. Heide (Holstein) 1990.

Gründer, Horst. *Welteroberung und Christentum: Ein Handbuch zur Geschichte der Neuzeit*. Gütersloh 1992.

Gruys, Jan Albert. „De reeks Republieken van de Elzeviers en Joannes de Laet.“ *Boekverkopers van Europa: Het 17de-eeuwse Nederlandse uitgevershuis* Elzevier, eds. Berry P.M. Dongelmans et al. Zuphen 2000, 77–106.

Günther, S[iegmund]. *Varenius* (Klassiker der Naturwissenschaften.4). Leipzig 1905. Nachdruck Amsterdam 1970.

Guhrauer, G.E. *Joachim Jungius und sein Zeitalter*. Stuttgart und Tübingen 1850.

Haase, Carl. „Bildung und Wissenschaft von der Reformation bis 1803.“ *Geschichte Niedersachsens*, Hg. Hans Patze. 3. Bd., Teil 2, Hildesheim 1983, 261–493.

Hall, A.R. "Newton's first book." *Archives Internationales d'Histoire des Sciences* 13 (1960): 55–61.

Harbsmeier, Michael. *Wilde Völkerkunde: Andere Welten in deutschen Reiseberichten der Frühen Neuzeit*. Frankfurt (M.) 1994.

——. Hg. *Stimmen aus dem äußersten Norden: Wie die Grönländer Europa für sich entdeckten* (Fremde Kulturen in alten Berichten 11). Stuttgart 2001.

Harvey, P.D.A. *The History of Topographical Maps: Symbols, Pictures and Surveys*. London 1980.

Hay, John. *De rebus Iaponicis, Indicis, et Pervanis epistolæ recentiores*. Antwerpen 1605.

Haycock, David Boyd. *William Stukeley: Science, Religion and Archaeology in Eighteenth-Century England*. Woodbridge 2002.

Heinrich, Gerd. „Nova-Ithaka: Fürstliches Landleben und soziale Wirklichkeit im Herzogtum Dannenberg-Hitzacker zwischen 1605 und 1635.“ *Fruchtblätter: Freundesgabe für Alfred Kelletat*, Hg. Harald Hartung et al. Berlin 1977, 257–283.

Henze, Dietmar. *Enzyklopädie der Entdecker und Erforscher der Erde*. 5 Bde. Graz 1978–2004.

Hofmeister, Adolph, Hg. *Die Matrikel der Universität Rostock*. Bd. 3. Rostock 1895.

Hoffmeister, Johannes, Hg. *Wörterbuch der philosophischen Begriffe*. Hamburg 1955.

Hosking, Geoffrey. *Russia: People and Empire, 1552–1917*. London 1997.

Hughes, Lindsey. *Russia in the Age of Peter the Great*. New Haven: Yale University Press 1998.

——. *Peter the Great: A Biography*. New Haven: Yale University Press 2002.

Huissen, Michael J. "England Encounters Japan: English Knowledge of Japan in the Seventeenth Century." *Terrae incognitae* 5 (1973): 43–59.

Humboldt, Alexander von. *Kritische Untersuchungen über die historische Entwicklung der geographischen Erkenntnis von der Neuen Welt und die Fortschritte der nautischen Astronomie in dem 15ten und 16ten Jahrhundert*. Bd. 1. Berlin 1836.

——. *Kosmos: Entwurf einer physischen Weltbeschreibung*. Bd. 1. Stuttgart 1845 [Neuausgabe Hg. Ottmar Ette u. Oliver Lubrich. Frankfurt a. M. 2004].

——. *Cosmos: A Sketch of the Physical Description of the Universe: Volume 1*, transl. E.C. Otte, introd. Nicholaas Rupke. Baltimore 1997.

Israel, Jonathan.I. *The Dutch Republic: Its Rise, Greatness, and Fall 1477–1806*. Oxford 1995.

Iwai, Hirosato. "A Study of the Varied Names for 'Japan'." *Cultural Nippon* 8 (1940): 21–48, 113–144.

James, Preston E. "On the Origin and Persistence of Error in Geography." *Annals of the Association of American Geographers* 57 (1967): 1–24.

James, Preston E., and Geoffrey J. Martin. *All Possible Worlds: A History of Geographical Ideas*. New York 1980. [For 1993 ed. → G.J. Martin.]

Jarvis, Simon. *Scholars and Gentlemen: Shakespearian Textual Criticism and Representations of Scholarly Labour, 1725–1765*. Oxford 1995.

Jefferson, Thomas. *Notes on the State of Virginia*, ed. William Peden. Chapel Hill 1982.

——. *Writings*. New York 1984.

Jorißen, Engelbert. *Das Japanbild im „Traktat“ (1585) des Luís Fróis* (Portugiesische Forschungen der Görresgesellschaft: 2. Reihe, 7). Münster, Westf. 1988.

Jungius, Joachim. *Logica Hamburgensis*, Hg. Rudolf W. Meyer. Hamburg 1957.

——. *Praelectiones Physicae*. Historisch kritische Edition. Hg. Christoph Meinel. Göttingen 1982.

Jurin: *The Correspondence of James Jurin (1684–1750): Physician and Secretary to the Royal Society*, ed. Andrea A. Rusnock. Amsterdam 1996.

Kaempfer, Engelbert. *Amoenitatum exoticarum politico-physico-medicarum fasculi V*. Lemgo 1712.

——. *Geschichte und Beschreibung von Japan*, Hg. Christian Wilhelm Dohm. 2 Bde. Lemgo 1777/79. Nachdrucke: Stuttgart 1964; Berlin 1980.

——. *Werke*. Kritische Ausgabe in Einzelbänden, Hg. Detlef Haberland et al. Bd.1.1 u. 1.2: *Heutiges Japan*, Hg. Wolfgang Michel u. Barend J. Terwiel. München 2001.

——. *Werke*. Bd. 2: *Briefe 1683–1715*, Hg. Detlef Haberland. München 2001.

Kaminske, Volker u. Ulrich Staffhorst. „Die Geographia Generalis des Bernhard Varenius: Ein disziplingeschichtlicher Erklärungsansatz in der Geographie auf der Grundlage eines ehemals verwendeten Schulbuchs im Gymnasium illustre Karlsruhe." *Jahresbericht des Bismarck-Gymnasiums Karlsruhe*. Karlsruhe 1986/87, 93–100.

Kant, Immanuel. „Entwurf und Ankündigung eines Collegii der physischen Geographie (1757)." *Gesammelte Werke*, Hg. Ernst Cassirer. Bd. 2 Hg. Artur Buchenau. Berlin 1912 [Edition 1922, 1–13].

——. „Metaphysische Anfangsgründe der Naturwissenschaft (Riga 1786)." *Gesammelte Werke*, Hg. Ernst Cassirer. Bd. 4 Hg. E. Cassirer u. A. Buchenau. Berlin 1913. [Edition 1922, 367–478].

Kapitza, Peter. „Engelbert Kaempfer und die europäische Aufklärung: Zur Wirkungsgeschichte seines Japanwerkes im 18. Jahrhundert." *Engelbert Kaempfers Geschichte und Beschreibung von Japan. Beiträge und Kommentar*, Hg. Heinz Götze, Hans Hüls et al. Tokio u. Berlin 1980, 41–63.

——. *Japan in Europa: Texte und Bilddokumente zur europäische Japankenntnis von Marco Polo bis Wilhelm von Humboldt*. 3 Bde. München 1990.

Kastrop, Rainer. *Ideen über die Geographie und Ansatzpunkte für die moderne Geographie bei Varenius unter Berücksichtigung der Abhängigkeit des Varenius von den Vorstellungen seiner Zeit*. Diss. Saarbrücken 1972.

——. „Die Bedeutung des Varenius innerhalb der Entwicklung des geographischen Denkens in Deutschland". *Zur Entwicklung der Geographie vom Mittelalter bis Carl Ritter*, Hg. Manfred Büttner. (Abh. und Quellen zur Geschichte der Geographie und Kosmologie Bd. 3) Paderborn 1982, 79–95.

Keckermann, Bartholomäus. *Systema geographicum*. Hanau 1616.

Keller, Mechthild, Hg. *Russen und Russland aus deutscher Sicht 8.–17. Jahrhundert* (Westöstliche Spiegelungen: Reihe A, 1). München 1985.

Keller, Rudolf. *Die Konfus Augustana im theologischen Wirken des Rostocker Professor David Chytraeus (1530–1600)*. Göttingen 1994.

Kelsall, Malcolm. *Jefferson and the Iconography of Romanticism: Folk, Land, Culture and the Romantic Nation*. London & New York 1999.

Kenney, E.J. *The Classical Text: Aspects of Editing in the Age of the Printed Book*. Berkeley 1974.

Kießling, M. „Varenius und Eratosthenes." *Geographische Zeitschrift* 15 (1909): 12–28.

Kirsch, Peter. *Die Barbaren aus dem Süden: Europäer im alten Japan 1543–1854* (Expansion. Interaktion. Akkulturation 6). Wien 2004.

Kish, Georg, ed. *A Source Book in Geography*. Cambridge, Mass. 1978.

Kivelson, V. "Cartography, Autocracy and State Powerlessness: The Uses of Maps in Early Modern Russia." *Imago Mundi* 51 (1999): 83–105.

Köster, He(i)nrich Martin Gottfried. *Encyclopädie* (1783). Nachdruck: *Historische Encyclopädie: Gesammelte Artikel über Historik und Didaktik aus der* Deutschen Encyclopädie, Hg. Horst Walter Blanke u. Dirk Fleischer. Bd. 2 (Wissen und Kritik, Bd. 9.2). Waltrop 2003, 35–103.

Krabbe, Otto. *Die Universität Rostock im fünfzehnten und sechzehnten Jahrhundert*. Rostock 1854.

Krafft, Fritz. „... *denn Gott schafft nichts umsonst!": Das Bild der Naturwissenschaft vom Kosmos im historischen Kontext des Spannungsfeldes Gott—Mensch—Natur*. Münster 1999.

Kreiner, Josef. „Das Bild Japans in der europäischen Geistesgeschichte." *Japanstudien* 1 (1990): 13–42.

Krogt, Peter van der. "Commercial Cartography in the Netherlands with Particular Reference to Atlas Production (16th-18th centuries)." *La cartografia dels Països Baixos*. Barcelona 1995, 71–140.

Kuhn, Thomas S. *Die Struktur wissenschaftlicher Revolutionen*. 5. Aufl. Frankfurt (Main) 1981.

Lach, Donald F. *Asia in the Making of Europe*. 3 parts. Chicago 1965–1993.

Lange, Gottfried. „Varenius über die Grundfragen der Geographie: Ein Beitrag zur Problemgeschichte der geographischen Wissenschaft." *Petermanns Geographische Mitteilungen* 105 (1961): 274–83.

——. „Das Werk des Varenius: Eine kritische Gesamtbibliographie." *Erdkunde* 15 (1961): 1–18.

——. „Philipp Clüver (1580–1622)", *Geographisches Taschenbuch* 1962/1963, 251–258.

Larner, John. *Marco Polo and the Discovery of the World*. New Haven 1999.

Lehmann, Klaus. „Das kurze Leben des Bernhard Varenius (1622–1650)." *Bernhard Varenius: Der Beginn der modernen Geographie*, Hg. Wolfgang Griep u. Susanne Luber. Begleitband zur Ausstellung 2001 (Veröffentlichungen der Eutiner Landesbibliothek 5). Eutin 2001, 9–18.

Lewis and Clark: *The Journals of the Expedition of Capts. Lewis and Clark*, ed. Nicholas Biddle. 2 vols. New York 1962.

Lewitter, L.R. "Peter the Great and the Modern World." *Russia and Europe*, ed. Paul Dukes. London 1991, 92–107.

Lexikon zur Geschichte der Kartographie, bearb. v. Ingrid Kretschmer, J. Dörflinger u. F. Wawrik. 2 Bde. Wien 1986.

Lindberg, Sten G. „Queen Christina Bindings." *Queen Christina of Sweden: Documents and Studies*, ed. Magnus von Platen. Stockholm 1966.

Linschoten, Jan Huygen van. *Itinerario. Voyage ofte Schipvaert / van Jan Huygen van Linschoten naer Oost ofte Portugaels Indien 1579–1592*. 4 Bde. Amsterdam 1596. Nachdruck: Hg. Hendrik Kern et al. 5 Bde. Den Haag 1910–1957.

Lipking, Lawrence. "The Marginal Gloss." *Critical Inquiry* 3 (1977): 609–55.

Livingstone, David N. *The Geographical Tradition: Episodes in the History of a Contested Enterprise*. Oxford 1992.

Lukermann, Fred. "The *Praecognita* of Varenius: Seven Ways of Knowing." *Text and Image: Social Construction of Regional Knowledges*, eds. Anne Buttimer, Stanley D. Brunn and Ute Wardenga. Leipzig 1999, 7–27.

Lutz, Walter, Hg. *Japan mit den Augen des Westens gesehen: Gedruckte europäische Landkarten vom frühen 16. bis zum 19. Jahrhundert*. München u. New York 1993.

Maanen, J.A. van. *Facets of Seventeenth Century Mathematics in the Netherlands*. Diss. Utrecht 1987.

McKenzie, Donald. *The Cambridge University Press, 1696–1712: A Bibliographical Survey*. 2 vols. Cambridge 1966.

McKitterick, David. *A History of Cambridge University Press*. Vol. 1: *Printing and the Book Trade in Cambridge, 1534–1698*. Cambridge 1992.

——. *A History of Cambridge University Press*. Vol. 2: *Scholarship and Commerce, 1698–1872*. Cambridge 1998.

Maffeus, Giovan Pietro. "De Japonicis rebus epistolarum" (1571). *Opera omnia latine scripta*. Bd. 2. Dillingen 1747, 91–316.

Mager, Inge. „Reformator. Theologie u. Reformationsverständnis an der Univ. Helmstedt im 16. u. 17. Jh." *Jahrbuch der Gesellschaft für niedersächsische Kirchengeschichte* 74 (1976): 11–33.

Martin, Geoffrey J. *All Possible Worlds: A History of Geographical Ideas*. New York 1993.
——. "The Emergence and Development of Geographic Thought in New England." *Economic Geography* 74, special issue (1998): 1–13.
Matriculation Hamburg: → Sillem, C.H.W.
Matriculation Königsberg: → Erler, G.
Matriculation Leiden: → Rieu du, G.
Matriculation Rostock: → Hofmeister, Adolph
Mayhew, Robert. "The Character of English Geography, c.1660–1800: A Textual Approach." *Journal of Historical Geography* 24 (1998): 385–412.
Meinel, Christoph. *In physicis futurum saeculum respicio: Joachim Jungius und die Naturwissenschaftliche Revolution des 17. Jahrhunderts* (Veröffentlichung der Jungius-Gesellschaft 52). Göttingen 1984.
——. *Die Bibliothek des Joachim Jungius: Ein Beitrag zur Historia litteraria der frühen Neuzeit* (Veröffentlichung der Jungius-Gesellschaft 67). Göttingen 1992.
——. „Reine und angewandte Wissenschaft." Publ. by Wissenschaftszentrum Nordrhein-Westfalen in: *Das Magazin* (2000: no.1).
Melon y Ruiz de Gordejuela, Amando. "Esquema sobre los modeladores de la moderna ciencia geográfica." *Estudios geográficos* n. 20–21 (Aug./Nov. 1945): 393–442.
Mill, Hugh Robert. "The Present Problems of Geography." *The Geographical Journal* 25 (1905): 2–17.
Möhlmann, Gerd. *Geschlechter der Hansestadt Rostock im 13.–18. Jahrhundert*. Neustadt a.d. Aisch 1975.
Moran, Joseph. *The Japanese and the Jesuits: Alessandro Valignano in Sixteenth-Century Japan*. London 1993.
Morhof, Daniel Georg. *Polyhistor literarius, philosophicus et practicus*. Hg. Johann Joachim Schwabe. 3 Bde. 4. Aufl. Lübeck 1747. Nachdruck: Aalen 1970.
Mozai, Torao, u. Nakamura Koji. "The Lost Fleet of Kublai Khan." *National Geographic Magazine* (1982:11): 635–649.
Münkler, Marina. *Marco Polo: Leben und Legende*. München 1998.
——. *Erfahrung des Fremden: Die Beschreibung Ostasiens in den Augenzeugenberichten des 13. und 14. Jahrhunderts*. Berlin 2000.
Nachod, Oskar. *Die Beziehungen der Niederländischen Ostindischen Kompagnie zu Japan im siebzehnten Jahrhundert*. Leipzig 1897.
Newton, Isaac. *Correspondence*. Volume 1: *1661–1675*, ed. H.W. Turnbull. Cambridge: The Royal Society 1959.
——. *The Mathematical Papers of Isaac Newton*. Volume 2: *1667–1670*, ed. D.T. Whiteside. Cambridge Press 1968.
——. *The Mathematical Papers of Isaac Newton*. Volume 4: *1674–1684*, ed. D.T. Whiteside. Cambridge 1971.
Nieuhof, Jo[h]an. *Het Gezandtschap der Neêrlandtsche Oost-Indische Compagnie, aan den grooten Tartarischen Cham, den tegenwoordigen Keizer van China*. Amsterdam 1665.
Olearius, Adam. *Vermehrte Newe Beschreibung der Muscowitischen und Persischen Reyse*. Schleswig 1656. Nachdruck: Dieter Lohmeier, Hg. Tübingen 1971.
Olschki, Leonardo. *Marco Polo's Asia: An Introduction to His* Description of the World *Called* Il Milione. Berkeley 1960.
Partsch, J. *Philipp Clüver, der Begründer der historischen Länderkunde*. Wien 1891.
Pauly, Ulrich. *Seppuku: Ritueller Selbstmord in Japan*. Tokyo 1995.
Pelliot, Paul. *Notes on Marco Polo*. 3 Bde. Paris 1959–1973.
Pfeiffer, Rudolf. *A History of Classical Scholarship: From 1300 to 1850*. Oxford 1976.
Philippson, A. „Zwei Vorläufer des Varenius." *Ausland* 52 (1892).
Plomer, Henry. *A Dictionary of Printers and Booksellers at Work in England, Scotland and Ireland from 1668 to 1725*. Oxford 1922.
Plomer, Henry, G.H. Bushnell and E.R. McC. Dix. *A Dictionary of Printers and Booksellers who were at Work in England, Scotland and Ireland from 1726 to 1775*. Oxford 1932.

Poelgeest, L. van. "The stadholder-King William III and the University of Leiden". *Fabrics and Fabrication: The Myth and Making of William and* Mary, eds. Paul Hoftijzer and C.C. Barfoot. Amsterdam 1990, 97–134.

Polo, Marco. *Il Milione, a cura di Luigi Foscolo Benedetto*. Firenze 1928.

——. *Il Milione: Die Wunder der Welt*. Übers. u. Nachw. Elise Guignard. Zürich 1983.

Popkin, Richard H. "The Crisis of Polytheism and the Answers of Vossius, Cudworth, and Newton." *Essays on the Context, Nature, and Influence of Isaac Newton's Theology*, eds. James E. Force and Richard H. Popkin. Dordrecht 1990, 9–25.

——. "Polytheism, Deism, and Newton." *Essays on the Context, Nature, and Influence of Isaac Newton's Theology*, eds. James E. Force and Richard H. Popkin. Dordrecht 1990, 27–42.

Postnikov, A.V. *Razvitie krupnomasshtabnoi kartografii v Rossii*. Moscow 1989.

Rademaker, C.S.M. *Life and Work of Gerardus Joannes Vossius (1577–1649)*. Assen 1981.

Raikov, B.E. *Ocherk po istorii geliotsentricheskogo mirovozreniya v Rossii*. Moscow-Leningrad 1947.

Rainov, T. *Nauka v Rossii, XI–XVII veko*. Moscow-Leningrad 1940.

Rantzau, Heinrich. *Methodus describendi regiones, urbes et arces* (1587). Nachdruck in: → Rassem u. Stagl 1994, 160–181.

Rassem, Mohammed u. Justin Stagl, Hg. *Geschichte der Staatsbeschreibung: Ausgewählte Quellentexte 1456–1813*. Berlin 1994.

Ratkowitsch, Christine. *Descriptio picturae: Die literarische Funktion der Beschreibungen von Kunstwerken in der lateinischen Großdichtung des 12. Jahrhunderts*. Wien 1991

Ratzel, Friedrich. „Varenius". *Allgemeine Deutsche Biographie* 39 (1895): 487–90.

Rebok, Sandra. „Alejandro de Humboldt y el modelo de la *Historia Natural y Moral*." *Humboldt im Netz* 3 (2001): http://www.uni-potsdam.de/u/romanistik/humboldt/hin.

——. "Two Exponents of the Enlightenment: Transatlantic Communication by Thomas Jefferson and Alexander von Humboldt," *The Southern Quarterly* 43: 4 special issue *Imagining the Atlantic World* (2006): 126–152.

Rehm, Albert. „Nilschwemme." *Paulys Real-Encyclopädie der Classischen Altertumswissenschaft. Neue Bearbeitung*, Hg. Wilhelm Kroll. Bd. 17. Stuttgart 1937, 571–90.

Reichert, Folker. *Begegnungen mit China: Die Entdeckung Ostasiens im Mittelalter* (Beiträge zur Geschichte und Quellenkunde des Mittelalters 15). Sigmaringen 1992.

——. „Marco Polos Buch: Lesarten des Fremden." *Fiktion des Fremden: Erkundung kultureller Grenzen in Literatur und Publizistik*, Hg. Dietrich Harth. Frankfurt a. M. 1994, 180–202.

——. „Die Erfindung Amerikas durch die Kartographie." *Archiv für Kulturgeschichte* 78 (1996): 115–143.

——. *Erfahrung der Welt: Reisen und Kulturbegegnung im Späten Mittelalter*. Stuttgart 2001.

——. „Zipangu: Marco Polos Japan und das europäische Weltbild zwischen Mittelalter und Neuzeit." *Engelbert Kaempfer (1651–1716) und die kulturelle Begegnung zwischen Europa und Asien*, Hg. Sabine Klocke-Daffa, J. Scheffler u. G. Wilbertz (Lippische Studien 18). Lemgo 2003, 147–168.

——. „Mythische Inseln." *Mittelalter-Mythen*, Hg. Ulrich Müller u. W. Wunderlich. Bd. 5. St. Gallen 2007.

Reinhard, Wolfgang. *Geschichte der europäischen Expansion*. Bd. 1: *Die Alte Welt bis 1818*. Stuttgart 1983.

——. „Gelenkter Kulturwandel im 17. Jahrhundert: Akkulturation in den Jesuitenmissionen als universalhistorisches Problem." In: W. Reinhard. *Ausgewählte Abhandlungen* (Historische Forschungen 60). Berlin 1997, 347–399.

Reynolds, L.D. and N.G. Wilson. *Scribes and Scholars: A Guide to the Transmission of Greek and Latin Literature*. Second edition. Oxford 1974.

Rhein, Stefan. „Ein Gruß aus Wittenberg: David Chytraeus und die Hausschule Melanchthons." *David Chytraeus (1530–160): Norddeutscher Humanismus in Europa*, Hg. Karl-Heinz Glaser u. Steffen Stuth. Ubstadt-Weiher 2000, 13–17.

Richter, Frank. „Gibt es eine antipositivistische Strömung in der Entwicklung der bürgerlichen Wissenschaftstheorie?" *Deutsche Zeitschrift für Philosophie* 5 (1980): 624–628.

——. „Michail Wassiljewitsch Lomonossow: Der Einfluß Christian Wolffs auf die russische Aufklärung." *Forschungen zur Physikotheologie im Aufbruch*. Bd. I. *Naturwissenschaft, Theologie und Musik in der Aufklärung*, Hg. Frank Richter u. Manfred Büttner. Referate des Symposiums in Halle 1994 (Physikotheologie im historischen Kontext Bd. 1). Münster 1995, 225–252.

Rieu du, G., ed. *Album studiosorum academiae Lugduno-Bataviae, 1575–1875*. Hagae Comitum 1975.

Roberts, Elizabeth T. *The* Geographia Generalis *of Bernhard Varenius*. MA Thesis in History. U. of New Hampshire. 1994. Unpubl. Typescript.

Röhrich, Lutz. *Lexikon der sprichwörtlichen Redensarten*. 5 Bände. 5. Aufl. Freiburg 2001.

Rossabi, Morris. *Khubilai Khan: His Life and Times*. Berkeley 1988.

Rusnock: → Jurin.

Ryan, W.F. *Magic in Russia: The Bathhouse at Midnight*. Stroud 1999.

St Clair, William. *The Reading Nation in the Romantic Age*. Cambridge 2004.

Sammler, Fürst, Gelehrter: Herzog August zu Braunschweig Lüneburg 1579–1666. (Ausstellungskatalog Herzog August Bibliothek 27). Braunschweig 1979.

Scheltema, Jacob. *Rusland en de Nederlandenbeschouwd in derzelver wederkeerige betrekkingen*. 4 vols. Amsterdam 1817–1819.

Schlegelmilch, Ulrich. *Descriptio templi: Architektur und Fest in der lateinischen Dichtung des konfessionellen Zeitalters*. Regensburg 2003.

Schmidt-Biggemann, Wilhelm. *Topica universalis: Eine Modellgeschichte humanistischer und barocker Wissenschaft* (Paradeigmata.1). Hamburg 1983.

——. „Polyhistoria/Polymathie." *Historisches Wörterbuch der Philosophie*, Hg. Joachim Ritter u. Karlfried Gründer. Bd.7. Basel 1989.

Schmitt, Eberhard, Hg. *Wirtschaft und Handel der Kolonialreiche* (Dokumente zur Geschichte der europäischen Expansion 4). München 1988.

——, Thomas Schleich u. Th. Beck, Hg. *Kaufleute als Kolonialherren: Die Handelswelt der Niederländer vom Kap der Guten Hoffnung bis Nagasaki 1600–1800*. Bamberg 1988.

Schmitthüsen, Josef. *Geschichte der geographischen Wissenschaft von den ersten Anfängen bis zum Ende des 18. Jahrhunderts*. Mannheim 1970.

Schomerus, Justus Christoph. *Die Gewissens-Freudigkeit im Tode* [Leichenpredigt August Varenius]. Rostock 1684.

Schuchard, Margret. *John Ogilby 1600–1676: Lebensbild eines Gentleman mit vielen Karrieren*. Hamburg 1973.

——. *A Descriptive Bibliography of the Works of John Ogilby and William Morgan*. Frankfurt 1975.

Schütte, Joseph. *Valignanos Missionsgrundsätze für Japan* (Storia e letteratura 36). 2 Bde. Rom 1951–1958.

Schurhammer, Georg. *Franz Xaver: Sein Leben und seine Zeit*. 2 Bde. Freiburg 1955–1973.

Schwabe → *Allgemeine Historie*.

Schwarz, Ingo. "From Alexander von Humboldt's Correspondence with Thomas Jefferson and Albert Gallatin." *Berliner Manuskripte zur Alexander-von-Humboldt-Forschung* 2 (1991): 1–20.

——. "Alexander von Humboldt's Visit to Washington and Philadelphia: His Friendship with Jefferson, and His Fascination with the United States." *Alexander von Humboldt's Natural History Legacy and Its Relevance for Today*. In: *Northeastern Naturalist* 8, special issue no. 1 (2001): 43–56.

——. *Alexander von Humboldt und die Vereinigten Staaten von Amerika*. Berlin 2004.

Schwerdfeger, Josef. „Bernhard Varenius und die morphologischen Capitel seiner *Geographia generalis* (Amsterdam 1650): Ein Beitrag zur Geschichte der Geographie." Sonderdruck aus *Jahresberichte des K.K. Staatsgymnasium in Troppau*, 1897/1898.

Schwind, Martin. „Die älteste Japanbeschreibung in europäischer Sprache: *Descriptio Regni Japoniae* von Bernhard Varenius, 1649." *Chirigaku Hyoron* [Geographical Review of Japan] 46 (1973): 81–91.

——. „Die wissenschaftliche Stellung der *Descriptio Regni Japoniae*," *Descriptio regni Japoniae: Beschreibung des japanischen Reiches*, Hg. Martin Schwind u. Horst Hammitzsch, Übers. Ernst-Christian Volkmann. Darmstadt 1974, XVII–XXXIX.

Seneca, L. *Annaei Senecae Naturalium qaestionum libri viii*, Hg. Alfred Gercke. Stuttgart 1970.

Shaw, D.J.B. "Geographical Practice and Its Significance in Peter the Great's Russia." *Journal of Historical Geography* 22 (1996): 160–176.

——. "A Strong and Prosperous Condition: The Geography of State Building and Social Reform in Peter the Great's Russia." *Political Geography* 18 (1999): 991–1015.

——. "Mapmaking, Science and State Building in Russia Before Peter the Great." *Journal of Historical Geography* 31 (2005): 409–429.

——. "The Translation and Publication of Bernhard Varenius' *Geographia Generalis* in Russia, 1718." *Study Group on Eighteenth Century Russia Newsletter* 34 (2006): 9–13.

Sillem, C.H.W. *Die Matrikel des Akademischen Gymnasiums in Hamburg, 1613–1883*. Hamburg 1891. Nachdruck Nendeln, Liechtenstein 1980.

Sitters, M.H. "Sybrandt Hansz. Cardinael (1578–1647): Een eigenzinnig meetkundige tussen Stevin en Huygens". *Gewina. Tijdschrift voor de geschiedenis der geneeskunde, natuurwetenschappen, wiskunde en techniek* 27 (2004): 14–32.

Sitwell, O.F.G. *Four Centuries of Special Geography*. Vancouver 1991.

Sowerby, E. Millicent, ed. *Catalogue of the Library of Thomas Jefferson*. 5 vols. Charlottesville 1983.

Staden, Hans. *Warhaftige Historia vnd beschreibung eyner Landschafft der Wilden / Nacketen / Grimmigen Menschenfresser Leuthen / in der Newenwelt America gelegen*. Marburg 1557. Nachdruck Kassel 1978.

Stagl, Justin. „Vom Dialog zum Fragebogen: Miszellen zur Geschichte der Umfrage." *Kölner Zeitschrift für Soziologie und Sozialpsychologie* 31 (1979): 611–38.

——. „Die Apodemik oder ‚Reisekunst' als Methodik der Sozialforschung vom Humanismus bis zur Aufklärung." *Statistik und Staatsbeschreibung in der Neuzeit vornehmlich im 16.–18. Jahrhundert*, Hg. Mohammed Rassem u. Justin Stagl. Paderborn 1980, 131–204.

Steiner, Arpad. "A Mirror for Scholars of the Baroque." *Journal of the History of Ideas* 1 (1940): 321–34.

Steinmeyer, Rudolf. *Die Chronik der Herzog-Ernst-Schule, der früheren Lateinschule in Uelzen*. Uelzen 1964.

Terra, Helmut de. "Alexander von Humboldt's Correspondence with Thomas Jefferson." *Proceedings of the American Philosophical Society* 103 (1959): 783–806.

Thüringer, Walter. „David Chytraeus (1530–1600): Aus dem Leben eines dem Kraichgau verbundenen Theologen und Historikers." *Kraichgau* 6 (1979): 161–173.

Tütken, Johannes. *Höhere und mittlere Schulen des Herzogtums Braunschweig-Wolfenbüttel, der Herrschaft Dannenberg und der Grafschaft Blankenburg im Spiegel der Visitationsprotokolle des Generalschulinspektors Christoph Schrader (1650–1666)*. Wiesbaden 1997.

—— et al. *Glaubenslehre, Bildung, Qualifikation: 450 Jahre Große Schule in Wolfenbüttel—ein Beitrag zur Geschichte des evangelischen Gymnasiums in Norddeutschland* (Ausstellungskatalog der Herzog August Bibliothek 69). Berlin 1993.

Überweg, Friedrich. *Grundriss der Geschichte der Philosophie der Neuzeit: Vom Aufblühen der Altertumsstudien bis auf die Gegenwart*, Hg. Max Heinze. Berlin 1880.

——. *Geschichte der Philosophie: Die Philosophie der Neuzeit bis zum Ende des 18. Jahrhunderts*. Darmstadt 1958.

Unno, Kazutaka. *Chizu ni miru Nippon—Japan in Maps: Wakoku, Zipangu, Dainippon*. Tokyo 1999 [japan.].

Vanpaemel, G.H.W. "Gemengde wiskunde en experimentele fysica." *De zeventiende eeuw* 7 (1991): 49–57.

Varenius, Bernhardus. *Disputatio medica inauguralis: De Febri in Genere*. Leiden 1649. [Übers. Carl Rohrbach, in: *Bernhard Varenius (1622–1650): Der Beginn der modernen Geographie*, Hg. Wolfgang Griep u. Susanne Luber. Begleitband zur Ausstellung 2001 (Veröffentlichungen der Eutiner Landesbibliothek 5). Eutin 2001, 70–75.]

——. *Descriptio regni Iaponiae cum quibusdam affinis materiae. Ex variis auctoribus collecta et in ordinem redacta*. Amsterdam 1649 [also 1673].

——. *Tractatus in quo agitur de Iapone religione*. Amsterdam 1649. [Usually bound with *Descriptio regni Iaponiae*. Amsterdam 1649 [also 1673].

——. *Geographia Generalis, in qua affectiones generales Telluris explicantur*. Amsterdam 1650 [also 1664, 1671].

——. *Geographia Generalis*, ed. Isaac Newton. Cambridge 1672 [also 1681]. [Complete digital facsimile in Wolfenbütteler Digitale Bibliothek maintained by Herzog August Bibliothek: http://diglib.hab.de/wdb.php?dir=drucke/ne-183.]

——. *Geographia Generalis*, ed. James Jurin. Cambridge 1712.

——. *A Compleat System of General Geography*, eds. Newton, Jurin, and Shaw, transl. Dugdale. 2 vols., London 1733 [also 1734, 1736, 1765].

——. (Bernhard Varen). *Descriptio regni Japoniae: Beschreibung des japanischen Reiches*, Hg. Martin Schwind u. Horst Hammitzsch, Übers. Ernst-Christian Volkmann. Darmstadt 1974. Nachdruck München 2000.

Varenius, Heinrich. *Trifolium Sacrum: Ein geistliches Kleeblätlein* [...] *Das ist ein Prob. Valet. und Antritspredigte*. Lüneburg 1628.

Verbeek, Theo. *Descartes and the Dutch: Early Reactions to Cartesian Philosophy 1637–1650*. Carbondale 1992.

Vermij, Rienk H. "Mathematics at Leiden: Stevin, Snellius, Scaliger." *Der "mathematicus": Zur Entwicklung und Bedeutung einer neuen Berufsgruppe in der Zeit Gerhard* Mercators, ed. Irmgard Hantsche. Bochum 1996, 75–92.

——. *The Calvinist Copernicans. The Reception of the New Astronomy in the Dutch Republic*. Amsterdam 2002.

Vidal de la Blache, Paul. "Le Principe de la Géographie Générale." *Annales de Géographie* 5 (1895–6): 129–42.

Vos, Ken. „Dejima und die Handelsbeziehungen zwischen den Niederlanden und dem vormodernen Japan." *Japan und Europa 1543–1929*, Hg. Doris Croissant u. Lothar Ledderose. Berlin 1993, S. 72–82.

Vries, Jan de and Ad M. van der Woude. *Nederland 1500–1815. De eerste ronde van moderne economische groei*. Amsterdam 1995. Engl. trsl.: *The First Modern Economy: Success, Failure, and Perseverance of the Dutch Economy, 1500–1815*. Cambridge 1997.

Vrieze, F.S. de. "Academic Relations between Sweden and Holland." *Leiden University in the Seventeenth Century: An Exchange of Learning*, eds. Th. Lunsingh Scheurleer et al. Leiden 1975.

Walsh, Marcus. *Shakespeare, Milton and Eighteenth Century Literary Editing: The Beginnings of Interpretative Scholarship*. Cambridge 1997.

Wandhoff, Haiko. *Ekphrasis: Kunstbeschreibungen und virtuelle Räume in der Literatur des Mittelalters*. Berlin 2003.

Warntz, William. *Geography Now and Then: Some Notes on the History of Academic Geography in the United States*. New York 1964.

——. "*Geographia Generalis* and the Early Development of American Academic Geography." *Origins of American Academic Geography*, ed. Brian Blouet. Hamden CT 1981, 245–263.

——. "Newton, the Newtonians, and the *Geographia Generalis Varenii*." *Annals of the Association of American Geographers* 79:2 (1989): 165–191.

Wasser, Hartmut. *Die große Vision: Thomas Jefferson und der amerikanische Westen*. Wiesbaden 2004.

Wendelinus, Markus Friedrich. *Wunder des Nil: Admiranda Nili 1623*. Faks.-Ausgabe. Hg Reinhard Düchting u. Boris Körkel. Heidelberg 2000.

Wiedemann, Conrad. „Polyhistors Glück und Ende: Von Daniel Georg Morhof zum jungen Lessing." *Festschrift Gottfried Weber*, Hg. Heinz Otto Burger u. Klaus von See (Frankfurter Beiträge zur Germanistik 1). Bad Homburg v.d.H. 1967, 215–35.

Wiegand, Hermann. *Hodoeporica: Studien zur neulateinischen Reisedichtung des deutschen Kulturraums im 16. Jahrhundert*. Baden-Baden 1984.

Wiesenfeldt, Gerhard. *Leerer Raum in Minervas Haus: Experimentelle Naturlehre an der Universität Leiden, 1675–1715*. Amsterdam 2002.

Willems, Alphonse. *Les Elzevier: histoire et annals typographiques*. Bruxelles 1880.

Wolff, Christian. *Cosmologia generalis*. *Gesammelte Werke*. Abt. 2, Bd. 4. Hildesheim 1964.

——. *Die Anfangsgründe aller mathematischen Wissenschaften*. 3. Teil. *Gesammelte Werke*, Hg. Jean Ecole. Abt. 1, Bd. 14. Hildesheim 1973.

——. *Mathematisches Lexicon* (1716). *Gesammelte Werke*. Abt. 1, Bd. 11, 2. Hildesheim 1978.

——. *Vernünftige Gedanken von den Würckungen der Natur*. *Gesammelte Werke*. Abt. 1, Bd. 6. Hildesheim 1981.

Wright, J.K. "Some British 'Grandfathers' of American Geography." *Geographical Essays in Memory of Alan G. Ogilvie*. London 1959.

Wurffbain, Johann Sigmund. *Vierzehen Jährige Ost-Indianische Krieg- und Ober-Kauffmanns-Dienste*, Hg. Johann Paul Wurffbain. Sulzbach Nürnberg 1686. Nachdruck als: *Reise nach den Molukken und Vorder-Indien 1632–1646*, Hg. R. Posthumus Meyjes. In: *Reisebeschreibungen von deutschen Beamten und Kriegsleuten im Dienst der Niederländischen West- und Ost-Indischen Kompagnien 1602–1797*, Hg. S.P. L'Honoré Naber. Bde. 8 u. 9. Den Haag 1931.

Xaverius, Francisus u. Orazio Torsellini. *Apostolisches Leben und Thaten des heiligen Francisci Xaveri*. München 1674.

——. *Gesammelte Briefe des heiligen Franciscus Xaverius*, Hg. Anton Eglauer. 3 Bde. Augsburg 1794.

Zeiller, Martin. *Fidus Achates, oder getreuer Raeßgefert*. Ulm 1651.

LIST OF CONTRIBUTORS

HORST WALTER BLANKE is a historian at the Kulturwissenschaftliches Institut Essen (Project "Einheit der Menschheit—Differenz der Kulturen—Chancen der Kommunikation") and at the University of Bielefeld. His research and his publications concentrate on the history of historiography, the German enlightenment, and travel literature.

REINHARD DÜCHTING, now retired, was Professor of Medieval and Early Modern Latin at the University of Heidelberg. He began his academic life with a degree in theology and German, but his doctoral dissertation was about Sedulius Scottus, and he has worked in the field of Neo-Latin ever since. He published on medieval poetry, also as the editor of texts, and on various sacred and secular themes.

KLAUS LEHMANN moved from being a history teacher at Bernhard-Varenius-Schule in Hitzacker (Elbe) to becoming keeper of the Museum Das Alte Zollhaus in Hitzacker (where the town archive is also kept, including research materials on Varenius). He has turned the museum into a centre of various activities and staged many successful exhibitions on local history (int. al. on Varenius and on Duke August of Braunschweig-Lüneburg).

ROBERT MAYHEW is Reader in Historical Geography at the University of Bristol, UK. He was educated at the University of Oxford and held a fellowship at Corpus Christi College, Cambridge. He is the author of two books, *Enlightenment Geography* (Macmillan: London, 2000) and *Landscape, Literature and English Religious Culture, 1660–1800* (Palgrave: London, 2004). His current research centres on early modern geographical writings as a print culture.

SANDRA REBOK studied Sociology and Anthropology in Heidelberg, Madrid, and Paris and wrote her Dr. phil. dissertation on "Alexander von Humboldt and Spain in the 19th Century: Analysis of the Reciprocal Perceptions." (Heidelberg University 2004). Since 1999 she has been a member of the History of Science Department of the Spanish Council for Scientific Research in Madrid, where her research on

Humboldt continues (work on Humboldt and Spain, a DVD edition of his works, a publication of his Mexican diaries in Spanish). Additional research includes the study of the perceptions of German travellers in America and the New World, transatlantic communication during the Enlightenment, and at present a three year project on German travellers in Spain in the 19th Century.

FOLKER REICHERT is Professor of Medieval History at the University of Stuttgart. He was educated in Würzburg and Heidelberg (Dr. phil. 1982, venia legendi 1990 in Heidelberg). He is the author of numerous publications on the history of travel, discoveries, and the world picture in the middle ages and the early modern period, among them *Begegnungen mit China: Die Entdeckung Ostasiens im Mittelalter* (Sigmaringen: Thorbecke 1994); *Erfahrung der Welt: Reisen und Kulturbegegnung im späten Mittelalter* (Stuttgart: Kohlhammer 2001; Japanese trsl. 2005).

FRANK RICHTER (1938–2003) was Professor of Philosophy at the Bergakademie Freiberg in East Germany until 1991 and taught Marxism–Leninism. After the end of the GDR, he remained a Marxist and attempted to assess Marxist philosophy in the Communist era critically. His research and publications concentrated on the history of science and scholarship and philosophical questions dealing with science and technology.

MARGRET SCHUCHARD is a retired senior lecturer (Akademische Direktorin) of the English Department of the University of Heidelberg where she taught English literature. She was educated in Hamburg, Cumberland Lodge (Windsor Park), Aberdeen and Heidelberg, where her subjects were English, German and History (Dr. phil. Heidelberg 1970). Her research interests and publications topics include biography, drama, and verse satire in the seventeenth and eighteenth centuries, literature of the 1930s and the history of scholarship.

DENIS J.B. SHAW is Reader in Russian Geography in the School of Geography, Earth and Environmental Sciences, and Associate Member of the Centre for Russian and East European Studies, University of Birmingham, UK. He is a specialist on the historical geography of late medieval and early modern Russia and on the history of Russian geographical and environmental thought and practice. He was educated at University College London, UK, and Moscow and Voronezh State

Universities, Russia, and has held numerous visiting professorships at universities in the USA, Canada, UK and Russia. He is the author of *Russia in the Modern World* (Blackwell, 1999); co-author of *Planning in the USSR* (Croom Helm, 1981); *Landscape and Settlement in Romanov Russia* (1613–1917) (OUP 1990); *Alternative Capitalisms: Geographies of Emerging Regions* (Arnold, 2003); and of many articles relating to geographical problems in Russia, including problems of historical geography.

Ulrich Staffhorst was educated in classics and history at the universities of Würzburg, Hamburg and Tübingen (Dr. phil Würzburg 1965). He was a teacher of Latin and History at the Bismarck-Gymnasium in Karlsruhe until he retired in 2003. Since 1999 he has taught courses in Ancient Greek at the University of Karlsruhe. His publications focus on his special subjects and school history.

Johann Anselm Steiger is Professor of Ecclesiastical and Dogmatics History at the University of Hamburg. He received his docorate in Heidelberg, his venia legendi in Leipzig. From 1995 to 2000 he was head of the research centre on Johann Gerhard studies in Heidelberg. His research interests include the reformation, Lutheran theology and the piety of the German baroque period, enlightenment, editorial matters, intersections between literature and theology, the relations between Jews and Christians throughout their history, and the history of exegesis.

Rienk H. Vermij obtained his PhD at the Institute for History of Science at the University of Utrecht (Netherlands) in 1991, has since worked at several universities in the Netherlands and Germany and is now at the History of Science Department of the University of Oklahoma. He has published several books and articles on the history of science in the sixteenth, seventeenth and eighteenth centuries.

Ernst-Christian Volkmann is a retired teacher of Latin and still teaches some Latin courses. He was educated at the universities of Göttingen and Tübingen and studied Latin, Greek and History. His thorough familiarity with Varenius' way of writing is owing to his translation of Varenius' *Descriptio Regni Japoniae: Beschreibung des japanischen Reiches* into German (Darmstadt 1974, reprinted 2000). It is the first complete translation of this book into a modern vernacular.

INDEX OF NAMES

Brill's Studies in Intellectual History

Series Editor: A.J. Vanderjagt

56. Pranger, M.B. *Bernard of Clairvaux and the Shape of Monastic Thought.* Broken Dreams. 1994. ISBN 90 04 10055 5
57. Van Deusen, N. *Theology and Music at the Early University.* The Case of Robert Grosseteste and Anonymous IV. 1994. ISBN 90 04 10059 8
58. Warneke, S. *Images of the Educational Traveller in Early Modern England.* 1994. ISBN 90 04 10126 8
59. Bietenholz, P.G. *Historia and Fabula.* Myths and Legends in Historical Thought from Antiquity to the Modern Age. 1994. ISBN 90 04 10063 6
60. Laursen, J.C. (ed.). *New Essays on the Political Thought of the Huguenots of the* Refuge. 1995. ISBN 90 04 09986 7
61. Drijvers, J.W. & A.A. MacDonald (eds.). *Centres of Learning.* Learning and Location in Pre-Modern Europe and the Near East. 1995. ISBN 90 04 10193 4
62. Jaumann, H. *Critica.* Untersuchungen zur Geschichte der Literaturkritik zwischen Quintilian und Thomasius. 1995. ISBN 90 04 10276 0
63. Heyd, M. *"Be Sober and Reasonable."* The Critique of Enthusiasm in the Seven-teenth and Early Eighteenth Centuries. 1995. ISBN 90 04 10118 7
64. Okenfuss, M.J. *The Rise and Fall of Latin Humanism in Early-Modern Russia.* Pagan Authors, Ukrainians, and the Resiliency of Muscovy. 1995. ISBN 90 04 10331 7
65. Dales, R.C. *The Problem of the Rational Soul in the Thirteenth Century.* 1995. ISBN 90 04 10296 5
66. Van Ruler, J.A. *The Crisis of Causality.* Voetius and Descartes on God, Nature and Change. 1995. ISBN 90 04 10371 6
67. Shehadi, F. *Philosophies of Music in Medieval Islam.* 1995. ISBN 90 04 10128 4
68. Gross-Diaz, T. *The Psalms Commentary of Gilbert of Poitiers.* From *Lectio Divina* to the Lecture Room. 1996. ISBN 90 04 10211 6
69. Van Bunge, W. & W. Klever (eds.). *Disguised and Overt Spinozism around 1700.* 1996. ISBN 90 04 10307 4
70. Floridi, L. *Scepticism and the Foundation of Epistemology.* A Study in the Meta-logical Fallacies. 1996. ISBN 90 04 10533 6
71. Fouke, D. *The Enthusiastical Concerns of Dr. Henry More.* Religious Meaning and the Psychology of Delusion. 1997. ISBN 90 04 10600 6
72. Ramelow, T. *Gott, Freiheit, Weltenwahl.* Der Ursprung des Begriffes der besten aller möglichen Welten in der Metaphysik der Willensfreiheit zwischen Antonio Perez S.J. (1599-1649) und G.W. Leibniz (1646-1716). 1997. ISBN 90 04 10641 3
73. Stone, H.S. *Vico's Cultural History.* The Production and Transmission of Ideas in Naples, 1685-1750. 1997. ISBN 90 04 10650 2
74. Stroll, M. *The Medieval Abbey of Farfa.* Target of Papal and Imperial Ambitions. 1997. ISBN 90 04 10704 5
75. Hyatte, R. *The Prophet of Islam in Old French:* The Romance of Muhammad *(1258) and* The Book of Muhammad's Ladder *(1264).* English Translations, With an Introduction. 1997. ISBN 90 04 10709 2
76. Jestice, P.G. *Wayward Monks and the Religious Revolution of the Eleventh Century.* 1997. ISBN 90 04 10722 3
77. Van der Poel, M. *Cornelius Agrippa, The Humanist Theologian and His Declamations.* 1997. ISBN 90 04 10756 8
78. Sylla, E. & M. McVaugh (eds.). *Texts and Contexts in Ancient and Medieval Science.* Studies on the Occasion of John E. Murdoch's Seventieth Birthday. 1997. ISBN 90 04 10823 8
79. Binkley, P. (ed.). *Pre-Modern Encyclopaedic Texts.* 1997. ISBN 90 04 10830 0

80. Klaver, J.M.I. *Geology and Religious Sentiment.* The Effect of Geological Discoveries on English Society and Literature between 1829 and 1859. 1997. ISBN 90 04 10882 3
81. Inglis, J. *Spheres of Philosophical Inquiry and the Historiography of Medieval Philosophy.* 1998. ISBN 90 04 10843 2
82. McCalla, A. *A Romantic Historiosophy.* The Philosophy of History of Pierre-Simon Ballanche. 1998. ISBN 90 04 10967 6
83. Veenstra, J.R. *Magic and Divination at the Courts of Burgundy and France.* Text and Context of Laurens Pignon's *Contre les devineurs* (1411). 1998. ISBN 90 04 10925 0
84. Westerman, P.C. *The Disintegration of Natural Law Theory.* Aquinas to Finnis. 1998. ISBN 90 04 10999 4
85. Gouwens, K. *Remembering the Renaissance.* Humanist Narratives of the Sack of Rome. 1998. ISBN 90 04 10969 2
86. Schott, H. & J. Zinguer (Hrsg.). *Paracelsus und seine internationale Rezeption in der frühen Neuzeit.* Beiträge zur Geschichte des Paracelsismus. 1998. ISBN 90 04 10974 9
87. Åkerman, S. *Rose Cross over the Baltic.* The Spread of Rosicrucianism in Northern Europe. 1998. ISBN 90 04 11030 5
88. Dickson, D.R. *The Tessera of Antilia.* Utopian Brotherhoods & Secret Societies in the Early Seventeenth Century. 1998. ISBN 90 04 11032 1
89. Nouhuys, T. VAN. *The Two-Faced Janus.* The Comets of 1577 and 1618 and the Decline of the Aristotelian World View in the Netherlands. 1998. ISBN 90 04 11204 9
90. Muessig, C. (ed.). *Medieval Monastic Preaching.* 1998. ISBN 90 04 10883 1
91. Force, J.E. & D.S. Katz (eds.). *"Everything Connects": In Conference with Richard H. Popkin.* Essays in His Honor. 1999. ISBN 90 04 110984
92. Dekker, K. *The Origins of Old Germanic Studies in the Low Countries.* 1999. ISBN 90 04 11031 3
93. Rouhi, L. *Mediation and Love.* A Study of the Medieval Go-Between in Key Romance and Near-Eastern Texts. 1999. ISBN 90 04 11268 5
94. Akkerman, F., A. Vanderjagt & A. Van der Laan (eds.). *Northern Humanism between 1469 and 1625.* 1999. ISBN 90 04 11314 2
95. Truman, R.W. *Spanish Treatises on Government, Society and Religion in the Time of Philip II.* The 'de regimine principum' and Associated Traditions. 1999. ISBN 90 04 11379 7
96. Nauta, L. & A. Vanderjagt (eds.) *Demonstration and Imagination.* Essays in the History of Science and Philosophy Presented to John D. North. 1999. ISBN 90 04 11468 8
97. Bryson, D. *Queen Jeanne and the Promised Land.* Dynasty, Homeland, Religion and Violence in Sixteenth-Century France. 1999. ISBN 90 04 11378 9
98. Goudriaan, A. *Philosophische Gotteserkenntnis bei Suárez und Descartes im Zusammenhang mit der niederländischen reformierten Theologie und Philosophie des 17. Jahrhunderts.* 1999. ISBN 90 04 11627 3
99. Heitsch, D.B. *Practising Reform in Montaigne's* Essais. 2000. ISBN 90 04 11630 3
100. Kardaun, M. & J. Spruyt (eds.). *The Winged Chariot.* Collected Essays on Plato and Platonism in Honour of L.M. de Rijk. 2000. ISBN 90 04 11480 7
101. Whitman, J. (ed.), *Interpretation and Allegory:* Antiquity to the Modern Period. 2000. ISBN 90 04 11039 9
102. Jacquette, D., *David Hume's Critique of Infinity.* 2000. ISBN 90 04 11649 4
103. Bunge, W. VAN. *From Stevin to Spinoza.* An Essay on Philosophy in the Seventeenth-Century Dutch Republic. 2001. ISBN 90 04 12217 6
104. Gianotti, T., *Al-Ghazālī's Unspeakable Doctrine of the Soul.* Unveiling the Esoteric Psychology and Eschatology of the Iḥyā. 2001. ISBN 90 04 12083 1
105. Saygin, S., *Humphrey, Duke of Gloucester (1390-1447) and the Italian Humanists.* 2002. ISBN 90 04 12015 7
106. Bejczy, I.P., *Erasmus and the Middle Ages.* The Historical Consciousness of a Christian Humanist. 2001. ISBN 90 04 12218 4
107. Brann, N.L. *The Debate over the Origin of Genius during the Italian Renaissance.* The Theories of Supernatural Frenzy and Natural Melancholy in Accord and in Conflict on the Threshold of the Scientific Revolution. 2002. ISBN 90 04 12362 8
108. Allen, M.J.B. & V. Rees with M. Davies. (eds.), *Marsilio Ficino: His Theology, His Philosophy, His Legacy.* 2002. ISBN 90 04 11855 1
109. Sandy, G., *The Classical Heritage in France.* 2002. ISBN 90 04 11916 7

110. Schuchard, M.K., *Restoring the Temple of Vision.* Cabalistic Freemasonry and Stuart Culture. 2002. ISBN 90 04 12489 6
111. Eijnatten, J. van. *Liberty and Concord in the United Provinces.* Religious Toleration and the Public in the Eighteenth-Century Netherlands. 2003. ISBN 90 04 12843 3
112. Bos, A.P. *The Soul and Its Instrumental Body.* A Reinterpretation of Aristotle's Philosophy of Living Nature. 2003. ISBN 90 04 13016 0
113. Laursen, J.C. & J. van der Zande (eds.). *Early French and German Defenses of Liberty of the Press.* Elie Luzac's *Essay on Freedom of Expression* (1749) and Carl Friedrich Bahrdt's *On Liberty of the Press and its Limits* (1787) *in English Translation.* 2003. ISBN 90 04 13017 9
114. Pott, S., M. Mulsow & L. Danneberg (eds.). *The Berlin Refuge 1680-1780.* Learning and Science in European Context. 2003. ISBN 90 04 12561 2
115. Gersh, S. & B. Roest (eds.). *Medieval and Renaissance Humanism.* Rhetoric, Repre-sentation and Reform. 2003. ISBN 90 04 13274 0
116. Lennon, T.M. (ed.). *Cartesian Views.* Papers presented to Richard A. Watson. 2003. ISBN 90 04 13299 6
117. Von Martels, Z. & A. Vanderjagt (eds.). *Pius II – 'El Più Expeditivo Pontefice'.* Selected Studies on Aeneas Silvius Piccolomini (1405-1464). 2003. ISBN 90 04 13190 6
118. Gosman, M., A. Macdonald & A. Vanderjagt (eds.). *Princes and Princely Culture 1450–1650.* Volume One. 2003. ISBN 90 04 13572 3
119. Lehrich, C.I. *The Language of Demons and Angels.* Cornelius Agrippa's Occult Philosophy. 2003. ISBN 90 04 13574 X
120. Bunge, W. van (ed.). *The Early Enlightenment in the Dutch Republic, 1650–1750.* Selected Papers of a Conference held at the Herzog August Bibliothek, Wolfenbüttel 22–23 March 2001. 2003. ISBN 90 04 13587 1
121. Romburgh, S. van, *"For My Worthy Freind Mr Franciscus Junius."* An Edition of the Correspondence of Francis Junius F.F. (1591-1677). 2004. ISBN 90 04 12880 8
122. Mulsow, M. & R.H. Popkin (eds.). *Secret Conversions to Judaism in Early Modern Europe.* 2004. ISBN 90 04 12883 2
123. Goudriaan, K., J. van Moolenbroek & A. Tervoort (eds.). *Educa-tion and Learning in the Netherlands, 1400-1600.* 2004. ISBN 90 04 13644 4
124. Petrina, A. *Cultural Politics in Fifteenth-Century England: The Case of Humphrey, Duke of Gloucester.* 2004. ISBN 90 04 13713 0
125. Schuurman, P. *Ideas, Mental Faculties and Method.* The Logic of Ideas of Descartes and Locke and Its Reception in the Dutch Republic, 1630–1750. 2004. ISBN 90 04 13716 5
126. Bocken, I. *Conflict and Reconciliation: Perspectives on Nicholas of Cusa.* 2004. ISBN 90 04 13826 9
127. Otten, W. *From Paradise to Paradigm.* A Study of Twelfth-Century Humanism. 2004. ISBN 90 04 14061 1
128. Visser, A.S.Q. *Joannes Sambucus and the Learned Image.* The Use of the Emblem in Late-Renaissance Humanism. 2005. ISBN 90 04 13866 8
129. Mooij, J.J.A. *Time and Mind.* History of a Philosophical Problem. 2005. ISBN 90 04 14152 9
130. Bejczy, I.P. & R.G. Newhauser (eds.). *Virtue and Ethics in the Twelfth Century.* 2005. ISBN 90 04 14327 0
131. Fisher, S. *Pierre Gassendi's Philosophy and Science.* Atomism for Empiricists. 2005. ISBN 90 04 11996 5
132. Wilson, S.A. *Virtue Reformed.* Rereading Jonathan Edwards's Ethics. 2005. ISBN 90 04 14300 9
133. Kircher, T. *The Poet's Wisdom.* The Humanists, the Church, and the Formation of Philosophy in the Early Renaissance. 2005. ISBN 90 04 14637 7
134. Mulsow, M. & J. Rohls (eds.). *Socinianism and Arminianism.* Antitrinitarians, Calvinists and Cultural Exchange in Seventeenth-Century Europe. 2005. ISBN 90 04 14715 2
135. Rietbergen, P. *Power and Religion in Baroque Rome.* Barberini Cultural Policies. 2006. ISBN 90 04 14893 0
136. Celenza, C. & K. Gouwens (eds.). *Humanism and Creativity in the Renaissance.* Essays in Honor of Ronald G. Witt. 2006. ISBN 90 04 14907 4

137. Akkerman, F. & P. Steenbakkers (eds.). *Spinoza to the Letter*. Studies in Words, Texts and Books. 2005. ISBN 90 04 14946 5
138. Finkelstein, A. *The Grammar of Profit: The Price Revolution in Intellectual Context*. 2006. ISBN 90 04 14958 9
139. Ittersum, M.J. van. *Profit and Principle*. Hugo Grotius, Natural Rights Theories and the Rise of Dutch Power in the East Indies, 1595-1615. 2006. ISBN 90 04 14979 1
140. Klaver, J.M.I. *The Apostle of the Flesh: A Critical Life of Charles Kingsley*. 2006. ISBN-13: 978 90 04 15128 4, ISBN-10: 90 04 15128 1
141. Hirvonen, V., T.J. Holopainen & M. Tuominen (eds.). *Mind and Modality*. Studies in the History of Philosophy in Honour of Simo Knuuttila. 2006. ISBN-13: 978 90 04 15144 4, ISBN-10: 90 04 15144 3
142. Davenport, A.A. *Descartes's Theory of Action*. 2006. ISBN-13: 978 90 04 15205 2, ISBN-10: 90 04 15205 9
143. Mazzocco, A. *Interpretations of Renaissance Humanism*. 2006. ISBN-13: 978 90 04 15244 1, ISBN-10: 90 04 15244 X
144. Verbaal, W., Y. Maes & J. Papy (eds.). *Latinitas Perennis, Volume I: The Continuity of Latin Literature*. 2007. ISBN-13: 978 90 04 15327 1, ISBN-10: 90 04 15327 6
145. Boulton, D'Arcy J.D. & J.R. Veenstra (eds.). *The Ideology of Burgundy*. The Promotion of National Consciousness, 1364-1565. 2006. ISBN-13: 978 90 04 15359 2, ISBN-10: 90 04 15359 4
146. Ruler, H. van, A. Uhlmann & M. Wilson (eds.). *Arnold Geulincx* Ethics. With Samuel Beckett's Notes. Translated by M. Wilson. 2006. ISBN-13: 978 90 04 15467 4, ISBN-10: 90 04 15467 1 (Published as Vol. 1 in the subseries *Brill's Texts and Sources in Intellectual History*)
147. Radding, C.M. & A. Ciaralli (eds.). *The* Corpus Iuris Civilis *in the Middle Ages*. Manuscripts and Transmission from the Sixth Century to the Juristic Revival. 2007. ISBN-13: 978 90 04 15499 5, ISBN-10: 90 04 15499 X
148. Birkedal Bruun, M. *Parables: Bernard of Clairvaux's Mapping of Spiritual Topography*. 2007. ISBN-13: 978 90 04 15503 9, ISBN-10: 90 04 15503 1
149. Lehner, U. *Kants Vorsehungskonzept auf dem Hintergrund der deutschen Schulphilosophie und -theologie*. 2007. ISBN 978 90 04 15607 4
150. Warnar, G. *Ruusbroec*. Literature and Mysticism in the Fourteenth Century. Translated by D. Webb. 2007. ISBN 978 90 04 15869 6
151. Treschow, M., W. Otten & W. Hannam (eds.). *Divine Creation in Ancient, Medieval, and Early Modern Thought*. Essays Presented to the Rev'd Dr Robert D. Crouse. 2007. ISBN 978 90 04 15619 7
152. Juste, D. *Les* Alchandreana *primitifs*. Étude sur les plus anciens traités astrologiques latins d'origine arabe (Xe siècle). 2007. ISBN 978 90 04 15827 6 (Published as Vol. 2 in the subseries *Brill's Texts and Sources in Intellectual History*)
153. Chardonnens, L.S. *Anglo-Saxon Prognostics, 900-1100*. Study and Texts. 2007. ISBN 978 90 04 15829 0 (Published as Vol. 3 in the subseries *Brill's Texts and Sources in Intellectual History*)
154. Melve, L. *Inventing the Public Sphere*. The Public Debate during the Investiture Contest (c. 1030–1122). 2007. ISBN 978 90 04 15884 9
155. Velema, W.R.E. *Republicans*. Essays on Eighteenth-Century Dutch Political Thought. 2007. ISBN 978 90 04 16191 7
156. Boone, R.A. *War, Domination, and the* Monarchy of France, *Claude de Seyssel and the Language of Politics in the Renaissance*. 2007. ISBN 978 90 04 16214 3
157. Smith, P.J. *Dispositio*: Problematic Ordering in French Renaissance Literature. 2007. ISBN 978 90 04 16305 8
158. Heddle, D.C. *John Stewart of Baldynneis'* Roland Furious, A Scots Poem in its European Context. 2007. ISBN 978 90 04 16318 8
159. Schuchard, M. (ed.). *Bernhard Varenius (1622-1650)*. 2007. ISBN 978 90 04 16363 8
160. Bejczy, I.P. (ed.). *Virtue Ethics in the Middle Ages*, Commentaries on Aristotle's *Nicomachean Ethics*, 1200-1500. 2008. ISBN 978 90 04 16316 4

161. Stern-Gillet, S. & K. Corrigan (eds.). *Reading Ancient Texts. Volume I: Presocratics and Plato.* Essays in Honour of Denis O'Brien. 2008. ISBN 978 90 04 16509 0
162. Stern-Gillet, S. & K. Corrigan (eds.). *Reading Ancient Texts. Volume II: Aristotle and Neo-Platonism.* Essays in Honour of Denis O'Brien. 2008. ISBN 978 90 04 16512 0

www.ingramcontent.com/pod-product-compliance
Lightning Source LLC
Chambersburg PA
CBHW071332230926
51849CB00002B/378
9789004163638